Gene Mapping
in
Laboratory Mammals

Part B

By the same author

Gene Mapping
in
Laboratory Mammals
Part B

by
Roy Robinson

ℙ PLENUM PRESS · London—New York · 1972

Plenum Publishing Company Ltd.
Davis House
8 Scrubs Lane
Harlesden
London NW10 6SE
Tel: 01-969 4727

U.S. Edition published by
Plenum Publishing Corporation
227 West 17th Street
New York New York 10011

ISBN 978-1-4684-7229-5   ISBN 978-1-4684-7227-1 (eBook)
DOI 10.1007/978-1-4684-7227-1

Library of Congress Catalog Card Number: 73-178776

Preface

The present work is an attempt to provide a systematic treatment of genetic linkage in diploid heredity. Part A presents a general account of statistical methods which can be brought to bear on the problem. The primary emphasis is on the practical aspects of estimation. A large proportion, if not the majority, of mutant genes fail to match up to 'textbook' genes—with faultless segregation ratios and expression—yet, these are the materials with which the practical researcher has to cope. For this reason, it is important to know how to deal with the assortment of genes which may display significant deviations from expectation.

Part B examines the accumulated data on linkage for most of the laboratory mammals and provides a comprehensive and up-to-date survey. The need for a critical review has often been expressed and it is hoped that the present analysis will fill the gap. The volume of material is probably the most important in the animal kingdom other than that for *Drosophila* species.

September, 1971 Roy Robinson

Acknowledgements

I should like to record my appreciation of the many people who have encouraged and assisted me in the writing of this book. I am indebted to Drs. C. J. Cooksey, D. S. Falconer, P. W. Lane, J. L. Southard and M. E. Wallace for specific acts of assistance. I am obliged to the many investigators who kindly sent me reprints of their published work and, particularly, to those who have generously granted permission to cite unpublished material. I am grateful to Miss M. B. Newcombe for computational assistance.

Contents of Part B

Contents of Part A: The Biometrical Approach

Introduction · Maximum Likelihood Estimation · Inviability, Impenetrance and Linkage Detection · Estimation with Normal Gene Ratios · Estimation with Inviability · Estimation with Impenetrance · Estimation with Inviability and Impenetrance · Scoring · Multi-point Crosses · Mapping Functions · General Bibliography · Index to Part A

Part B

LINKAGE IN MAMMALIAN SPECIES

CHAPTER 11

House Mouse

Mus musculus ($n = 19 + X + Y$)

The house mouse has emerged as the laboratory animal *par excellence* for mammalian genetics. The main reason for this is technical but there is also the fact that the mouse 'got in first' in those early days when the variations present in fanciers' stocks were eagerly exploited for confirmation and extension of Mendelian inheritance. This early lead has never been lost and the accumulative information on the species is now so extensive that no other mammalian species is likely to supplant it. This is not to imply that the house mouse is ideal for all types of research, nor that the comparative genetics of other species should be neglected. Far from it in fact. Excessive concentration on one species is to be deplored.

The mouse is well served with reviews on genetics and descriptions of known mutants. The most important of the older works is that of Gruneberg (1952) and is still valuable in several respects. Two useful catalogues are those of Gruneberg (1956) and Sidman, Green, and Appel (1965), the latter having the added appeal of being a specialist account of behavioural and neurological mutants. The assignment of genes to the known linkage groups has received attention. Dickie (1954b), Green and Dickie (1959), E. L. Green (1966), and M. C. Green (1963, 1966, 1968), have each compiled reviews. The fact of these successive compilations is a reflection of the progressive increase of information on linkage in the mouse.

The mouse has far outstripped other laboratory animals in the number of known mutant genes/loci. Some loci in fact possess an impressive series of mutant alleles. A recent estimate places the number of known loci in the region of 400. It is almost certain at this time that this number has been exceeded. Table 11.1 lists those loci whose alleles have featured in tests for linkage. The investigation of linkage is an important aspect of mouse genetics and this work has resulted in a substantial proportion of known loci being assigned to linkage groups.

Although no clear evidence has yet emerged for the existence of partial sex-linkage (i.e.. crossingover between the X and Y), crossingover regularly occurs for the X chromosome in the female.

TABLE 11.1. Genes/loci of the house mouse which have featured in studies on linkage: assigned linkage group, symbol, and conventional designation.

Linkage Group	Symbol	Designation	Prime Characteristic
V	*a*	Non-agouti	Coat colour
	ad	Adipose	Body size
	ag	Agitans	Behaviour
VII	*Al*	Alopecia	Hypotrichosis
	Amy-1	Amylase, Salivary	Electrophoretic variant
	Amy-2	Amylase, Pancreatic	Electrophoretic variant
VIII	*an*	Anaemia	Blood
V	A^s	Agouti-suppressor	Coat colour
VIII	*asp*	Audiogenic seizure	Behaviour
IV	*av*	Ames waltzer	Behaviour
XV	*ax*	Ataxia	Behaviour
VIII	*b*	Brown	Coat colour
XVII	*bf*	Buff	Coat colour
XIV	*bg*	Beige	Coat colour
	bh	Brain hernia	Physiology
XVII	*bl*	Blebbed	Physiology
XII	*bm*	Brachymorphic	Skeleton
XX	*Bn*	Bent-tail	Tail
V	*bp*	Brachypodism	Skeleton
VI	*bt*	Belted	White spotting
I	*c*	Albino	Coat colour
VI	*Ca*	Caracul	Coat texture
	Cat	Dominant cataract	Eyes
XI	*Cd*	Crooked	Tail
XX	*Cg*	Controlling element	Physiology
XIV	*ch*	Congenital hydrocephalus	Physiology
	cl	Club-foot	Feet
V	*Cm*	Coloboma	Eyes
VII	*co*	Cocked	Behaviour
XIV	*cr*	Crinkled	Hypotrichosis
V	*Cs*	Catalase	Electrophoretic variant
II	*cw*	Curly-whiskers	Hair texture
II	*d*	Dilution	Coat colour
I	*da*	Dark	Coat colour
VIII	*db*	Diabetes	Physiology

Linkage Group	Symbol	Designation	Prime Characteristic
XII	*Dc*	Dancer	Behaviour
XVI	*de*	Droopy-ear	Ears
VIII	*dep*	Depilated	Hypotrichosis
VII	*df*	Ames dwarf	Body size
XIII	*Dh*	Dominant hemimelia	Feet
IV	*dl*	Downless	Hair texture
V	*dm*	Diminutive	Body size
XIII	*dr*	Dreher	Behaviour
III	*Ds*	Disorganization	Physiology
XIII	*dt*	Dystonia musculorum	Behaviour
II	*du*	Ducky	Behaviour
XI	*dw*	Dwarf	Body size
IV	*dy*	Dystrophia muscularis	Behaviour
XVIII	*e*	Extension	Coat colour
XVIII	*Ea-1*	Erythrocytic antigen-1	Immunogenetics
IV	*eb*	Eye-blebs	Physiology
	Ee-1	Erythrocytic acetylesterase	Electrophoretic variant
	Ee-2	Erythrocytic propionylesterase	Electrophoretic variant
XII	*ep*	Pale ears	Coat colour
XVIII	*Es-1*	Serum esterase-1	Electrophoretic variant
XVIII	*Es-2*	Serum esterase-2	Electrophoretic variant
VII	*Es-3*	Kidney esterase-3	Electrophoretic variant
XVIII	*Es-5*	Esterase-5	Electrophoretic variant
XVIII	*Es-6*	Esterase-6	Electrophoretic variant
I	*ex*	Ealier X-zone degeneration	Physiology
XIV	*f*	Flexed-tail	Blood, tail
	fa	Falter	Behaviour
V	*fi*	Fidget	Behaviour
III	*Fkl*	Freckled	Coat colour
I	*fr*	Frizzy	Hair texture
XIV	*fs*	Furless	Hypotrichosis
XVI	*ft*	Flaky tail	Physiology
IX	*Fu*	Fused	Tail
II	*Fv-2*	Friend virus-2	Immunogenetics
XIII	*fz*	Fuzzy	Coat texture
XVII	*g*	Glucuronidase level	Physiology
IV	*gl*	Grey-lethal	Coat colour
XVII	*go*	Angora	Hair texture
VIII	*Gpd-1*	Glucose-6-phosphate dehydrogenase	Electrophoretic variant
I	*Gpi-1*	Glucophosphate isomerase	Electrophoretic variant
X	*gr*	Grizzled	Coat colour
XX	*Gs*	Greasy	Hair texture
XX	*Gy*	Gyro	Behaviour

Linkage Group	Symbol	Designation	Prime Characteristic
I	*H-1*	Histocompatibility-1	Immunogenetics
IX	*H-2*	Histocompatibility-1	Immunogenetics
V	*H-3*	Histocompatibility-3	Immunogenetics
I	*H-4*	Histocompatibility-4	Immunogenetics
	H-5	Histocompatibility-5	Immunogenetics
V	*H-6*	Histocompatibility-6	Immunogenetics
V	*H-13*	Histocompatibility-13	Immunogenetics
H	*H-14*	Histocompatibility-14	Immunogenetics
	H-Y	*Y* Histocompatibility	Immunogenetics
	Hba	Haemoglobulin α chain	Electrophoretic variant
I	*Hbb*	Haemoglobulin β chain	Electrophoretic variant
	Hc	Haemolytic complement	Immunogenetics
XI	*Hd*	Hypodactyly	Feet
I	*hf*	Hepatic fusion	Physiology
XVIII	*Hk*	Hook	Tail
VI	*hl*	Hairloss	Hypotrichosis
XVII	*Hm*	Hammer-toe	Feet
III	*hr*	Hairless	Hair texture
VI	*Ht*	High-tail	Tail
	hy-1	Hydrocephalus-1	Physiology
XVIII	*hy-3*	Hydrocephalus-3	Physiology
XIII	*Id-1*	Isocitrate dehydrogenase	Electrophoretic variant
	Ig-1	Immunoglobulin-1	Immunogenetics
	Ig-2	Immunoglobulin-2	Immunogenetics
	Ig-3	Immunoglobulin-3	Immunogenetics
	Ig-4	Immunoglobulin-4	Immunogenetics
IX	*Ir-1*	Immune response-1	Immunogenetics
V	*Ir-2*	Immune response-2	Immunogenetics
IV	*jc*	Jackson circler	Behaviour
	je	Jerker	Behaviour
XVII	*jg*	Jagged-tail	Tail
X	*ji*	Jittery	Behaviour
XX	*jp*	Jimpy	Behaviour
	jt	Joined-toes	Feet
XX	*k*	Phosphorylase kinase	Physiology
X	*kd*	Kidney disease	Physiology
IX	*Ki*	Kinky	Tail
V	*kr*	Kreisler	Behaviour
XVIII	*la*	Leaner	Behaviour
XI	*Lc*	Lurcher	Behaviour
V	*ld*	Limb deformity	Skeleton
XI	*Ldr-1*	Lactate dehydrogenase regulator	Electrophoretic variant
XVII	*le*	Light ears	Coat colour
XIII	*ln*	Leaden	Coat colour

Linkage Group	Symbol	Designation	Prime Characteristic
IX	*Low*	Low-ratio	Physiology
XIII	*Lp*	Loop-tail	Tail
	lr	Lens rupture	Eyes
V	*ls*	Lethal spotting	Coat colour
V	*lst*	Strong's luxoid	Skeleton
VII	*lt*	Lustrous	Hair texture
II	*lu*	Luxoid	Skeleton
VIII	*Lv*	δ-aminolevulinate dehydrogenase	Electrophoretic variant
XVII	*lx*	Luxate	Skeleton
	Ly-A	Lymphocyte antigen	Immunogenetics
	Ly-B	Lymphocyte antigen	Immunogenetics
VIII	*m*	Misty	Coat colour
XVI	*ma*	Matted	Hypotrichosis
XVII	*mc*	Marcelled	Coat texture
II	*Mdh-1*	Malate dehydrogenase	Electrophoretic variant
VI	*med*	Motor end-plate disease	Physiology
V	*mg*	Mahogany	Coat colour
XI	*Mi*	Microphthalmia	Eyes
	mk	Microcytic anaemia	Blood
VI	*mn*	Miniature	Body size
XX	*Mo*	Mottled	Coat colour
XIV	*mu*	Muted	Coat colour
VIII	*Mup-1*	Major urinary protein	Electrophoretic variant
	my	Myelencephalic blebs	Physiology
VI	*N*	Naked	Hypotrichosis
I	*Nil*	Neonatal intestinal lipoidosis	Physiology
XVIII	*nr*	Nervous	Behaviour
VII	*nu*	Nude	Hypotrichosis
I	*nv*	Nijmegen waltzer	Behaviour
XI	*ob*	Obese	Body size/growth
VII	*oe*	Open eyelids	Eyes
I	*ol*	Oligodactyly	Feet
XX	*Op*	Osteopetrotic	Skeleton
XVIII	*Os*	Oligosyndactylia	Feet
I	*p*	Pink eye	Coat colour
V	*pa*	Pallid	Coat colour
XIV	*pe*	Pearl	Coat colour
VIII	*pf*	Pupoid foetus	Physiology
IV	*pg*	Pygmy	Body size
XVII	*Pgm-1*	Phosphoglucomutase	Electrophoretic variant
XVII	*Ph*	Patch	Coat colour
XVII	*pi*	Pirouette	Behaviour
III	*pn*	Pugnose	Skeleton
	Pre	Prealbumin component	Electrophoretic variant

Linkage Group	Symbol	Designation	Prime Characteristic
	Pro-1	Erythrocyte protein	Electrophoretic variant
VIII	*Po*	Polysyndactyly	Feet
VIII	*Pt*	Pin-tail	Tail
I	*pu*	Pudgy	Skeleton
XI	*px*	Postaxial hemimelia	Feet
XIII	*py*	Polydactyly	Feet
XVIII	*Q*	Quinky	Tail
IX	*qk*	Quaking	Behaviour
I	*qv*	Quivering	Behaviour
IV	*r*	Rodless retina	Eyes
V	*Ra*	Ragged	Hypotrichosis
XVII	*rd*	Retina degeneration	Eyes
VII	*Re*	Rex	Hair texture
IX	*rgv-1*	Gross leukaemia virus susceptibility	Physiology
	rgv-2	Gross leukaemia virus susceptibility	Physiology
V	*rh*	Rachiterata	Skeleton
XVII	*rl*	Reeler	Behaviour
V	*ro*	Rough	Hair texture
XII	*ru*	Ruby-eye	Coat colour
I	*ru-2*	Ruby-eye-2	Coat colour
XVII	*Rw*	Rump white	White spotting
III	*s*	Piebald	White spotting
XIV	*sa*	Satin	Hair texture
	Sas-1	Antigenic serum substance	Immunogenetics
V	*Sd*	Danford's short-tail	Tail
II	*se*	Short-ears	Ears
	Sey	Small eye	Eyes
XX	*sf*	Scurfy	Physiology
II	*sg*	Staggerer	Behaviour
I	*sh-1*	Shaker-1	Behaviour
VII	*sh-2*	Shaker-2	Behaviour
VI	*Sha*	Shaven	Hypotrichosis
VII	*shm*	Shambling	Behaviour
IV	*si*	Silver	Coat colour
XI	*Sig*	Sightless	Eye
IV	*Sl*	Steel	Blood, coat colour
XX	*Sla*	Sex-linked anaemia	Blood
IX	*Slp*	Sex-limited protein	Immunogenetics
	slt	Slate	Coat colour
VIII	*sno*	Snubnose	Skeleton
XIII	*Sp*	Splotch	White spotting
XVI	*spa*	Spastic	Behaviour
XX	*Spf*	Sparse-fur	Hair texture

Linkage Group	Symbol	Designation	Prime Characteristic
	sr	Spinner	Behaviour
IX	*Ss*	Serological serum	Immunogenetics
V	*Stb*	Stubby	Skeleton
XX	*Str*	Striated	Hair texture
	Sut	Short undulated tail	Tail
II	*sv*	Snell's waltzer	Behaviour
V	*Svp*	Seminal vesicle protein	Electrophoretic variant
VI	*sw*	Swaying	Behaviour
IX	*T*	Brachyury	Tail
XX	*Ta*	Tabby	Hair texture
XIII	*tb*	Tumbler	Behaviour
XI	*tc*	Truncate	Tail
IX	*tf*	Tufted	Hypotrichosis
XX	*Tfm*	Testicular feminization	Physiology
XVIII	*tg*	Tottering	Behaviour
XIII	*th*	Tilted head	Behaviour
VII	*ti*	Tipsy	Behaviour
II	*tk*	Tail kinks	Tail
IX	*Tla*	Thymus leukemia antigen	Immunogenetics
VII	*tn*	Teetering	Behaviour
I	*tp*	Taupe	Coat colour
VII	*Tr*	Trembler	Behaviour
II	*Trf*	Transferrin	Electrophoretic variant
XV	*Tw*	Twirler	Behaviour
	U	Umbrous	Coat colour
V	*un*	Undulated	Tail
VI	*uw*	Underwhite	Coat colour
X	*v*	Waltzer	Behaviour
XVI	*Va*	Varitint-waddler	Coat-colour, behaviour
VII	*vb*	Vibrator	Behaviour
VIII	*vc*	Vacillans	Behaviour
VI	*Ve*	Velvet	Coat texture
XIII	*vl*	Vacuolated lens	Eyes
VII	*vt*	Vestigial-tail	Tail
XVII	*W*	Dominant spotting	Coat colour, blood
XI	*wa-1*	Waved-1	Hair texture
VII	*wa-2*	Waved-2	Hair texture
VIII	*wd*	Waddler	Behaviour
V	*we*	Wellhaarig	Coat texture
VIII	*wi*	Whirler	Behaviour
III	*wl*	Wabbler-lethal	Behaviour
II	*wy*	Wavy	Coat texture
XIV	*Xt*	Extra-toes	Feet
XX	*Ym*	Yellow mottling	Coat colour
	'*θ*'	Thymocyte antigen	Immunogenetics

It will be noted that the microphthalmia locus is symbolized by Mi instead of the more conventional mi. This former symbol is more convenient for present purposes because the allele employed in the overwhelming proportion of linkage tests is Mi^{wh}. The use of Mi simplifies a rather cumbersome symbol and signifies that a dominant mutant gene was employed in most crosses. Also, mainly for convenience, the symbol Fu is used for fused rather than Ki^{fu}. Fu has been employed in the majority of crosses whereas Ki has featured only rarely. It is appreciated that Fu is either closely linked to Ki or is probably an allele (Dunn and Gluecksohn-Waelsch, 1954).

A number of communications, dealing with aspects of independent assortment or even with linkage, could not be considered in the sections which follow. There are several reasons for this. (1) The gene could be of a relatively minor character. (2) A gene has been discarded without provision for its preservation by at least one laboratory. This has often meant that the gene cannot be readily connected up with present-day mutants of similar phenotype. (3) The discovery of a new gene may be so recent that its standing with comparable genes cannot be properly evaluated. It is unlikely that the following cases constitute a full listing (especially of the older literature) but these may give some idea of the difficulties which have been encountered.

A kinky tail is described by Plate (1910) to be associated with a pink-eyed gene (presumably p). However, the data are inconclusive. Gates (1926) presents evidence that a kinky tail gene is linked to se. He also showed that the tail condition is independent of v. Keeler (1927) featured a kinky tail gene linked to se but independent of r. These two latter reports are interesting but the relationships of this kinky tail gene to present-day tail mutants is unclear.

Fisher and Mather (1936b) give details of the independent segregation of a minor spotting gene ('light head') from a, b, d, wa-1 and sex. The gene displays interaction with s but yields no evidence of linkage. Gruneberg (1936c) has described the independent heredity of a minor tail spotting gene from that of a.

In 1940, Mather and North described a dominant gene, under the name of 'umbrous', which produces a suffusion of melanism in the agouti phenotype. The gene assorted independently of a.

A similar gene is stated by Carter (1951c) to be inherited independently of *lx*. Yet another gene of the type is described by Robinson (1959), dominant in inheritance and independent of *a* and sex. At this time, several 'umbrous' or agouti darkening genes are known, both dominant and recessive, and it is difficult to assess how these relate to those reported earlier.

There is doubt whether or not rodless retina (*r*) is still in existence (Sidman and Green, 1965). However, Keeler (1966) is of the opinion that the currently available retinal degeneration (*rd*) gene is identical to *r*. It is possible that this ambiguity will only be settled in a negative manner (i.e., by the demonstration of the non-identicalness of *r* and *rd*). Sidman and Green entertain the possibility that *r* may yet be found in an obscure mouse stock since the presence of an eye defect of this nature could pass unnoticed unless a specific examination is made. It is primarily for this reason that details of assortment of *r* are tabulated in this book.

An association of a polydactylous condition with anophthalmia is outlined by Beck (1963). The polydactyly is ascribed to a recessive gene with marked impenetrance. The association is clear but it is doubtful if genetic linkage is involved.

Shreffler (1966) has outlined the discovery of an erythrocyte antigen which he considers to be new, because its pattern of incidence between inbred strains and its presence or absence in various body tissues, do not correspond to those of previously described antigens. The antigen (unsymbolized as yet) has failed to show linkage with *a, b, c, d,* and *Hbb*. Two other reports have given details of erythrocyte antigens which may or may not be identical to the above. Egorov (1965) has described an erythrocyte antigen denoted as lambda (λ). The antigen is inherited as an autosomal dominant and assorted independently of *H-2* in several small samples. Similarly, Klein and Martinkova (1968) have described an automsomal erythrocyte antigen which is inherited independently of *b, d,* and *H-2* in a testcross of 28 individuals. Stimpfling and Snell (1968) have isolated an autosomal erythrocyte antigen which is inherited independently of *c* and *H-1*.

The two major disturbances of normal assortment are partial penetrance and reduced viability. To these may be added affinity. The latter has been reported in the mouse and is the subject of a subsequent section. Its magnitude is perhaps not large (in the main) and its generality has yet to be properly evaluated. Some chromosomes (hence linkage groups) may exhibit the phenomenon either more generally or more strongly than others. It is uncertain if whole chromosomes or merely regions may show affinity. However, it is a factor not easy to allow for in the statistical analysis of segregations but can probably be mitigated by balanced experiments. At this time, the assortment of genes in linkage groups II, V, and VII are likely to show affinity. Fuller details may be found in the section on the phenomenon.

LINKAGE GROUPS

Nineteen groups are known for the mouse, just one short of the haploid number of chromosomes. It may be wondered if this could mean that most of the chromosomes are tagged by mutant loci. In all probability this is so, since it seems unlikely that many of the chromosomes are long enough to accommodate more than one linkage group. Of course, should the number of groups come to exceed the chromosome number, it will be obvious that at least one chromosome is capable of supporting two or more quasi-independent groups. However, as linkage research progresses and the known groups extend in length, together with the likelihood of accurate positioning of the centromere for each group, the present vagueness will disappear. Eventually, it should be possible to relate the linkage group to the actual chromosome in which it resides by a combined genetical and karyological analysis of gene tagged translocations. Already, a start has been made in this direction.

Sick and Nielsen (1964) have postulated the existence of two closely linked loci to explain variation of amylase in the saliva and pancreas. The two loci are designated as *Amy-1* and *Amy-2*, respectively. Crossingover between the loci have been tentatively calculated as $2/208 = 0.96 \pm 0.68$. The loci have yet to be assigned to a linkage group.

A series of immunoglobulins known symbolically as *Ig-1*, *Ig-2*, *Ig-3*, and *Ig-4* (and sometimes under other designates; e.g.,

Asa for *Ig-1*) have been tested for independent assortment but with negative results. Backcross matings of *Ig-1* and *Ig-2* have not given recombinants (Herzenberg, 1964; Lieberman and Potter, 1966). Intercrosses of *Ig-1* and *Ig-3* (Dray *et al.*, 1965; Lieberman *et al.*, 1965) and *Ig-1* and *Ig-4* (Herzenberg *et al.*, 1967; Minna *et al.*, 1967) failed to produce recombinant types. These results suggest a complex locus, although the extent and nature has yet to be evaluated. The complex has yet to be assigned to a linkage group.

Group I (c, da, ex, fr, Gpi-1, H-1, H-4, Hbb, Hf, Nil, nv, ol, p, pu, qv, ru-2, sh-1, tp)
Group I may be said to have come into existence as a result of the comment by J. B. S. Haldane that genes *c* and *p* were displaying 'reduplication' in Darbishire's (1904) experiments (Haldane *et al.*, 1915). These two genes are now the most extensively investigated pair in mouse linkage literature. The group as a whole possesses an impressive range of loci and the results to-date are summarized by the trigon of Fig. 11.1. Given reasonably consistent data, the correct linear order of the loci will emerge as the sequences of crossover values increase steadily towards the nadir of the trigon, reading diagonally either from the left or right. The individual data which culminate in the trigon are documented in Table 11.3.

Many of the gene pairs have data of various segregational types and from different investigations. These have been combined where appropriate and have yielded the following mean values:

c - fr	17.1 ± 2.6	*fr - p*	29.3 ± 4.4
c - H-1	8.7 ± 1.4	*fr - sh-1*	15.9 ± 2.3
c - Hbb	4.8 ± 0.8	*Gpi-1 - Hbb*	33.3 ± 3.6
c - Nil	5.7 ± 1.1	*Hbb - p*	17.9 ± 1.0
c - ol	12.7 ± 2.3	*Hbb - sh-1*	1.6 ± 0.4
c - p	15.1 ± 0.3	*Nil - p*	2.9 ± 0.9
c - pu	31.3 ± 1.3	*p - pu*	16.2 ± 1.0
c - qv	28.3 ± 3.2	*p - ru-2*	2.7 ± 1.0
c - sh-1	3.7 ± 2.4	*p - sh-1*	14.8 ± 1.2

Several genes are known to be in the group but cannot be

mapped with precision because of the inadequate data. No crossingover has been recorded for *H-1 - p* but the considerations put forward by Snell and Stevens (1961) make it very likely that the order is *p - c - H-1*. *H-4* is closely linked to *p* but which side is unknown. The *ru-2* loci is and *ol* are both linked to *c* but the order is unknown. The *ru-2* loci is close to *p* but so far no explicit tests of *ru-2* have been reported with other group I genes. However, Eicher (1970a) has shown that the order is probably *ru-2 - p - c* on the basis of relative frequency of expected single and double crossovers. Gene *tp* is closely linked to *c* and while Fielder (1952) failed to observe crossingover between *p* and *tp*, his R11 data were only large enough to exclude a crossover frequency greater than 39. Hence, although *tp* may lie between *c* and *p*, this is not certain.

Separate determinations of crossingover according to the sex of the heterozygous parent have been made for 16 gene pairs. These are listed in Table 11.4. Six of the comparisons show a greater frequency of crossingover in the male and ten show a greater frequency in the female. None of the greater differences for the male are significant, whereas, five of the female differences are significant. These are for the intercepts *c - Hbb* (3.7 ± 1.0), *c - Nil* (4.9 ± 2.3), *c - p* (4.6 ± 0.7), *c - sh-1* (1.5 ± 0.5) and *Hbb - p* (4.7 ± 2.0). The significance of the *c - Nil* difference is only just beyond the five per cent level.

Bunker (1959) states that the following sex differences were observed for the crossover value in his data: intercept *c - hf*, 4 versus 2.6, *c - p*, 16 versus 17.1, and *hf - p*, 12 versus 14.5 for male versus female gametogenesis, respectively. No original figures are presented, hence the significance of the difference cannot be assessed. Presumably, all were non-significant in these samples but *c - p* has emerged as possessing a clear sex difference in other observations.

Both Castle and Wachter (1924) and Gruneberg (1936b) examined their data for variation of crossingover with age for the *c - p* interval but with negative results. Castle and Wachter (1924) also considered briefly the possible effect of seasonal variation of age for the same pair of genes but again with negative results.

Snell (1946) has recorded crossingover in a translocated chromosome for the intercepts *c - p*, *c - sh-1*, and *p - sh-1*. The translation is T(1;?)12Sn and the break is very close to c. Some interference is evident for c - p and c - sh-1 but only attaining significance for female gametogenesis (Table 11.16). The female differences are c - p (5.0 ± 2.0) and c - sh-1 (2.6 ± 1.1), the translocated chromosome displaying a lower frequency in each case. It is possible that the sexually differentiated decrease is related to the notably higher frequency of crossingover in the female for these intercepts. Male gametogenesis showed a non-significant increase in frequency for c - p (3.6 ± 3.4) but a non-significant decrease for c - sh-1 (1.5 ± 1.9). The differences in crossingover for p - sh-1 were non-significant for both sexes although in the direction of a decrease (5.0 ± 3.0) in the female. The break probably has to be close to the loci concerned before interference is marked. The translocated data for c - p and c - sh-1 have not been amalgamated with those for the normal chromosome but the data for p - sh-1 have been.

Investigation of four radiation-induced deficiencies involving the c locus by Erickson et al. (1968) revealed that the inviable homozygotes, in addition to lacking pigment, possessed anomalies of the thymus and kidneys and a deficiency of glucose-6-phosphatase. It is conceivable that the deficient chromosomal region contains loci controlling basic pigment formation, thymus and kidney morphogenesis and production of glucose-6-phophatase. However, Gluecksohn-Waelsch and Cori (1970) are of the opinion that this might be too naive a

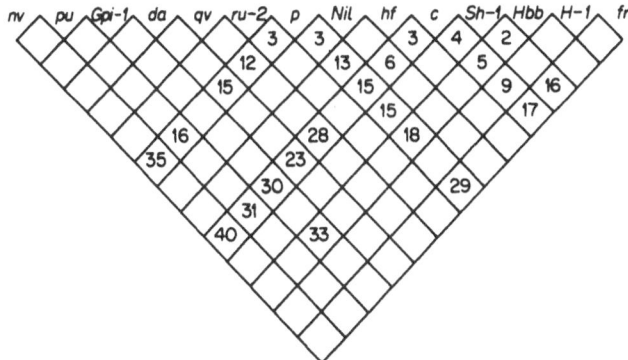

Fig. 11.1. Linkage group I of the house mouse.

picture and later observations suggest a complex of effects resulting from a single fundamental change.

Chang and Hildemann (1964) propose that one of a pair of genes determining susceptibility to a runting syndrome following inoculation with polyoma virus is linked to c. The observations are suggestive but the evidence does not conform to a 'formal' linkage experiment and is best accepted on a provisional basis.

Gilman and Smithies (1968) found a close association between a notable difference of foetal haemoglobin and heterozygosity at the *Hbb* locus controlling components of adult haemoglobin (zero crossovers in a backcross of 56 individuals). It is not clear if two linked loci are involved or if the differences are due to modes of expression of *Hbb* in foetal and adult blood.

The possibility of linkage between c and *bh*, reported by M. E. Wallace (1963; *MNL* 29, 22), turned out to be inconclusive and Wallace (1970; personal communication) now regards the suspected linkage 'as not all certain'.

The position of the centromere has been discussed by Eicher (1970b). The evidence is indirect but compatible with the hypothesis that the centromere is at the *fr* end of the group.

Group II (cw, d, du, Fv-2, lu, Mdh-1, se, sg, sv, tk, Trf, wy)

Group II was brought into being by the independent discovery of exceedingly close linkage between *d* and *se* by Gates (1927a, 1928a) and Snell (1928). Many other genes have since been located in the group. The accompanying trigon summarizes most of the work on the loci involved and provides an indication of the linear order. It is obvious that *d* and *se* could be interchanged without materially altering the map. This remark also applies to *sg* and *sv*.

The following gene pairs have crossover values based upon more than one assortment, resulting in the means:

cw - se	34.9 ± 1.5	d - tk	3.3 ± 1.0
cw - tk	43.2 ± 2.1	d - Trf	12.7 ± 1.9
d - du	20.3 ± 1.7	dse - lu	16.9 ± 2.0
d - Fv-2	15.6 ± 4.0	dse - sg	4.5 ± 8.3
d - se	0.13 ± 0.05	du - lu	44.7 ± 5.7

du - se	19.9 ± 3.6	*lu - tk*	23.8 ± 3.1
du - tk	13.4 ± 1.3	*lu - Trf*	22.2 ± 3.7
lu - Mdh-1	12.0 ± 6.5	*se - sv*	2.5 ± 0.6
lu - se	16.3 ± 3.7	*se - tk*	5.4 ± 0.9
lu - sv	21.8 ± 4.2		

The close linkage of *d* and *se* aroused interest when it was first discovered. By repulsion intercross matings, no crossovers were observed among 4355 young (Gates, 1927a, 1928a; Snell, 1928, 1931). However, by CII test matings of the form *+-sese* × *dd+-*, Bates and Snell found one crossover out of 736 mice tested, giving the estimate of 0.068 ± 0.068. This value is a little outside the permitted limits for the final estimate of the crossover value but the testing method could lead to underestimation. A minimum of five progeny was raised from each tested individual and the failure to detect even one heterozygote would make a lot of difference. The mean estimate for the *d - se* intercept is 0.129 ± 0.048. By Steven's table, the five per cent limits are 0.384 and 0.079, computed on the basis of an equivalent number of 5666 testcross young for the total observed information.

When two loci are exceedingly closely linked, it is often difficult to determine their relative order to a third. Snell (1955) has speculated that the order may be *d - se - du,* rather than the reverse, on the basis of a single crossover between *d* and *se* which separated *d* from *du.* Russell (1967), on the basis of analysis of deficiencies involving the *dse* region, concluded that the loci sequence is *d - se - sv.* Incidental observations of this nature are often the only practical means of solving the problem and it may be assumed that the order is *d - se - sv - du.* The *lu* gene had rather low penetrance in the crosses of Hutton and Roderick (1970) and this has probably biased the crossover value for the genes *lu* and *Mdh-1.* The value for the whole of the first batch of data, allowing for impenetrance, is 21.1 ± 10.1. An estimate based on *lulu* only, and supplemented by additonal data, gives the value 12.0 ± 6.5. This latter determination gives the better overall picture in relation to the values found for *d - lu* (16.9 ± 2.0) and *d - Mdh-1* (10.1 ± 2.9) but not necessarily with complete satisfaction.

It is not possible to position the genes *Fv-2* and *wy* in the map because each has only been tested with one member at this time. It may be noted that *cw* shows a crossover value of 35.0 ± 1.5 with *se* while *wy* shows a value of 25.3 ± 1.9. These two genes both affect vibrissae and coat texture and it may be wondered if they are allelic. The difference in crossover values (9.7 ± 2.4) is significant and this is reasonably conclusive that they are not but a direct test for allelism might be desirable.

Russell (1967) has examined the question whether the *dse* locus may be complex and has given an affirmative answer. Her data suggest that the components or units may be roughly classified as dilute colour, opisthotonic convulsions, short-ear, and waltzer; together with various pre- and post-natal lethals. The waltzer may be *sv* which is closely linked to *dse*. Most of the lethals may be radiation products *per se* (e.g., deficiencies) and, strictly, not be part of the normal complex. In fact, the evidence — at present — for complexity is not as decisive as for *a* (group V), *H-2* and *T* (group IX).

Information is available on crossingover between the sexes for ten pairs of genes. In eight of these, the female has the greater frequency of crossingover. In most cases, the difference is not significant but it is so for the *d* - *Trf* intercept (8.5 ± 2.6 versus 16.9 ± 2.6). Neither of the two cases approach significance where the male has the greater frequency of crossingover.

Some of the data on the intercepts *cw* - *se, cw* - *tk, d* - *tk,* and *se* - *tk* have involved translocated chromosomes. On the evidence, none of the translocations had any influence on the rate of crossingover. Accordingly, these data have been incorporated with such other that may be available. The translocated data for *d* - *tk* was derived from the studies of Green and Lane (1967) and for *cw* - *se, cw* - *tk,* and *se* - *tk* from the studies of Lyon *et al.* (1968). Fuller details may be found in Table 11.16.

The centromere has been located at the *cw* end of the linkage group by M. F. Lyon and S. Hawkes (1970: *MNL*, 42 27). This is a general indication but it usefully confirms early studies in which the centromere was shown to be closely linked to *cw* (Lyon *et al.*, 1968). The disadvantage of the earlier work was that these involved a metacentric translocation apparently produced by fusion of two centromeres or the joining up of unequal chromosome arms. The doubtful aspect is that it is unknown if the metacentric was formed by simple fusion or if other structural changes might have occurred. If simple fusion or breakage had produced the metacentric, then it is a reasonably safe conjecture that the *cw* locus lies towards one end of the normal chromosome. The later results indicate that this is the likely situation and the *cw* locus is probably located close to the centromere.

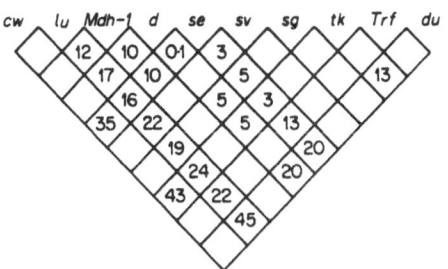

Fig. 11.2. Linkage group II of the house mouse.

Both Gates (1926) and Keeler (1927) have reported linkage between *se* and a recessive tail kinking gene. Gates' data reveal a crossover value of 10.0 ± 1.8 and Keeler's a value of 29.3 ± 6.6. The two values differ significantly; although Keeler's intercross sample is not large. It is possible that Gates' gene may have been an early occurrence of *tk* although the crossover values for the two genes with *se* differ significantly. These two instances of linkage have not been incorporated into the group II map, primarily because the genes probably have been irretrievably lost. A new tail mutant showing a similar crossover value with *se* could be a recurrence, of course, but for practical purposes it will have to be regarded as a distinct gene.

Group III (Ds, Fkl, hr, pn, s, wl)

The present group has a relatively small membership, particularly when it is remembered that the group was established as long ago as 1931 by Snell. At one time, however, it was thought to consist of some 13 loci until Lane (1967) demonstrated that several of these displayed independent assortment. The outcome was that seven loci were hived off to form group XVII. The existing situation is shown by the trigon of Fig. 11.3. The loci *Ds* and *wl* could be interchanged and still produce a creditable map. Gene *hr* could be interchanged with *Ds* and *wl* although K. P. Hummel and D. B. Chapman (1966; *MNL*, 34, 31) state that 'a three-point backcross shows the order to be *Ds - hr - s*'. No numerical data appear to have been published so far.

Two of the above pairs of loci have information on differences of crossingover between the sexes. In two cases, the male has the greater frequency while, in the other, the female. The higher frequency in the male for the *Fkl - s* interval is significant (Table 11.4).

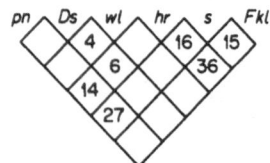

Fig. 11.3. Linkage group III of the house mouse.

The centromere is located at the *pn* end of the map, although its distance from *pn* has yet to be determined (Green, 1970; *MNL*, 43, 32).

The loci *ag* and *hr* have been listed as linked but this seems doubtful on the data available (Hoecker *et al.*, 1954). The relevant data consist of a repulsion intercross of 65 animals. This is a small sample to detect linkage in a R11 but, even so, no double recessives are reported. However, the single gene ratios of both *ag* and *hr* are significantly deficient from expectation due to inviability and the absence of *agaghrhr* animals could be due to this alone or to a viability interaction of the genes. This seems to be the most likely explanation. If linkage is assumed, the small number of animals examined merely excludes very loose linkage. Using Stevens' method, only a linkage value greater than 47 per cent is rendered unlikely.

Group IV (av, dl, dy, eb, gl, je, pg, r, si, Sl)

Group IV was initiated by Keeler (1930) by the discovery of linkage between *r* and *si*. It is remarkable that although many other genes have since been added, almost all of the work is comparatively recent and has yet to be published in detail. The accompanying trigon summarizes the results and presents a consistent picture.

Those gene pairs which have more than one type of assortment data or have been investigated at different times have the following mean crossover values:

av - dy	26.9 ± 1.75	*r - si*	12.5± 8.8
av - si	32.7 ± 2.1	*gl - Sl*	29.8 ± 4.6
av - Sl	14.7 ± 1.2		

The genes *je, r,* and *si* cannot be inserted with confidence into the trigon because each has only shown linkage with one other gene. On present showing, both *je* and *dy* may prove to be sited close to one another and evidently well away from the centrally placed *Sl.*

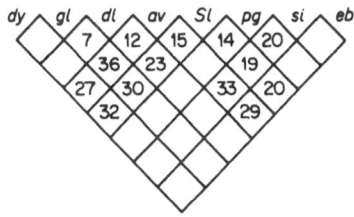

Fig. 11.4. Linkage group IV of the house mouse.

Only two of the above pairs of genes have data on sex differences of crossingover. These are for the intercept *av - si* and *gl - Sl* and the differences are not significant. The male sample is very small for *gl - Sl* and the low crossover value for the sex could be far from reliable.

The *r* gene is listed here because of the early studies of Keeler (1930), despite the possibility that it may be identical to *rd* (group XVII). Sidman and Green (1965) propose that the two genes are different but Keeler (1966) feels that they are identical. Genetically, the crucial point is that *r* failed to display linkage (41.9 ± 7.6) with *W* in Keeler's (1930) material whereas *rd* does so (13.4 ± 1.7; Sidman and Green, 1965). Gene *r* shows linkage with *si* (13.4 ± 1.7) whereas *rd* does not (50.7 ± 2.5; DiPaolo and Noell, 1962). Keeler has argued that the *si* gene with which he worked originated in his stock and may not be the same gene as the one commonly taken to be *si*. The failure to detect linkage between *r* and *W* is attributed to chance. The problem is discussed in the section on group XVII.

Group V (a, A^s, bp, Cm, Cs, dm, fi, H-3, H-6, H-13, Ir-2, kr, ld, ls, lst, mg, pa, Ra, rh, ro, Sd, stb, Sut, Svp, un, we)
The initiation of Group V results from the discovery of linkage between *a* and *we* by Hertwig (1942). Subsequently, the group has grown progressively to become the largest at this time. The large membership is due to the high frequency of *a* among mouse strains and stocks (Fig. 11.5). Although not a necessary consequence, the result has produced the longest map (approximately 79 - 85 units according to route) of the 19 groups. The trigon shows the most probable linear order of the loci and the extent to which their relationships have been investigated. A reasonably consistent picture has emerged although further work may demand some changes. Unfortunately, rather a large number of loci owe their position to weak linkages and these tend to be unreliable.

Mean crossover values have been determined for those intercepts which have assortments of different types and phases and/or observations by different workers and are as under:

a - A^s	0.44 ± 0.15	*dm - mg*	6.0 ± 1.4
a - bp	0.34 ± 0.18	*fi - pa*	25.8 ± 1.4
a - dm	3.5 ± 1.1	*fi - Sd*	19.7 ± 1.1
a - fi	31.7 ±1.2	*H-3 - un*	12.1 ± 1.5
a - H-3	16.7 ± 1.8	*H-3 - we*	2.0 ± 0.7
a - H-6	25.8 ± 1.5	*H-6 - pa*	7.9 ± 1.1
a - ld	28.0 ± 5.2	*ld - mg*	6.1 ± 2.0
a - lst	22.4 ± 2.7	*lst - Ra*	42.7 ± 3.5
a - mg	11.8 ± 0.7	*mg - Ra*	27.5 ± 1.5
a - pa	16.4 ± 0.4	*pa - Ra*	41.4 ± 4.6
a - Ra	22.0 ± 0.7	*pa - un*	9.6 ± 4.7
a - Sd	43.9 ± 1.2	*pa - we*	3.0 ± 0.2
a - stb	39.5 ± 2.6	*Ra - sd*	41.0 ± 3.3
a - Sut	4.4 ± 1.1	*Ra - we*	33.0 ± 1.2
a - un	4.7 ± 0.2	*ro - we*	3.5 ± 0.8
a - we	11.8 ± 0.3	*Sd - stb*	11.2 ± 1.7
bp - un	5.0 ± 0.8	*un - we*	6.0 ± 0.3

A number of loci have data on only a single linkage with other members of the group. These cannot be positioned with any degree of accuracy. Genes *a* and *A^s* are closely linked and indeed may be part of a complex. Genes *kr*, *Cm*, *Ir-2*, *Sut*, and *Svp* display various degrees of linkage with *a*. Fisher (1949) has remarked that *kr* has been found to lie on the side of *a* away from *un* but without presenting any supporting data. He indicates that his estimate of 3 crossingover for the *a - kr* interval is very rough. It may be noted that Lane (1959; appendix) has found much closer linkage than this (0.4 ± 0.6).

Gene *rh* has been inserted in the trigon on the strength of the statement that the gene shows a crossover value of 35 with *a* but no sign of linkage with *Ra* (L. C. Stevens, personal communication 1970). Apart from the entries in the trigon, loci *fi - stb* have data which exclude a crossover value greater than nine (Lane and Dickie, 1968) and this is in keeping with their respective values in conjunction with *a*. Gasser (1969) has considered the possibility that *Ir-2* might be part of a complex involving either *H-3* or *H-6*; on the analogy of the demonstrated close association of *Ir-1* and *H-2*. The crossover value for *a - Ir-2* is comparable for those either of *a - H-3* or *a - H-6*.

Information on crossingover between the sexes is available for 25 pairs of genes. Of these, nine pairs show a higher frequency of crossingover in spermatogenesis. The difference is only significant for one intercept, however, namely *a - Ra*

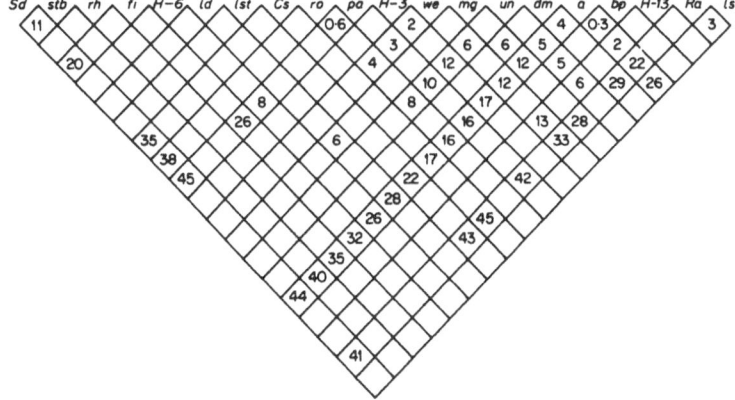

Fig. 11.5. Linkage group V of the house mouse.

(3.9 ± 1.5). The other 16 pairs display a higher frequency of crossingover in oogenesis and, out of these, seven show significant differences. The pairs are: a-fi (8.9 ± 2.8), a-mg (3.3 ± 1.5), a-we (3.9 ± 0.7), H-13-un (4.2 ± 1.1), pa-un (4.9 ± 0.8), pa-we (1,7 ± 0.5), and un-we (2.7 ± 0.6). Most of these intercepts span a definite map region, except for a-fi, which spans most of the map. It is doubtful if much importance can be attached to the former because of the detection of sex differences which have not been studied systematically (direction, magnitude, sequences of intercepts) over the whole map.

One of the clearest demonstrations of a significant change in crossingover with age is provided by the analysis of the a-un intercept by Fisher (1949). A summary of the results is given in Table 11.2. The drop is about 77 per cent for the age range and shows no significant departure from linearity. Both sexes behaved similarly. The decline with age is noteworthy but only future comparative research can decide just how general such large declines may be. The trend is in keeping with a fall in chiasma frequency with age but this latter can only be regarded as a general indication. The position of the intercept in the chromosome may turn out to be the prime variable.

TABLE 11.2.
The trend in crossover value with age for the a-un intercept.

Age (Mths.)	Male		Female		Combined	
	No of Animals	Crossover Value	No. of Animals	Crossover Value	No. of Animals	Crossover Value
0 - 3	288	7.0	314	6.4	542	6.6
4 - 5	506	5.1	525	3.6	1031	4.4
6 - 7	434	4.6	522	5.6	956	5.1
8 - 9	342	5.3	267	4.5	609	4.9
10 -11	522	4.1	81	3.7	203	2.6
11 - 13	105	1.9	-	-	133	1.5

Bodner (1961b) has also observed a significant decline in the crossover value with age. Data are tabulated from female heterozygotes for the fi-pa intercept, about 10 crossover units from a-un. The change in frequency varied from 32 units at two months of age to 20 units for animals of 11 months and older. There was no significant departure from a linear trend. Over the age range, the fall is 40 per cent, not as great as that recorded for *a* - *un*.

On the other hand, Wallace (1957a) could not find clear evidence of a decline in her data for the intercepts *a* - *fi*, *fi* - *Sd*, and *a* - *Sd*. The age range was the same as that examined by Fisher and Bodner. The data for oogenesis show a decrease with age but somewhat erratically and non-significantly. The data for spermatogenesis on the contrary display an increase with age, again erratically but significantly for *a* - *fi*. This intercept is the only one of the three to show a significant sex difference and it may be wondered if the male data had given unusually low crossover values for part of the range. Wallace also cites unpublished observations of Owen and Darrock for the *a* - *un* intercept which fail to exhibit a significant trend. This is the same intercept as studied by Fisher and the discrepancy between the two experiments is intriguing, particularly because of the large number of animals involved (3,500 and 5,000 respectively). The data for all three experiments show a small dip in the curve of crossover values for the 3 - 5 age interval as if the trend may in fact be U shaped.

That the *a* locus may be complex has long been suspected, especially after Pincus (1929) and Keeler (1931b) gave the concept substance. The four alleles A^w, A, a^t, and a form a pattern of phenotypes to be expected either from two closely linked genes or from two intra-locus mutational sites. These are agouti versus non-agouti and dark versus light belly. Two experiments to discover crossovers between the linked genes were negative (Morgan, 1914; Keeler, 1931b). Many other alleles or pseudo-alleles have since been described and, at the phenotypic level, the evidence for complexity is overwhelming (Wallace, 1965). Among the components which can be separated are: development versus suppression of eumelanism, full versus partial development of agouti hairs, dark versus light belly, and presence versus absence of light pinnae and nipple hairs. Dominance relationships are irregular, with some of the components showing dominance, recessiveness or interaction in the same heterozygous compound. A *cis-trans* effect completes the argument at least for two presumed pseudo-alleles (Phillips, 1966).

GMLM Pt B—2

Crossingover within the complex would be expected to be low and, in many instances, indistinguishable from mutation. Wallace (1954) has described such an instance and has reviewed earlier cases. Russell *et al.* (1963) have estimated that crossingover between A^y and a^x to be of the order of 0.5 per cent. Further, the order of the units relative to kr and un may be kr - a^x - A^y - un. Phillips (1966) has presented details of crossingover between a and A^s (A^s was originally symbolized as As, an implied independent gene, but has now been designated as A^s, part of the a complex). The crossover value for a - A^s is 0.4 ± 0.2. An intriguing aspect of A^s is that the allele appears to enhance crossingover between a and bp to 2.3 ± 0.9 (Phillips, 1968; *MNL*, 39, 25) but to suppress it between A^s, un, we, and pa while allowing some crossingover between these genes and fi (Phillips, 1970; *MNL*, 42, 27). No data are given but it is proposed that A^s may be associated with an inversion.

It is of interest that I. K. Egorov and Z. K. Blandova (personal communication 1970) found a marked reduction in crossingover for the intercept A^s - Sut (0.16 ± 0.11) as compared with a - Sut (4.4 ± 1.1). These observations confirm those of Phillips. Furthermore, since Phillips has found that crossingover is either suppressed or reduced only on the un side of a, this is suggestive that Sut is the same side. This gives rise to the possibility that Sut is an allele of un (or part of a complex) in view of the closely similar crossover values of the two genes with a.

Several of the gene pairs have segregations involving a translocation. In some cases, there is evidence that the frequency of crossingover is affected by the presence of the translocation; in other instances, there is no apparent effect (Table 11.16). The prime factor is distance of intercept from the break point. A decrease in the amount of crossingover is the rule and this is evident for the intercept a - pa, for one of Carters *et al.* (1956) translocations (T(5;?)2Ca) but not for another (T(5;11)7Ca), a - we (Carter *et al.*, 1956) and probably pa - Ra (Carter *et al.*, 1956). The decrease for the latter is not formally significant but of such a size that some effect cannot be discounted. The intercept a - Ra curiously showed an increase in the amount (Carter *et al.*, 1956). On the other hand, the amount of crossingover for a - pa, in the case of Snell's

(1946) experiments and for one of Carter's *et al.* (1956) translocations, and for *fi - Sd* (Carter *et al.,* 1956) showed no signs of disturbance. These have been incorporated into the general analysis.

The centromere position for group V has been determined by two different workers employing three different translocations (Snell 1946, Searle, 1968; *MNL* 39, 25; Searle and Beechey, 1970; *MNL,* 43, 29). All three experiments indicate that the centromere is located at or beyond the *Sd* end of the group. It is not possible to be more precise at present.

The gene *hy-1* (now apparently extinct) is sometimes assigned to group V although the evidence is slender. Genes *hy-1* and *pa* show linkage of 33.4 ± 6.8 in a single RII segregation (Clark, 1936). This segregation is suspect because both genes display sub-viability and it is possible for the double homozygote *hy-1hy-1papa* to be deficient due to a viability interaction. If so, then a linkage distribution of class frequencies would be simulated. Without a balancing CII segregation, the matter cannot be resolved. The *hy-1* gene fails to show linkage with *a* but this is inconclusive. The intercept *a - pa* is about 16 units long and if *hy-1* is linked to *pa* on the opposite side to *a,* the sum of the two crossover values would approximate that for independent assortment.

Group VI (bt, Ca, hl, Ht, med, mn, N, Sha, sw, uw, Ve)

This group was formed by the discovery of linkage between *Ca* and *N* (Cooper 1939). Additions to the group have not been as rapid as for some others and the present situation is shown by the trigon of Fig. 11.6. The number of entries is barely able to sustain the ordering of the loci. Despite this drawback, the arrangement is reasonably consistent. The group appears to be characterized by low crossover values between the member loci.

Those pairs of loci which have segregations of more than one type or phase or with observations by more than one worker have given the following mean crossover values:

bt - Ca	7.0 ± 0.3	*bt - uw*	40.6 ± 1.7
bt - hl	7.9 ± 0.4	*bt - Ve*	9.0 ± 1.0
bt - Ht	6.4 ± 0.5	*Ca - hl*	1.9 ± 1.9
bt - N	6.8 ± 0.4	*Ca - Ht*	2.5 ± 0.3

Ca - N	0.59 ± 0.08	N - Sha	1.8 ± 0.2
Ca - uw	43.9 ± 3.0	N - Ve	1.8 ± 0.7

Thus far, in experiments with Sha, a direct estimate of the crossover value has only been obtained between N and Sha. Three points tests by Flanagan and Isaacson (1967) involving Ca, N, and Sha gave no crossovers between Ca and Sha but 12 crossovers between N versus Ca or Sha. This implies that N cannot lie between Ca or Sha but either of the orders Ca - Sha - N or Sha - Ca - N are feasible. The absence of crossingover between Ca and Sha implies that Sha is closer to Ca than to N.

Only tentative estimates are available for several loci. Searle (1970; appendix) remarks that Ca and med show close linkage. Lane (1970; appendix) similarly notes that there is close linkage between Ca and sw. Mallyon and Wallace (1970; appendix) conclude that mn is about 20 crossover units from Ca and N.

Separate determinations have been made of crossingover between the sexes for twelve intercepts. In nine, the male revealed a higher rate of crossingover, in two, the female showed a slightly higher rate; while, in the tenth, there was a tie. Out of the nine cases where the male has the greater amount of crossingover, five of these are significant. The loci concerned are fairly closely linked and are as follows: bt - Ca (6.9 ± 0.6), bt - N (8.6 ± 0.8), ca - hl (3.3 ± 0.5), Ca - Ht (3.2 ± 0.7), and Ca - N (0.93 ± 0.18). These are remarkable and, at first blush, would appear to be in opposition to the general tendency for crossingover to be greater in the female. Apart from the general consideration that this tendency may not hold for all chromosomes (or parts thereof), it may be remembered that Carter's (1954) analysis of chiasmata distribution and interference in the mouse has revealed the possibility that

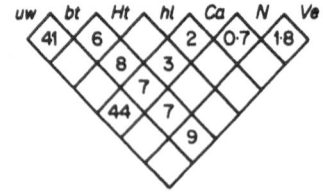

Fig. 11.6. Linkage group VI of the house mouse.

crossingover in the male may exceed that of the female for a short region of each chromosome. The region is about 30 crossover units from the centromere and extends for about 20 units. The bt - N intercept of about seven units could easily be accommodated in such a region. Carter cited the results for bt - Ca - N as possible verification of his analysis. The addition of three loci (hl, Ht, and uw) have not weakened the tendency for the region and Carter's remarks are reinforced. In fact, the intercept bt - uw, which may be held to be outside of the region, shows a greater (though insignificant) frequency of crossingover in the female.

Group VII (Al, co, df, Es-3, lt, nu, oe, Re, sh-2, shm, ti, tn, Tr, rb, vt, wa-2)

Linkage between the genes $sh-2$ and $wa-2$ laid the foundation for the group (Snell and Law, 1939). The trigon of Fig. 11.7 shows the probable linear order of the loci although there are several inconsistencies in the diagonal sequences of crossover values. Some of these are doubtless due to chance fluctuation (e.g., for the closely linked genes $sh-2$ and vt in comparison with Re) but others could be indicative that some of the loci may require interchanging. Further data are needed for clarification in the latter event, especially in view of the irregularities of assortment to be discussed anon. Re is closely linked to shm (1.5 ± 7.0) but which side is unknown. Similarly, the genes df, $Es-3$, and lt are linked to Re but the side is unknown. The co mutant has recently been found to exhibit linkage with oe (A. C. Peterson, F. G. Biddle, N. S. Virgo, and J. R. Miller, personal communication 1970).

The following mean crossover values were derived for those loci with segregations consisting of more than one linkage type or which have been investigated by more than one worker:

Al - Re	16.0 ± 1.9	oe - Re	16.6 ± 1.5
Al - $sh-2$	26.6 ± 1.4	oe - $sh-2$·	17.1 ± 1.2
Al - vt	44.8 ± 5.3	oe - tn	42.8 ± 3.9
Al - $wa-2$	39.5 ± 2.3	oe - $wa-2$	36.1 ± 2.3
df - Re	22.5 ± 2.2	Re - $sh-2$	23.3 ± 2.1
nu - Re	15.0 ± 3.1	Re - shm	1.5 ± 7.0
nu - Tr	7.4 ± 3.6	Re - ti	19.8 ± 1.5

Re - tn	22.7 ± 2.5	*sh-2 - vb*	35.0 ± 6.2
Re - Tr	21.5 ± 2.1	*sh-2 - vt*	17.0 ± 1.7
Re - vb	18.0 ± 2.6	*sh-2 - wa-2*	25.0 ± 0.7
Re - vt	20.1 ± 1.8	*ti - vt*	8.9 ± 1.6
Re - wa-2	42.7 ± 1.7	*ti - wa-2*	29.4 ± 1.9
sh-2 - tn	44.4 ± 4.9	*Tr - wa-2*	33.3 ± 5.4
sh-2 - Tr	2.9 ± 2.0	*vt - wa-2*	24.8 ± 0.8

Several of the gene pairs have displayed notable irregularities of assortment. In some cases, this has led to heterogeneity of the data with respect to estimation of the crossover value. Just how far this may have biased the final estimate remains to be seen. In general, the bias is probably negligible but, in others, the effect may not be insignificant. Data on the assortment of *Al* and *Re* constitutes a pertinent case. Dickie (1955) reported the following crossover values for three different backcrosses in repulsion 11.4 ± 4.8, 30.4 ± 3.9, and 45.1 ± 7.0, and Lane (1970; personal communication) a crossover value of 6.7 ± 2.5. The mean is 16.0 ± 1.9, seemingly the 'best' value which can be devised but not necessarily a good approximation of the true value. Examination of the trigon suggests that the value of 16.0 is probably too large. Lane's estimate may turn out to be the most reliable. Dickie has proposed that the affinity phenomenon may be at the root of her curiously contradictory segregations. In all, *Al* behaved in an associative manner with six other genes belonging to five linkage groups. The assortment of *df - Re* displayed a significant difference of recombination between coupling and repulsion phases (34.4 ± 4.4 versus 18.3 ± 2.6; Bartke, 1970; appendix). Affinity could be an

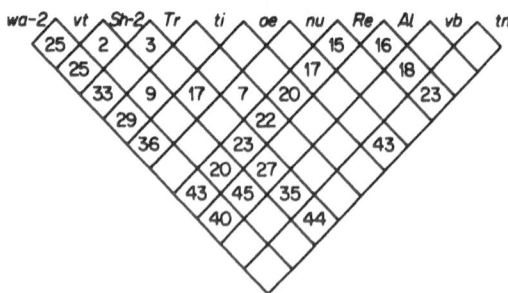

Fig. 11.7. Linkage group VII of the house mouse.

explanation for these differences. Yet, while *Re* has been cited on more than one occasion for displaying an affinity effect, the gene has not actually featured so prominently in this connection as several others.

Michie (1955a) analysed his data on the *Re - vt* intercept for a possible trend with age in the amount of crossingover. A slight trend towards a decrease with age was indeed noticeable but the amount was statistically insignificant.

Crossover values for each sex have been reported for eight intercepts. In four, the male heterozygotes exhibited the higher frequency while, in four, the female had the higher. Only one (*sh-2 - wa-2*) of the male cases was significant and the difference (5.8 ± 1.7) was less than those cases where the female had the higher frequency. Three of the four female cases are formally significant; namely, *Re - sh-2*, (8.7 ± 3.1), *Re - tn* (7.2 ± 2.1), and *Re - vt* (10.7 ± 3.3).

The data of Carter *et al.* (1956) for the *sh-2 - wa-2* interval were obtained from experiments with a translocated chromosome. However, the crossover value obtained (23.3 ± 1.4 from female heterozygotes) is so close to the mean (23.6 ± 0.9) exclusive of Carter's *et al.* data, that it is improbable that the presence of the translocation had any effect. Carter's *et al.* results were therefore incorporated into the computations for the female and general means.

Heston (1941) reported an association between the susceptibility to induced pulmonary tumours and the *sh-2* and *wa-2* loci. Later, Heston *et al.* (1952) pin-pointed the association to *sh-2* but concluded that the susceptibility was probably due to a side-effect (reduced body weight) of the gene itself rather than to linkage between it and a 'susceptibility' gene. The *wa-2* did not seem to be involved *per se;* the association arising from the linkage of *sh-2* and *wa-2*. In 1953, Heston concluded that he had detected true linkage between a major locus for pulmonary tumour susceptibility and *vt*. Tatchell (1961a) has reported evidence for a significant association between the susceptibility and *vt* and *wa-2*. The linkage was loose and of a similar magnitude for each gene (44.6 ± 2.4 and 43.8 ± 2.4, respectively), upon the assumption of the segregation of a major susceptibility gene. However, Bloom (1961) is dubious that the evidence will bear such a precise interpretation. The susceptibility to tumours is basically polygenic and the apparent segregation of a major gene may have been a 'freak' effect, peculiar to the experiment. Tatchell (1961b) maintained that the existence of a major locus has been *prima facie* demonstrated but that a more exacting analysis is certainly required.

In 1947b, Wright published the finding that both *sh-2* and *wa-2* show a weak disassociation with sex of about 56 per cent. The data were well balanced and sufficiently extensive to establish the significance of the excess recombination. However, Carter and Phillips (1953) failed to confirm the partial sex-linkage in any straightforward sense. Some sex association was discovered but the effect was noted in female heterozygotes as well as in males. This fact alone would negate true sex-linkage. Subsequently, Carter *et al.* (1954) found that the VII linkage group was involved in the T(1;7)8Ca translocation and Slizynski (1954) observed karyologically that the translocation configuration was independent of the *XY* bivalent. Carter and Phillips (1953) speak briefly of 'pseudo-sex-linkage' in connection with Wright's and their own observations but could not offer any worthwhile explanation. The design of the experiment tended to eliminate affinity as a possible explanation. Michie (1955c) feels that affinity could explain at least some of Wright's results but agrees that affinity is a doubtful explanation for those of Carter and Phillips. Tatchell (1961a) failed to detect an association with sex for the genes *Re, vt,* or *wa-2*.

Group VIII (an, asp, b, db, dep, Gpd-1, Lv, m, Mup-1, pf, Pt, sno, vc, wd, wi)
Group VIII was founded in 1942 by Hertwig with the genes *an* .and *b* displaying a mean crossover value of 5.0 ± 1.2. Since then, many other loci have been added to the group, particularly *b* and *m* which have proved to be most useful for tying in new mutant genes. The trigon provides a convenient summary for most of this work. It also gives the most probable order consistent with the present data although, for the usual reasons, some of the adjoining loci may yet have to be interchanged.

The following gene pairs have crossover values estimated

from more than one assortment and have the means:

an - b	6.7 ± 1.1	*b - sno*	16.8 ± 0.8
asp - b	41.6 ± 1.3	*b - ve*	6.7 ± 1.2
b - db	9.9 ± 2.5	*b - wd*	32.9 ± 2.4
b - Gpd-1	31.7 ± 3.1	*b - wi*	3.8 ± 0.8
b - m	7.5 ± 0.6	*m - Ps*	3.5 ± 3.7
b - Mup-1	5.3 ± 1.5	*m - Pt*	3.5 ± 0.5
b - Ps	8.2 ± 1.5	*Pf - Pt*	15.7 ± 3.8
b - Pt	4.6 ± 3.1	*Pt - wi*	9.0 ± 0.8

It is impossible to place one gene in the map. This is the *dep* gene which displays linkage with *Pt* (6.5 ± 2.4). The chromosome map is long enough to accommodate an intercept length of this size on either side of *Pt*, even on the principle of minimizing the total length.

Eight of the gene pairs have data on crossingover between the sexes. In six cases, the female has the greater amount of crossingover. In the case of *Pt - wi* (6.1 ± 1.3 versus 11.0 ± 1.1), the difference is significant.

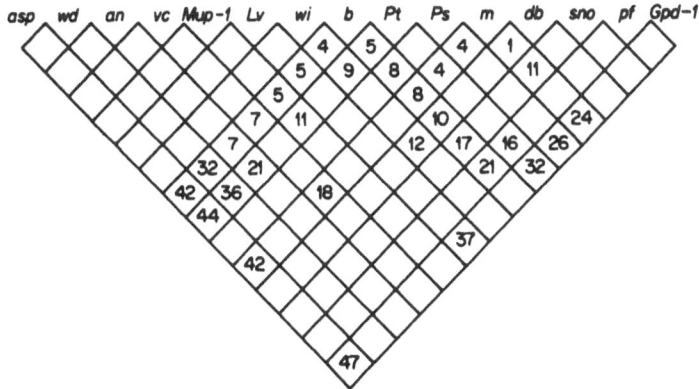

Fig. 11.8. Linkage group VIII of the house mouse.

It is of interest that Yoon (1961) bred a series of litters from females kept in a temperature of about 10°C above normal and obtained a crossover value of 46.9 ± 5.1 for the intercept *b - wd*. This compares with 32.5 ± 3.7 for females of the same phase bred under normal conditions. The difference (15 ± 5.6) is significant.

The observations of Snell (1946) were from segregations involving a translocation of the VIII chromosome. However, the break was sufficiently distant (about 29 units from *b*) that no obvious disturbance is apparent for the *b* - *m* intercept. The data have been included in the general analysis.

Strong (1946) has concluded that a locus concerned with susceptibility or resistance to methylcholanthrene-induced fibrosarcoma is,linked to *b*.

The centromere has been located at the *asp* end of the group (Snell, 1946; Ford *et al.*, 1956; Searle 1968; *MNL*, 39, 25). At present, it has not been possible to pin-point the location more closely.

Group IX (Fu, H-2, Ir-1, Ki, Low, qk, Slp, Ss, T, tf, Tla)
This group is of interest because it contains two loci (namely, *H-2* and *T*), situated moderately close in terms of crossover units but possessing different functions, in which the intra-locus complexity is being subjected to intensive analysis. The trigon of Fig. 11.9 shows the probable order for the map, as determined by crossingover values and special observations by individual investigators. These latter consist of observations which point towards the linear order but are not of such a nature (or are not precise enough) to be expressed in crossover units. These feature prominently for the group because of the peculiar complexity of *H-2*. The gene *qk* has been shown to be linked to *T* but the side has yet to be decided.

Those gene pairs with more than one type of assortment data and contributions from different investigators have mean crossover values as follows:

Fu - *H-2*	4.6 ± 1.1	*Ki* - *T*	3.3 ± 0.7
H-2 - *Ki*	7.6 ± 1.6	*Ki* - *tf*	0.56 ± 0.23
H-2 - *T*	10.8 ± 0.6	*T* - *tf*	5.5 ± 0.3
H-2 - *tf*	3.6 ± 0.6		

Genes *Fu* and *Ki* are either alleles or are extremely closely linked. Dunn and Caspari (1945) did not observe any crossovers in 105 RBB progeny from male heterozygotes but five possibles out of 140 RBB progeny from female heterozygotes. This gives a combined crossover value of 2.0 ± 0.9. However, these observations are not regarded as reliable and the five possibles could be normal overlapping of either *Fu* or *Ki* (Dunn and Gluecksohn-Waelsch, 1954). A subsequent and meticulous investigation by these authors (which entailed progeny testing of all suspected crossovers) failed to uncover a single crossover among 971 testcross young. The upper limit at the five per cent level for this sample size is 0.38 (including the negative results from the male heterozygotes of Dunn and Caspari, the upper limit is 0.09 per cent). Dunn and Gluecksohn-Waelsch concluded that the two genes are alleles. This is a reasonable conclusion although it must be recognized that *Fu* and *Ki* could be part of a gene complex which includes the closely linked gene *T*. It should be noted that the trigon shows a pattern of entries to be expected if *Fu* and *Ki* are closely linked but separable. However, the pattern may be fortuitous.

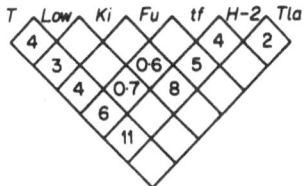

Fig. 11.9. Linkage group IX of the house mouse.

The complex nature of *H-2* was suspected shortly after the existence of the locus had been placed on a firm basis. This was indicated by the interrelated antigenic specifities of the alleles, as shown for example by Snell *et al.* (1953). For technical reasons, progress was slow at first but has recently accelerated, as witness the comprehensive reviews of Snell *et al.* (1964) and Shreffler and Snell (1969). Eighteen different alleles are now recognized and at least 25 different antigenic specifities.

The first detectable crossover types within the *H-2* complex date from Allen (1955b), Amos *et al.* (1955), and Gorer and Mikulska (1959). Subsequent reports not only confirmed the earlier conclusions but have extended the number of recombinations between different specifities. Thus, the fine structure of *H-2* is steadily being laid bare, although the genetic aspects will inevitably lag behind the immunological, if only because of the labour which is required to accumulate the quantitative data. The first tentative map for *H-2* was presented by Gorer and Mikulska (1959). Even then, four regions could be defined (viz., 4, 13(D), 3(C), 22(V), 11(K)). Initially, letters were used to denote the specifities but numbers are now employed; letters are now used to denote the main regions (Snell *et al.*, 1964).

Estimation of the amount of recombination poses a problem of numbers. Indeed, sensible estimates can only be derived by summing the contribution of many workers. The greatest amount of data is available for the D and K regions which, on current thinking, appear to constitute the extremes of the complex. A summation of the data gives the crossover value of 0.40 ± 0.09 for the complexity based upon 5264 testcross offspring (Allen, 1955b; Gorer and Mikulska, 1959; Pizarro *et al.*, 1961; Boyse *et al.*, 1964; Shreffler, 1964b, 1965; Stimpfling and Richardson, 1965; Shreffler *et al.*, 1966; Snell *et al.*, 1967; and Pizarro and Dunn, 1970). The five per limits are 0.61 and 0.25 per Steven's table. A sex difference is apparent; since crossingover for the male is 0.27 ± 0.09, compared with 0.62 ± 0.15. The difference (0.35 ± 0.20) is not significant although in the expected direction of more frequent crossingover for the female.

Stimpfling and Richardson (1965) have published carefully determined data not only on the D-K regions but also for the intervening regions D-C and C-K. The respective crossover values are: males, 0.09 ± 0.19 and 0.06 ± 0.02, and females, 0.25 ± 0.17 and 0.05 ± 0.05 (means for both sexes, 0.21 ± 0.09 and 0.05 ± 0.04, respectively). The data for the D-K interval gave the values 0.19 ± 0.14 and 0.55 ± 0.221 approximately in line with the overall means given above. Shreffler (1965) has observed crossingover for the D-K interval and also for two others. These are D-E with 0.16 ± 0.11 of crossingover and E-K with 0.20 ± 0.14 of crossingover, E being a fifth defined region of specifities. The accepted order for the regions are D, C, E, and K (Fig. 11.10). V is not shown in the figure but is currently placed between C and E.

Shreffler and Owen (1963) initially established that the *Ss* locus was either identical with or part of the *H-2* complex. Later, Shreffler (1964b, 1965) found that *Ss* is located within the E-K region. Crossingover was observed between the D - *Ss*,

E - *Ss*, and K - *Ss* intervals to the order of 0.18 ±b 0.13, 0.10 ± 0.10, and 0.18 ± 0.13 respectively. ·Allen (1955b) has speculated briefly on the possible order of the D and K components relative to the *Fu* and *T* loci but her data could not discriminate between the alternatives. However, Pizarro and Dunn's (1970) observations point towards the order *T* - *tf* - K - D. All of the details of the preceding few paragraphs, together with a few relevant items contributed by Shreffler (1967), lead up to the depicture of Fig. 11.10.

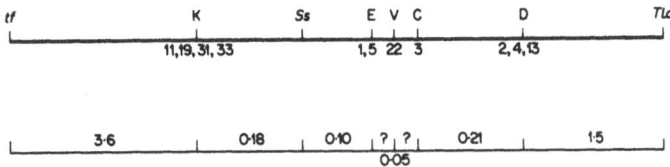

Fig. 11.10. Diagram of the *H-2* region of linkage group IX of the house mouse.

The discovery that *Ss* lies within the *H-2* complex is of unusual interest since, unless the function of *Ss* can be related to the histocompatibility and immunological properties of *H-2*, it means that the region is not a simple series of antigenic sites. Possibly this would be too simple a view in any case, for there is evidence that the region is genetically versatile. Passmore and Shreffler (1968, 1970) and Klein *et al.* (1969) have described a sex-limited serum protein variant (expressed only in males). The gene (*Slp*) is closely linked to *H-2*. In fact, the only crossovers observed have been intra-*H-2* and indicate that *Tlp* is probably located close to *Ss*. Also, McDevitt and Tyan (1968; Herzenberg *et al.*, 1968) found that an immune response represented as *Ir-1* is linked to *H-2*, with only the dubious possibility of crossingover in an (as yet) not very large sample.

One of the two genes (*rgv-1*), proposed by Lilly (1966) to be involved in the susceptibility to the Gross leukaemia virus, has been found to be closely linked to, or be part of, the *H-2* complex. Lilly pointed out that *rgv-1* was associated with the K component in his material; rather similarly, in fact, to the distributional association of *Ss* and K in the investigations of Shreffler and Owen (1963). He speculated very briefly that *Ss* and *rgv-1* might be interconnected or possibly identical.

In view of recent developments, it seems advisable to

consider the suggestion that the D and K components are in need of recasting as histocompatibility loci in their own right (Shreffler, 1967). Crossingover between D and K is low but comparable in frequency with other pairs of closely linked genes (such as *d* and *se*) which are accepted òn phenotypic grounds as belonging to distinct loci. This aspect is tacitly recognized in effect if the symbolism *H-2.4* and *H-2.11* becomes universally adopted as representative of the D and K regions.

The *T* locus may be regarded as complex because of the genetic behaviour and apparent differential composition of its *t* type alleles. However, a series of mutants to *T* type alleles have been treated as if they are repeats or true alleles of *T* on parsimonial grounds. These are *T^f* (Fitch's brachury; Dunn *et al.*, 1962), *T^h* (T-Harwell; Lyon 1959), and *T^c* (curtailed; Searle, 1966). The many and remarkable *t* alleles have been frequently analysed. A thorough-going review is that of Dunn (1956) while a recent summary is provided by Dunn *et al.* (1962). The complex nature of the *t* alleles has been analysed by Lyon (1960b, 1961c), Dunn *et al.* (1962) and Lyon and Meredith (1964a, b). Seven features are ascribed to the alleles as a group:

(1) Modifying effect on *T* (short-tailed to taillessness, usually).
(2) Homozygous lethality.
(3) Male sterility.
(4) Unusual male segregation ratio.
(5) High mutation rate to other *t* alleles.
(6) Suppression of crossingover between *T* and *tf*.
(7) Complementation effects between many of the alleles.

The above is suggestive that the locus could be a complex, with the implication provided by the lethality that the chromosomal region could be abnormal. The most developed conception (Lyon and Meredith, 1964b) is that of a region extending from *T* to at least *tf*, this being the distance primarily affected by the suppressor effect. Evidence is brought forward to indicate that the following effects could be recombined between alleles: absence or presence of modification of the *T* phenotype, lethality versus viability, variation in male segregation ratio, and amount of crossover suppression. The assumed high mutation rate could be better explained by crossingover between or rearrangement of units of the complex. On this criterion, the lethal factor is probably distal from T, possibly located beyond tf, although this is tentative at present. The suppression of crossingover is indicative of an inversion or a deficiency which, in turn, also finds expression as the lethality.

In calculating the crossover values between T and other loci, only data from segregations with T (or similar alleles) are utilized, because the t alleles tend either to suppress or distort crossingover. However, there are exceptions. Dunn and Gluecksohn-Waelsch (1953b) found that t^3 did not suppress crossingover since the crossover value for the Ki - t^3 intercept was 5.3 ± 3.0. Lyon and Phillips (1959) found that t^{13} apparently did not suppress crossingover in the female (11.0 ± 2.0) but did so in the male (2.9 ± 1.4) for the T - tf intercept. Pizarro and Dunn (1970) may have found another allele (t^{w35}) with similar properties. Dunn et al. (1962) and Lyon and Meredith (1964a, b) have since described numerous t alleles which allow either partial or normal crossingover between T and tf. Summing the results in these reports for 20 such alleles gives the values $129/2889 = 4.5 \pm 0.4$ for the male and $184/2012 = 9.2 \pm 0.6$ for the female; with a mean of 5.7 ± 0.3. These compare remarkably well with the means based on the T alleles of 4.3 ± 0.4, 9.7 ± 0.7, and 5.5 ± 0.3, respectively. An example of absence of crossingover in an adequate sample is provided by results with the allele t^{w8} which have no crossovers among 474 testcross offspring (Dunn et al., 1962).

In many cases, the expression of crossingover is associated with viability of t but, not completely, for Lyon and Meredith (1964b) have isolated two lethal alleles which did not suppress crossingover. This combinational feature of the properties displayed by the t alleles, together with examination of the sequence by which the various t alleles have arisen from one another, forms the basis for the concept of the abnormal region. Finally, it is by no means uncertain that some of the alleles may have arisen by unequal crossingover. This could be due to the disturbance introduced by the postulated inversion or deficiency. If the latter, it may be an academic point which came first, the deficiency or the unequal crossingover.

A karyological investigation by Geyer-Dusynska (1964) has

contributed evidence for deficiencies associated with certain of the t alleles. No abnormal configurations were shown. by chromosomes bearing either T or t^{w1} or, if so, were too minute to be detected. The deficiencies shown by t^{w6} and t^{w18} are depicted as being smaller than that shown by t^o. The t^{12} allele is depicted as being associated with either a small instial deficiency or a large terminal deficiency. Some of the configurations are suggestive of duplicative regions which exaggerated the asynapsis. However, Geyer-Dusynska points out that the assumption of deficiencies of various sizes, according to the particular t allele, would be sufficient to produce non-homologous regions, without the additional although complementary postulation of duplication.

Latterly, Dunn and Bennett (1970, personal communication), have found recombination between T and a sex-limited factor *Low*. *Low* reduces the transmission of T and *tf* in the male to about 13 per cent (instead of the normal 50 per cent) and appears to be sited close to *tf*. The crossover values for *Low* given in the trigon are based on Dunn and Bennett's observations, which define the order of the genes and *Low* but do not give precise values of the distance. The factor reduces crossingover for the T - *tf* intercept to 2.7 ± 0.3 in the male and 6.5 ± 1.7 in the female. It will be interesting to discover if the low ratio and reduction of crossingover are related to the same mechanisms which produce modified transmission ratios and reduced crossingover for the t alleles and, as a corollary, whether or not the various effects can be separated from *Low*.

Six of the intercepts have information on the frequency of crossingover between the sexes. In each instance, crossingover is greater in the female. The differences are significant for *H-2* - T and T - *tf* and probably would be for a few other gene pairs had the sample sizes been larger.

The centromere is almost certainly located at or beyond the T end of the linkage group (Lyon et al., 1968: Lyon and Hawkes, 1970; *MNL*, 42, 27).

Group X (gr, ji, kd, v)

Snell (1945) observed linkage between *ji* and *v* and, as a consequence, brought the present group into being. The mean linkage value is 15.0 ± 3.8 but this is not based upon particularly homogeneous data. A limited amount of testcross data gives the slightly lower, but possibly more reliable, value of 12.4 ± 4.2.

Bloom and Falconer (1966) found linkage between *gr* and *v* of 14.9 ± 2.6 while Green and Woodworth (1970, appendix), reported a linkage value of 8.1 ± 5.2 between *gr* and *ji*. The order *gr - v - ji* is made improbable on these data and, of the other two alternatives, the order *gr - ji - v* is marginally preferable to *ji - gr - v* but a decisive choice must wait upon further data.

The gene *kd* is linked to *gr* with a crossover value of 16.4 ± 5.0 or 24.1 ± 5.7, depending upon whether the curious high value of 37.6 ± 23.9 for one segregation is rejected or accepted (Hulse and Lyon 1970, appendix).

The data of Snell (1945) on the *ji* and *v* interval shows a large sex difference (29.8 ± 16.1) in crossingover but the sample numbers are too small for the result to have any worthwhile significance. The high value for female gametogenesis has the appearance of being anomalous.

Group XI (Cd, dw, Hd, Lc, Ldr-1, Mi, ob, px, Sig, tc, wa-1)

This group came into being with the discovery of linkage between *Mi* and *wa-1* by Bunker and Snell (1948). Few other loci were assigned to the group until quite recently. In fact, much of this work has yet to be published in detail. The accompanying trigon shows the probable order of the loci. Not necessarily the most probable because at least one pair of loci could be interchanged without materially destroying the internal consistency of the sequencies of crossover values.

Only a few of the crossover values rest upon more than one segregation and those that do have the following means:

Lc - Mi	11.2 ± 0.7	*Mi - wa-1*	4.7 ± 6.7
Lc - wa-1	6.9 ± 0.9	*Mi - tc*	5.8 ± 0.8
Mi - ob	25.3 ± 4.4		

The genes *Cd* and *Ldr-1* have displayed linkage with *Mi*, with crossover values of 11.5 ± 4.4 and 28.7 ± 4.4, respectively. As yet, these genes should not be included in the map but, on the principle of minimizing the length, *Cd* could provisionally be inserted between *Hd* and *Lc*, and *Ldr-1* likewise between *dw* and *ob*. Genes *Lc* and *px* are subject to impenetrance. An allowance is made for this in the estimation but some bias may still remain for those segregations involving one or the other gene. P. W. Lane (personal communication 1970) reports close linkage between *tc* and *wa-1* but gives no numerical data.

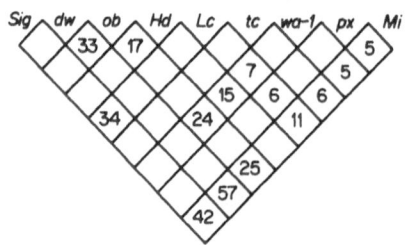

Fig. 11.11. Linkage group XI of the house mouse.

Genes *Lc* and *Mi* have been involved in the study by Carter *et al.* (1956) to localize the position of translocation breaks. Repulsion test-crosses between the two genes give a frequency of crossingover of 6.82 ± 2.7 for the male and zero crossingover for a progeny of 13 for the female. This rate of crossingover is not significantly different from the mean value for male heterozygotes but, since a disturbance could easily be present due to the proximity of the break in the chromosome, these data have not been used.

Separate determinations of the frequency of crossingover between the sexes have been reported for five pairs of genes. Of these, four show a greater frequency for female heterozygotes while the fifth shows no difference.

A gene producing the general condition of 'eyes open at birth' has been described by Bennett and Gresham (1956). The gene is linked to *wa-1*, with a crossover value of 9.22 ± 2.0 for males and 5.34 ± 1.4 in females. Some doubt exists on the status of this gene. The condition is briefly mentioned by Green (1966) but not as a monogenic trait in its own right. The linkage with *wa-1* is indirectly mentioned. Bennett and Gresham denote the gene as *eo* but this symbol has been used by Gower (see Green, 1966) for a mutant form designated as 'eye opacity'. The position evidently requires clarification; not only genetically but also as symbol nomenclature if one or both mutants are extant. At this time, eyes open at birth will not be formally listed as a gene in group XI, although a *prima facie* case has been made.

The centromere has been shown to be at the *Sig* end of the linkage map by the work of M. F. Lyon and S. Hawkes (1970; *MNL*, 42, 27).

Group XII (bm, Dc, ep, ru)

The first linkage for this group was between *Dc - ep* with a crossover value of 34.7 ± 4.9 (Deol and Lane, 1966). The gene pairs *ep - ru* and *bm - ep* were soon added, with values of 1.6 ± 0.3 and 8.6 ± 3.5, respectively (Lane and Green, 1967; Lane and Dickie, 1968). The data are too meagre for an obvious order to be apparent; *bm* may lie between *Dc* and *ep* but this can only be regarded as a pointer for future studies.

Separate information is available for crossingover between the sexes for *ep - ru*. For the male, the crossover value is 2.6 ± 0.6 versus 1.2 ± 0.4 for the female. The difference (1.4 ± 0.7) is formally significant but only just so. The difference is of interest because it is in the opposite direction (greater amount of crossingover in the male) to that normally expected. However, in view of the bare significance of the difference, it may be wise to wait for confirmatory data before too much is made of the result.

Linkage group XII was first proclaimed by Fisher and Snell (1948) for presumptive weak linkage of 44.7 ± 1.9 between *je* and *ru*. However, Wallace (1958a) showed that the data were definitely ambiguous. Affinity emerged as an equal, if not better, explanation. Lane and Green (1967) have contributed data which show independent segregation of *ep* and *je*. There is, therefore, no justification of retaining *je* in the group at this time.

Group XIII (Dh, dr, dt, fz, Id-1, ln, Lp, py, Sp, tb, th, vl)

This group owes its inception to the notification of linkage between *fz* and *ln* and the possibility of linkage between *fz* and *Sp* by Dickie and Woolley (1950). The present situation is shown by Fig. 11.12. The usual cautions apply, however, particularly for the genes *dr*, *py*, and *Lp*. The available information on these genes is not as precise as would be desirable and it is possible that new data may require some interchanges.

The following pairs of genes have data of different types and phases, contributed in many instances by different workers, and these give the mean crossover values:

Dh - fz	41.8 ± 1.0	fz - Sp	36.3 ± 0.6
Dh - ln	3.3 ± 0.4	ln - Lp	32.5 ± 4.2
dr - ln	32.4 ± 2.5	ln - py	31.8 ± 0.6
dt - fz	23.0 ± 4.6	ln - Sp	4.7 ± 0.1
dt - ln	32.5 ± 3.5	ln - th	1.9 ± 0.4
fz - ln	39.2 ± 3,5	py - Sp	34.6 ± 0.7
fz - Lp	52.4 ± 3.7	Sp - th	5.7 ± 0.8
fz - py	49.4 ± 0.7		

A suggestion of between data heterogeneity is evident for *dt - fz*. Falconer and Isaacson (1965; appendix) reported a crossover value of 15.6 ± 5.6 whereas Kelton (1965; appendix) reported a crossover value of 35.6 ± 7.8. The difference (20.0 ± 9.6) is just formally significant. The two experiments were both three-point repulsion intercrosses involving the genes *dt, fz,* and *ln.* Crossover values for *dt - ln* and *fz - ln* differed between the two crosses but not significantly, and, in the case of *fz - ln,* within the range reported by other investigators. It seems likely that the rather large difference for the *dt - fz* is due to chance. If so, then the value from Kelton's data is probably spuriously high. The trigon implies that a lower value would produce a better fit.

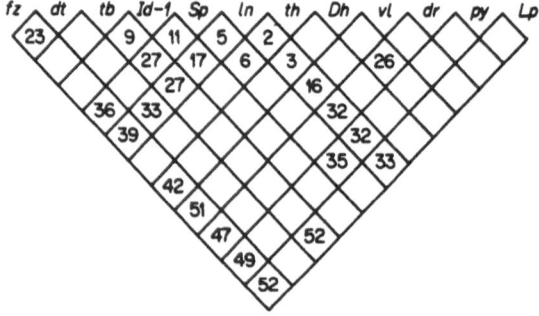

Fig. 11.12. Linkage group XIII of the house mouse.

Variation of the crossover value with age of parent for the *fz - ln* intercept has been examined by Reid and Parsons (1963). The results do not display a significant trend. That for the male shows a slight increase for the 3 to 15 months period while that for the female shows little change except for the last 11 to 15 month sub-period.

The *fz - ln* interval has also featured in the analysis of translocations performed by Carter *et al.* (1956). In only one of the three crosses involving the translocations is there significant evidence that the translocation interfered with crossingover between the two genes (Table 11.16; translocation T(5;13)83Ca). This datum, therefore, has not been incorporated into the computations to produce mean crossover values, although those for the other two translocated chromosomes have.

Data on the sex difference of crossingover has been reported for twelve of the gene pairs. Three of these show the male as possessing the higher frequency but, in each case, the difference is insignificant. The female shows the higher frequency in the other nine; in four cases significantly so. These four cases are *fz - ln* (7.1 ± 1.0), *fz - Sp* (9.0 ± 1.2), *ln - py* (14.0 ± 1.3), and *py - Sp* (12.8 ± 1.4). The last two intercepts, incidentally, display two of the largest significant sex differences of crossingover for any chromosome.

The relative position of the centromere has been located for the group. Searle (1961; *MNL*, **40**, 25) has shown that the centromere almost certainly lies beyond the *fz* end. The distance is unknown at present.

Group XIV (bg, ch, cr, f, fs, mu, pe, sa, Xt)
This group was discovered almost simultaneously by King (1956) for the genes *cr* and *f* and by Phillips (1956) for *ch* and *f*. Many other loci have since been located in the group. The accompanying trigon summarizes the results to date and shows the most probable gene order.

Most of the intercepts have data of different linkage phases and contributions by various workers. These have been combined to give the following mean crossover values:

bg - f   26.8 ± 1.9   *bg - fs*   31.9 ± 2.0

bg - sa	7.2 ± 0.7	*f - pe*	22.1 ± 1.5
bg - Xt	0.52 ±0.34	*f - Xt*	20.1 ± 1.4
ch - f	14.4 ± 3.0	*fs - pe*	13.8 ± 1.0
cr - f	22.0 ± 1.4	*mu - pe*	26.4 ± 2.1
cr - pe	44.9 ± 2.9	*mu - Xt*	12.2 ± 1.4
cr - Xt	1.0 ± 0.4	*pe - Xt*	36.0 ± 1.3
f - fs	10.3 ± 0.9	*sa - Xt*	5.2 ± 1.1

Heterogeneity is manifested between the data of King (1956) and of Lyon *et al.* (1967) for the *cr - f* intercept. The mean crossover value for the former is 33.8 ± 2.8 versus 15.3 ± 1.8 for the latter. The difference (18.6 ± 3.1) is highly significant and for no outstanding reason. The King contribution consists of coupling and repulsion data, both phases yielding similar crossover values. The Lyon's *et al.* results were drawn from male and female heterozygotes, each sex yielding a similar crossover value. King's data were obtained wholly from females and, while the female portion of Lyon's data gave a higher crossover frequency than the male, the value was still much lower than that reported by King. The data of Phillips (1956) is too small (and not entirely reliable) to be decisive although these do point towards an intermediate value to those observed by King and Lyon *et al.* It is of interest that the mean of 22.0 ± 1.4 is not greatly in conflict with the sequence of values shown by the trigon. Perhaps the deviations have roughly cancelled to produce a mean not too far removed from the true value. For this reason, this mean has been used in the trigon, despite the heterogeneity.

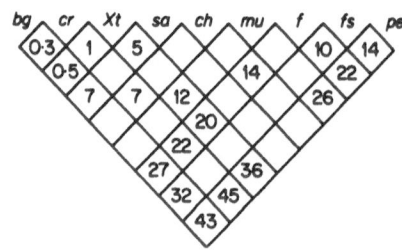

Fig. 11.13. Linkage group XIV of the house mouse.

Crossover values between the sexes have been computed for ten pairs of genes. With one exception, all of the comparisons show a higher level of crossingover in the female. Three of the differences for the latter are statistically significant; namely, bg - sa (3.3 ± 1.3), mu - Xt (10.1 ± 3.2), and pe - Xt (10.3 ± 2.7). In the case of cr - f, the sex difference is based upon the results of Lyon et $al.$ (1967), in which somewhat similar crossover values were observed for both sexes. The data of King (1956) for this pair of genes were obtained from female heterozygotes and the crossover value which these give (33.8 ± 2.8) is significantly greater than those of Lyon et $al.$ However, it is doubtful if the difference (18.6 ± 3.1) can be attributed wholly to sex and some other factor may be assumed to be complicating the comparison.

The work of Searle (1969; MNL, 41, 28) indicates that the centromere lies at or beyond the bg end of the group. At this time, it is not possible to be more definite.

Group XV (ax, Tw)

Lyon (1958) has *prima facie* established the XV group by the discovery of close linkage between ax and Tw. No worthwhile crossover value can be derived since no crossovers were observed in two crosses of 21 CII and 223 CIB young, respectively.

It is possible for the two genes to be remotely situated on a chromosome from another group. However, both ax and Tw have been extensively tested with a variety of genes from other groups with satisfactorily negative results in most cases (Lyon 1955, 1958).

Group XVI (de, ft, ma, Spa, Va)

The first linkage in this group was found by Curry (1959) for the genes de and Va. The crossover value is 27.8 ± 3.2 and is probably based on pooled data from both sexes. Three other genes have since been added with the following relationships: ft - ma (4.4 ± 0.8), ft - Va (25.6 ± 2.5), ma - Va (25.7 ± 2.7), and Spa - Va (25.6 ± 2.7) (Lane 1970, appendix). It is impossible on this data to arrange the loci in serial order with any sense of conviction. The relationships with Va would suggest that at least two of the genes are extremely closely linked but the size of the linkage between ft and ma suggests that this may not necessarily be so.

Crossingover between the sexes has been recorded for the

intercepts *ft - ma, ft - Va,* and *ma - Va* but, in each case, the difference is non-significant.

Group XVII (bf, bl, g, go, Hm, jg, le, lx, mc, Pgm-1, Ph, pi, rd, rl, Rw, W)
This group was initiated by Lane (1967) who demonstrated that the loci formerly comprising group III actually consisted of two mutually independent groups. One of these was left to compose group III but the other was reassigned to a new group. At that time, this new group was made up of the loci *g, le, lx, pi, rd, rl,* and *W.* Since then the number of loci in the group has more than doubled. Much of this work is new and has yet to be described in detail. The accompanying trigon summarizes the present picture.

The following gene pairs have mean crossover values based upon more than one batch of data:

g - rd	11.4 ± 1.8	*le - W*	13.0 ± 0.7
go - Hm	29.5 ± 0.8	*lx - pi*	12.9 ± 2.6
go - W	7.7 ± 1.3	*lx - rl*	15.7 ± 3.4
Hm - jg	40.3 ± 4.7	*lx - W*	17.0 ± 1.1
Hm - W	20.2 ± 1.0	*rl - W*	29.4 ± 3.4
jg - le	4.7 ± 1.6	*Pgm-1 - W*	2.7 ± 1.6
jg - W	19.4 ± 2.0	*pi - W*	3.5 ± 5.9
le - lx	30.0 ± 3.6	*rd - W*	12.6 ± 1.4
le - pi	18.5 ± 0.9		

It is impossible at present to insert the genes *bl, Pgm-1,* and *Ph* into the map because each have been tested only against *W.* The *W* locus is centrally placed in the group and the three genes could lie on either side of it and still be within the known map length.

The close linkage of genes *Ph, Rw,* and *W* are of interest because, taken together, they display several common or inter-relating features. A single crossover has been recorded between *Ph - W* among 1302 RBB animals (Gruneberg and Truslove, 1960) but none between *Ph - Rw* (257 RBB animals) nor between *Rw - W* (1403 RBB animals; Searle and Truslove, 1970). Searle and Truslove have given good reasons for thinking that the genes are not true alleles. They also tend to discount pseudo-allelism yet accept the likelihood that the three genes could have arisen by repeated duplication. The upper limit to the amount of crossingover which would be compatible with the number of animals examined is 1.44 per cent for *Ph - Rw* and 0.74 for *Ph - W* by Steven's method.

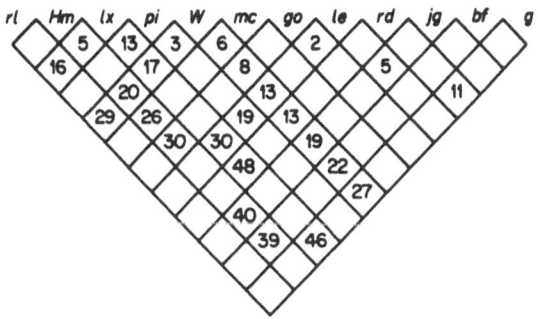

Fig. 11.14. Linkage group XVII of the house mouse.

Eight of the gene pairs have had data separately recorded for the sexes. Of these, three showed a higher frequency of crossingover for the male and five for the female. In two instances, the difference has attained significance. The female shows the greater amount of crossingover (9.3 ± 1.5) for the intercept *le - W* and rests upon ample and consistent data. The male shows the greater amount of crossingover (5.7 ± 2.7) for the intercept *pi - W* but rests upon inconclusive data. Dickie and Woolley (1946) state that crossingover in their male data is 11.4 per cent (sample of 114 animals), compared with 5.3 per cent (94 animals). These figures are internally inconsistent but can be made consistent by assuming that male crossingover is of the order of 8.8 per cent. Even this value is significantly higher than for the female data (3.0 ± 06) of Lane (1967) or of the combined data (3.1 ± 0.6) but only just so. Lane (1967) records the exceptionally low value of 0.63 ± 0.63 for male data but these are suspect because of inviability or impenetrance.

Sidman and Green (1965) have considered the question whether or not the rodless retina gene (*r*) of Keeler (1930) may be identical or allelic to their gene *rd* for retinal degeneration. The two anomalies have features in common, both developmentally and histologically, as shown by the discussions of Sidman and Green (*op. cit.*) and Keeler (1966). The main objection to acceptance for identicalness of the two genes resides in the linkage data. Gene *r* shows linkage with *si* (12.5 ± 8.8) but not with *W* (41.9 ± 7.6; Keeler, 1930) while *rd* does not show linkage with *si* (50.7 ± 2.5; DiPaolo and Noell, 1962), but does so with *W* (13.4 ± 1.7). Sidman and Green concluded that the two genes are not identical or allelic. Keeler (1966), on the other hand, has argued in favour of their being the same. He states that the *si* gene which he used originated in his own stock and could be different from the gene of DiPaolo and Noell. The absence of linkage between *r* and *W* is abscribed to chance. It now seems unlikely that a rodless mutant will be discovered in circumstances which are conclusive that *r* has been rediscovered; although a subsequent mutant linked to *si* but not to *W*, yet independent of *rd*, would be decisive to many people. In fine, an element of uncertainty must remain, despite the persuasiveness of Keeler's argument.

Group XVIII (e, Ea-1, Es-1, Es-2, Es-5, Es-6, Hk, hy-3, la, nr, Os, Q, tg)

This group was established by Green and Sidman (1962) and Green, Snell, and Lane (1963) with reports of linkage between the genes *Hk*, *Os*, and *tg*. Within recent years, the membership has swelled to 10 loci. The trigon summarizes the current situation. The trigon is internally consistent but it is clear that there is notable deficiency of crossover determinations for many of the allegedly adjacent loci. A few interchanges may be necessary when these have been determined. A major shift could involve *Es-1* and *Es-5* since these could be moved to beyond *tg* and still produce a consistent trigon. This example reveals the basic inadequacy of the present data.

The following gene pairs have data consisting of more than one segregation, giving the mean values:

e - Os	30.0 ± 1.1	Hk - Os	10.8 ± 1.8
e - Q	30.0 ± 2.7	nr - Os	24.6 ± 1.5
Es-1 - Es-2	10.2 ± 1.4	Os - Q	5.0 ± 1.0
Es-1 - Os	0.79 ± 0.56		

Linkage has been noted between the gene pairs *Es-2 - Es-6*,

Hk - Q, la - Os, and *Os - tg* but without precise evaluation being possible at this time. No crossovers of *Es-2* and *Es-6* were observed by Petras and Sinclair (1969) among 51 RII progeny but the small sample only effectively excludes very weak linkage. W. F. Hollander (personal communication 1970) states that *Hk - Q* are closely linked. The data of J. L. Southard (1970; appendix), imply fairly close linkage between *la* and *Os* but are not presented in sufficient detail for an assessment to be made of the permissible maximum. Green and Sidman (1962) and Green *et al.* (1963) failed to observe any double recessive animals among 880 RII offspring for *Os - tg.* By Steven's method, a sample of this size would exclude a crossover frequency of over 13 per cent at the 5 per cent level. This amount or less would be compatible with the arrangement of loci shown by the trigon. Foster *et al.* (1968) state that the *Ea-1* blood group locus displays linkage with *Es-1, Es-2,* and *Es-5.*

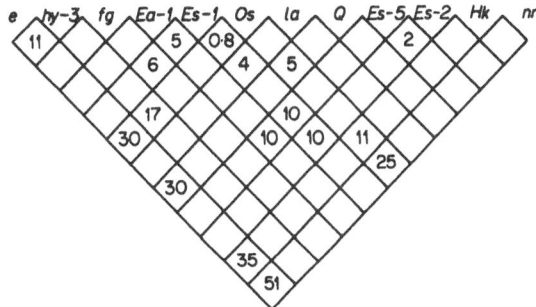

Fig. 11.15. Linkage group XVIII of the house mouse.

Six pairs of genes have data on crossingover between the sexes. In two cases, the male has the higher frequency, in two cases, the female has the higher frequency and in the last two cases, the frequency is identical for the sexes. None of the differences are significant.

Group XX (Bn, Cg, Gs, Gy, jp, k, Mo, Ms, Op, sf, sla, spf, Str, Ta, Tfm, Ym)

Group XX comprises the sex-linked *X* borne genes. Sound evidence for the transmission of a sex-linked lethal has been reported by Hauschka *et al.* (1951). However, the lethality could have been due to a deficiency, rather than to a point mutation. The first undoubted sex-linked mutants were discovered by Falconer (1952b) and Garber (1952). It may be briefly noted that *sf* was discovered even earlier (in 1949) but was not recognized as sex-linked until several years had elapsed (Russell *et al.,* 1959). At this time, at least 16 mutant loci are known for the *X* chromosome. Genes *k* (Lyon *et al.,* 1967), *Ms* (H. Krzanowska, 1966; *MNL,* 35, 35) and *Op* (M. M. Dickie 1969; *MNL,* 41, 30) have been shown to be sex-linked. This fact is sufficient to locate any gene in group *XX* whence it is solely a matter of determining its relative position.

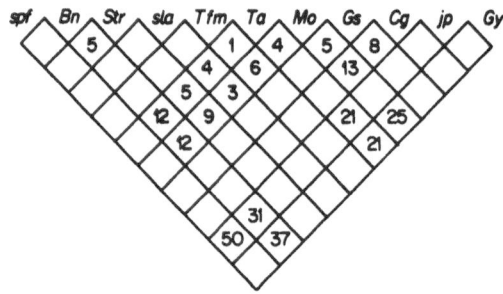

Fig. 11.16. Linkage group XX of the house mouse.

The linear order of the loci is shown by Fig. 11.16. The trigon displays a fairly consistent picture, in spite of impenetrance being a persistent (though varying) feature of some mutants. Some of the determinations may be biased in this respect. The worst offenders are *Bn* and *Str.* The crossover values suggest an ambiguity in the order of the *sla - Ta - Mo* loci but the observations of Falconer and Isaacson (1962) indicate that this is almost certainly correct. Little concrete data seems to be available in the genes *sf* and *spf.* Welshons and Russell (1959) state that *sf - Ta* shows a crossover value of approximately 44, while Cattanach (1966) gives data that demonstrated very weak linkage between *jp - spf* (50 ± 7). At the same time, he cites L. B. Russell as stating that *sf - spf* are closely linked. This information is sufficient to place the two genes on the map, albeit not precisely. *Ym* has been shown to be closely linked to *Mo* (2.6 ± 3.4). *Ta* has been shown to be linked (12.9 ± 3.2) to an unnamed gene which is briefly described as 'male lethal, female superficially tabby, not scalely when young' (Hunsicker 1966; *MNL*, 34, 41).

Those loci which have data consisting of different types and phases, and contributions by different workers, have given mean values as under:

Bn - jp	31.2 ± 3.7	*Mo - Str*	9.2 ± 1.6
Bn - Mo	12.1 ± 1.0	*Mo - Ta*	3.9 ± 0.4
Bn - Str	4.9 ± 2.0	*sla - Ta*	4.2 ± 0.9
Bn - Ta	11.5 ± 1.0	*Str - Ta*	5.1 ± 0.7
Cg - Mo	13.4 ± 1.3		

A noteworthy series of genes with similar or identical expression are *Blo, Br, Dp, Dp$_2$, Mo, To,* and *'26K'*. Although it has been suspected that all of these are either alleles or repeats, the lethality of the male for some of the genes has frustrated a direct approach. Similar crossover values with a common locus (usually *Ta*) has confirmed the suspicion. Recently, the comparative studies of Grahn *et al.* (1969) have produced good evidence in favour of multiple allelism. Here, it is assumed that all of the genes are alleles of *Mo.* It is possible that *Mo* may be a duplicative complex (in the style of *Ph - Rw - W* in group XVII), but the evidence for this is slight at present. Welshons and Russell (1959) state that *'26K'* and *Ta* show about four per cent of crossingover; while L. B. Russell (1960; *MNL*, 23, 58) has arrived at the conclusion that *'26K'* is a repeat of *Mo.* It seems probable that the sex-linked mutant *Ms* is a *Mo* type mutant (H. Krzanowska 1966; *MNL*, 35, 35, 1968; *MNL*, 38, 25). It certainly seems as if the *Mo* locus has a high rate of mutation. This itself is indicative of complexity since a series of duplicated loci, each with the usual rate of mutation, would *ipso facto* display an enhanced rate.

The location of the postulated centre(s), which are thought to be primarily responsible for the mosaic phenotype of females heterozygous for locally autonomous acting genes, have received tentative consideration. The concept has been particularly developed by Cattanach and colleagues (Cattanach and Isaacson, 1967; Cattanach *et al.*, 1969, 1970). The last report has conjectured that one inactivation centre (or controlling element, to use a more general term) may be close to the *Mo - Ta* region or close to the *Gy - jp* region. It is impossible to be more precise at this time but incidental data are suggestive that *Ta* is more closely associated with the centre than the other loci. On the other hand, Grahn *et al.* (1970: appendix; also *MNL*, 42, 16) have found evidence of a controlling element (symbolized as *Cg*) located between *Gs* and *jp*. The whole concept of *X* chromosome inactivation has been extensively reviewed by Eicher (1970b). She argues that, although all of the data can be accounted for by the postulation of two inactivation centres, it is probable that many centres exist, each one being responsible for inactivating a given region of chromosome.

The position of the centromere has yet to be finally settled. However, Eicher (1970b) has proposed that the available evidence is suggestive that it is at the *Gy* end of the group.

Bailey (1962, 1963a, 1964) has shown that the *X* chromosome can have antigenic properties. The elicited reactions tend to be weak and techniques specially designed to measure weak histocompatibility loci are necessary. This is some indication of variation of reaction and Bailey considers that this is more probably due to differences of *X* chromosome antigenicity than to differences in host reactivity. Bailey (1963b) has featured cases of partial rejection which he interprets as a facet of the complete or partial inactivation of one or the other *X* chromosome in different parts of the graft. Little is known at present of the nature of the *X* antigenicity in terms of number or location of histocompatibility loci. The existence of coat colour mutants on the *X* should facilitate an investigation into the problem of whether the *X* element *per se* is antigenic or if one or more loci can be pin-pointed.

Epstein (1969) has demonstrated that glucose-6-phosphate dehydrogenase (G6PD) activity is very probably governed in part, if not primarily, by the *X* chromosome. Epstein was able to accomplish this by comparison of the activity in *XX* and *XO* individuals although as yet no allelic differences are currently known for the postulated locus.

Little (1920) has given details of a stock of Japanese waltzing mice in which the sex ratio is approximately 50 per cent of normal. The lack of males is attributed to a sex-linked (X-borne) recessive lethal gene. Evidence is presented to show that the males of the stock produce a normal sex ratio while the females produced a low sex ratio when mated to unrelated laboratory mice. These results are consistent with the transmission of a postulated X borne lethal. However, the critical demonstration that half of the females from male depleted litters were capable of transmitting the low sex ratio and half were not, could not be undertaken.

A parallel situation was described by Hauschka, Goodwin, and Brown (1951) for a mouse strain in which the proportion of males is significantly less than normal. These authors successfully demonstrated that the abnormal sex ratio was not transmitted by the male and that the females were of two sorts: those which produced the normal 1 : 1 ratio and those which produced an abnormal 2 : 1 ratio. The inheritance of a lethal gene causing death of the hemizygous male during the first 10 days of gestation was proposed. This is a sufficient explanation but so is the alternative of the transmission of a small but lethal deficiency.

Chase (1946) initially claimed to have discovered a sex-linked polydactylia but subsequent breeding results failed to confirm the early results (Gruneberg 1952). The polydactylia was of variable expression, often as low as 25 per cent, and possibly may have exhibited sex-limited expression.

Y-linked group

It is commonly held that the Y chromosome must be sparsely populated with gene loci. Two reasons are usually proferred in support of this thesis. Firstly, the lack of known mutants and, secondly, the variability in size of the chromosome for certain species (e.g., man and rat). The implication is that material can be lost or added with no obvious correlated changes in phenotype. Yet, the fact that the Y persists indicates that it has some genetic function, even if the extent has yet to be evaluated.

Only within very recent years. has it been possible to contemplate that Y borne genes do exist. Welshons and Russell (1959) have demonstrated that the Y is necessary for the individual to be male since the XO karyotype is female. These authors speak of 'male determining factors' and this cautionary attitude is doubtless justified because (as far as the mouse is concerned) it is still unknown how many factors are involved or if the efficacy of the determination may vary. At present, the efficacy is absolute in the sense of an absence of intersexual states. These are so uncommon that one report of sex abnormalities within a stock deserves citation (Hollander, Gowen and Stadler, 1956). However, these seem to be hermaphrodites, rather than intersexes, and explanations other than an inefficacious Y may be more appropriate.

The presence of a Y linked histocompatibility locus was inherent in the observations of Eichwald and Silmser (1955), and explicitly proposed by Hauschka (1955) and Snell (1956). In essence, male tissue grafts are rejected by female hosts in inbred isogenic strains. The phenomenon has been confirmed by numerous experiments (Prehn and Main, 1956; Eichwald, Silmser, and Wheeler, 1957; Hirst, 1957; Short and Sobey, 1957; Bernstein, Silvers, and Silvers, 1958; Eichwald, Silmser, and Weissman, 1958; Feldman, 1958; Sachs and Heller, 1958; Zaalberg, 1959; Gittes and Russell, 1961; Bailey 1964, 1965; and Medvedev and Egorov, 1966). Further studies have shown that while there is good evidence of variation of response by the female to the Y antigen (Prehn and Main, 1956; Zaalberg, 1959; Klein and Linder, 1961; Silvers and Billingham, 1967), there is no clear evidence as yet that the Y chromosome displays variation of antigenicity (Billingham and Silvers, 1960; Celada and Welshons, 1963). The symbol H-Y has been proposed for the postulated Y histocompatibility locus.

The concept of a histocompatibility locus (or loci) on the Y chromosome has wide but not necessarily universal support. The proposal of Hauschka and Holdridge (1962) and Hauschka et al. (1961, 1962) that the Y chromosome itself may be the source of the antigenicity has not been formally disproved although Billingham and Silvers (1963) consider it to be unlikely. Should it turn out that the Y chromosome has several antigenic sites dotted along its length, then the proposal would seem to have some merit.

The absence of a marked reaction to the Y antigen by females of some strains has caused speculation. This 'differential capacity' may be mediated by autosomal genes (by raising or lowering a threshold of responsiveness, say). It has also been suggested that some strains may not react because these have an autosomal replica of the Y antigen (Michie and McLaren, 1958). Hauschka and Holdridge (1962) and Hauschka et al. (1961, 1962) have argued strongly in favour of autosomal translocation of small pieces of the Y chromosome (presumably carrying the postulated H-Y locus) as a possibility. However, Celada and Welshons (1962) feel that this suggestion cannot be generalized and that other mechanisms are seemingly operating.

It has been conjectured that the histocompatibility locus may actually be on an autosome but only manifests in the presence of the Y chromosome or in a male hormonal environment (Silvers, Billingham, and Sanford, 1968); or, the antigenicity of an autosomal locus is inhibited by two X chromosomes but not by one (Fox, 1958). Silvers et al. (1968) admit that their experiments are not conclusive but they argue that the results are at variance with the interpretation of an autosomal basis to the Y antigenicity. However, the work of Celada and Welshons (1963) is reasonably decisive in establishing the existence of Y histocompatibility.

Hints have been made that the male determining factor(s) and the Y histocompatibility may be identical. This is possible but whether a practical differentiation is feasible remains to be seen. It may be that only a portion of the Y differentiated segment is genetically active in the sense of possessing mutable loci. The X differential segment is proving to be more genetically active than many people would have seriously considered over two decades ago and it will be intriguing to observe what the future holds for the Y.

The possibility of a gene in the Y chromosome controlling development of the spermatozoan head piece.has been mooted by Krzanowska (1966, 1969). Sperm head morphology is a polygenic character in general but Krzanowska has adduced evidence that the percentage of abnormal sperms for a particular mouse stock is determined by a few genes. In fact, three are proposed, one of which is depicted as being on the Y. The evidence is indirect but none the less suggestive.

TABLE 11.3 Established linkage groups in the mouse

Group	Loci	Sex	Mating type	Crossover value	References
I	c - fr	♂	CBB	18.0 ± 4.9	Falconer and Snell (1952)
		♀	CBB	16.8 ± 3.1	„ „ „ „
	c - Gpi-1	♀	-BB	29.5 ± 5.2	Hutton and Roderick (1970)
	c - H-1	♂	-BB	5.4 ± 3.7	Snell (1958), Snell and Stevens (1961)
		♀	-BB	8.4 ± 2.1	Snell (1958), Snell and Stevens (1961)
		-	-II	7.8 ± 3.2	Snell and Stevens (1961)
			-II	10.9 ± 4.0	„ „ „ „
		?	-BB	12.5 ± 7.3	Stimpfling and Snell (1968)
		-	-II	12.8 ± 5.0	„ „ „ „
	c - Hbb	♂	-BB	2.5 ± 0.8	Popp and St. Amand (1960)
		♀	-BB	5.1 ± 0.9	„ „ „ „
		?	-BB	4.3 ± 2.4	Hutton et al. (1962)
		♂	-BB	3.6 ± 3.5	Wolfe et al. (1963)
		♀	-BB	9.8 ± 2.0	„ „ „ „
		♂	-BB	7.0 ± 2.5	Popp and St. Amand (1964)
		♀	-BB	10.1 ± 1.7	„ „ „ „
		♀	-BB	6.4 ± 2.8	Hutton and Roderick (1970)
	c - hf	?	CBB	2.9 ± 2.1	Bunker (1959)
	c - nv	?	CBB	39.5 ± 3.3	Abeelen and Kroon (1967)
	c - ol	♂	RBI	14.1 ± 3.2	Hertwig (1942)
		♀	RBI	10.3 ± 3.3	„ „
		-	RII	28.5 ± 3.6	„ „
	c - p	-	CII	19.6 ± 14.4	Darbishire (1904)
		-	CII	27.2 ± 6.0	Haldane et al. (1915)
		♂	RBB	13.0 ± 3.1	Dunn (1920a)
		♀	CBB	16.2 ± 1.5	„ „
		♂	RBB	15.6 ± 3.5	„ „
		-	CII	13.3 ± 5.8	„ „
		-	CII	16.7 ± 3.4	„ „
		-	CII	29.6 ± 8.2	„ „
		♂	CBB	14.0 ± 0.8	Castle and Wachter (1924)
		♀	CBB	19.1 ± 2.4	„ „ „
		♂	CBB	12.0 ± 1.7	Detlefsen and Clemente (1924)
		♀	CBB	20.4 ± 4.1	„ „ „
		♂	CBB	11.6 ± 3.0	Feldman (1924)
		♂	RBB	11.6 ± 1.9	„ „
		♀	CBB	20.0 ± 1.2	„ „
		♀	RBB	16.3 ± 2.1	„ „
		♂	CBB	10.2 ± 1.7	Detlefsen (1925)
		♂	RBB	10.4 ± 2.2	„ „
		♀	CBB	11.9 ± 1.8	„ „

Group	Loci	Sex	Mating type	Crossover value	References
		♀	RBB	18.3 ± 1.3	Detlefsen (1925)
		-	RII	36.2 ± 8.0	„ „
		♂	CBB	11.1 ± 1.3	Gruneberg (1935, 1936b)
		♂	RBB	13.1 ± 1.8	„ „
		♀	CBB	16.4 ± 0.9	„ „
		♀	RBB	14.2 ± 1.0	„ „
		♂	RBB	16.1 ± 3.4	Snell (1946)
		♀	RBB	11.8 ± 2.0	„ „
		♂	CBB	8.2 ± 3.5	Falconer and Snell (1952)
		♀	CBB	11.4 ± 2.6	„ „ „ „
		?	RBB	16.2 ± 4.5	Bunker (1959)
		♂	CBB	13.9 ± 1.9	Popp and St. Amand (1960)
		♀	CBB	17.4 ± 1.8	„ „ „ „ „
		♀	CBB	16.9 ± 4.7	Snell and Stevens (1961)
	c - qv	-	CII	28.8 ± 4.1	Yoon and Les (1957)
		-	RII	27.6 ± 5.0	„ „ „ „
	c - sh-1	♂	CBB	2.9 ± 1.1	Gates (1928b, 1929, 1931)
		♀	CBB	4.1 ± 0.5	„ „ „ „
		-	CII	5.0 ± 1.4	„ „ „ „
		♂	CBB	3.9 ± 0.8	Gruneberg (1935, 1936b)
		♀	RBB	1.7 ± 0.7	„ „ „
		♀	CBB	4.8 ± 0.5	„ „ „
		♀	RBB	3.2 ± 0.5	„ „ „
		♂	CBB	1.6 ± 1.6	Falconer and Snell (1952)
		♀	CBB	8.1 ± 2.2	„ „ „ „
	fr - p	♂	CBB	26.2 ± 5.6	„ „ „ „
		♀	CBB	34.9 ± 3.9	„ „ „ „
		♂	CBB	34.0 ± 6.9	Eicher (1967)
	fr - sh-1	♂	CBB	16.4 ± 4.7	Falconer and Snell (1952)
		♀	CBB	15.4 ± 3.0	„ „ „ „
		♂	CBB	17.0 ± 5.5	Eicher (1967)
	Gpi-1 - Hbb	♀	-BB	31.6 ± 4.7	Hutton (1969a, b)
		♀	-BB	35.9 ± 5.6	Hutton and Roderick (1970)
	H-4 - p	-	RBB	<4.2	Snell and Stevens (1961)
	Hbb - p	♂	RBB	15.9 ± 2.0	Popp and St. Amand (1960)
		♀	RBB	23.1 ± 2.0	„ „ „ „
		♂	RBB	15.3 ± 2.0	Popp (1962a)
		♀	RBB	16.6 ± 2.0	„ „
	Hbb - sh-1	♂	RBB	2.0 ± 0.5	„ „
		♀	RBB	1.4 ± 0.4	„ „
	hf - p	?	RBB	14.3 ± 3.8	Bunker (1959)
	p - qv	-	CII	12.4 ± 1.9	Yoon and Les (1957)
	p - ru-2	♂	CBB	1.7 ± 1.0	Eicher (1970a)
		♀	CBB	3.2 ± 1.2	„ „
	p - sh-1	?	RBB	14.7 ± 6.8	Gates (1931)

Group	Loci	Sex	Mating type	Crossover value	References
		♂	CBB	17.2±3.9	Snell (1946)
		♀	CBB	13.3±1.8	„ „
		♂	CBB	9.8±3.8	Falconer and Snell (1952)
		♀	CBB	19.5±2.8	„ „ „ „
		♂	CBB	13.3±1.2	Popp (1962a)
		♀	CBB	15.8±1.3	„ „
		♂	CBB	17.0±5.5	Eicher (1967)
	p - tp	-	RII	<39.2	Fielder (1952)
II	cw - se	♂	CBB	36.6±3.9	Falconer and Isaacson (1966)
		♀	CBB	40.1±2.7	„ „ „ „
		-	CII	31.7±2.4	„ „ „ „
		-	RII	36.2±5.6	„ „ „ „
		♂	CBB	35.5±6.1	Lyon et al. (1968)
		♀	RII	38.5±6.7	„ „ „ „
	cw - tk	♂	CBB	40.3±5.8	Falconer and Isaacson (1966)
		♂	RBB	44.9±5.6	„ „ „ „
		♀	CBB	42.6±4.0	„ „ „ „
		♀	RBB	43.8±3.8	„ „ „ „
		♂	CBB	38.7±6.2	Lyon et al. (1968)
		♀	CBB	50.0±6.9	„ „ „ „
	d - du	-	RII	12.8±6.3	Snell (1955)
		-	RII	19.0±3.2	Green and Lane (1967)
		-	CII	21.9±2.2	„ „ „ „
	d - Mdh-1	♀	-BB	10.1±2.9	Hutton (1969a), Hutton and Roderick (1970)
	d - se	-	RII	0.094±0.129	Snell (1928)
		♀	RBB	0·098±0·098	Snell (1931)
		♂	CBB	0.735±0.732	Castle et al. (1936)
		♂	CBB	0.182±0.09	Goodwins and Vincent (1955)
		♀	CBB	0.115±0.08	„ „ „ „
	d - Trf	♂	-BB	8.5±2.6	Shreffler (1963)
		♀	-BB	16.9±2.6	„ „
	t - tk	♂	CBB	3.0±1.2	Green and Lane (1967)
		♂	RBB	3.6±1.0	„ „ „ „
		♀	CBB	3.5±1.9	„ „ „ „
		♀	RBB	2.9±1.4	„ „ „ „
		-	CII	5.5±0.9	„ „ „ „
	dse - lu	♂	CBB	26.7±15.7	Green (1961)
		♂	RBB	12.1±5.3	„ „
		♀	CBB	17.3±4.7	„ „
		♀	RBB	22.0±7.7	„ „
		?	CBB	23.3±7.7	„ „
		-	CII	14.6±4.5	„ „
		-	RII	17.6±3.7	„ „

Group	Loci	Sex	Mating type	Crossover value	References
	dse - sg	-	CII	4.4±9.2	Green and Lane (1967)
		-	RII	5.0±2.0	,, ,, ,, ,,
	du - lu	?	RBB	44.7±5.7	Green (1961)
	du - se	♂	RBB	15.0±7.0	,, ,,
		♀	CBB	21.3±4.7	,, ,,
		♀	RBB	22.6±8.7	,, ,,
	du - tk	-	CII	12.4±1.6	Green and Lane (1967)
		-	RII	11.0±2.4	,, ,, ,, ,,
	lu - Mdk-1	♀	-BB	12.0±6.5	Hutton and Roderick (1970)
	lu - se	♀	RBB	16.3±3.7	Deol and Green (1966)
	lu - sg	♀	RBB	19.1±3.2	Green and Lane (1967)
	lu - sv	♀	CBB	16.4±5.0	Doel and Green (1966)
		♀	RBB	35.1±7.9	,, ,, ,, ,,
	lu - tk	-	CII	24.8±3.7	Green and Lane (1967)
		-	RII	21.5±5.5	,, ,, ,, ,,
	lu - Trf	♂	-BB	20.7±5.2	Shreffler (1963)
		♀	-BB	23.8±5.3	,, ,,
	se - sv	♂	CBB	1.8±1.7	Doel and Green (1966)
		♀	CBB	2.4±1.0	,, ,, ,, ,,
		♀	RBB	2.9±1.0	,, ,, ,, ,,
	se - tk	♂	CBB	6.9±3.0	Falconer and Isaacson (1966)
		♂	RBB	5.1±2.5	,, ,, ,, ,,
		♀	CBB	7.1±2.1	,, ,, ,, ,,
		♀	RBB	4.1±1.5	,, ,, ,, ,,
		-	RII	16.2±7.4	,, ,, ,, ,,
III	hr - pn	♀	CBB	13.0±1.3	Kidwell et al. (1966)
	hr - s	♂	CIB	2.3±7.7	Snell (1931)
		♀	CBB	9.8±2.2	,, ,,
		♀	RBB	10.0±9.4	,, ,,
		♀	CBB	11.9±1.2	Kidwell et al. (1966)
	hr - wl	-	CII	3.7±0.7	Lane and Dickie (1961)
		-	RII	8.7±4.6	,, ,, ,, ,,
	pn - s	♂	RBB	24.4±6.7	Kidwell et al. (1961)
		♀	RBB	30.1±2.2	,, ,, ,, ,,
		-	RII	27.9±3.0	,, ,, ,, ,,
		♀	CBB	27.1±1.3	Kidwell et al. (1966)
		♀	RBB	26.0±2.9	,, ,, ,, ,,
IV	av - Sl	?	?	16.9±3.0	Nash and Venier (1964)
	r - si	♂	CBB	14.3±13.2	Keeler (1930)
		?	RBB	11.1±11.7	Keeler (1930)
V	a - As	♂	CBB	0.35±0.25	Phillips (1966)
		♂	RBB	1.17±0.52	,, ,,

Group Loci	Sex	Mating type	Crossover value	References
	♀	CBB	0.66±0.38	Phillips (1966)
	♀	RBB	0.31±0.22	,, ,,
a - bp	♀	CBB	0.60±0.60	Runner (1959)
	♀	RBB	0.30±0.21	,, ,,
	?	CBB	0.38±0.39	Phillips (1966)
a - Cs	?	CBB	17.2±3.4	Dickerman et al. (1968)
a - fi	-	CII	35.1±2.5	Carter and Gruneberg (1950), Carter (1951a)
	-	RII	27.6±4.2	Carter and Gruneberg (1950), Carter (1951a)
	♂	CBB	30.3±3.8	Wallace (1957a)
	♂	RBB	24.0±2.6	,, ,,
	♀	CBB	36.3±2.9	,, ,,
	♀	RBB	35.4±2.9	,, ,,
a - H-3	♂	-BB	36.8±10.9	Snell and Bunker (1964)
	♀	-BB	21.8±3.2	,, ,, ,, ,,
	♂	-BB	10.7±3.4	Snell et al. (1967)
	♀	-BB	15.8±2.7	,, ,, ,, ,,
a - H-6	?	-BB	25.1±3.4	Lilly (1967)
	-	-II	25.5±2.7	,, ,,
a - H-13	♀	-BB	1.5±0.9	Snell et al. (1967)
a - Ir-2	?	CBB	20.0±6.0	Gasser (1969)
a - lst	♂	CBI	19.1±4.4	Forsthoefel and Shenk (1970)
	♀	CBI	27.9±5.9	,, ,, ,, ,,
a - mg	♂	CBB	9.9±1.8	Lane and Green (1960)
	♂	RBB	9.8±1.5	,, ,, ,, ,,
	♀	CBB	12.3±1.8	,, ,, ,, ,,
	♀	RBB	13.5±1.2	,, ,, ,, ,,
a - pa	♂	CBB	19.6±2.5	Roberts and Quisenberg (1935)
	♀	CBB	21.2±2.4	Roberts and Quisenberg (1935)
	♂	RBB	15.8±3.3	Snell (1946)
	♀	RBB	19.8±3.7	,, ,,
	?	?BB	16.6±2.5	Borger (1950)
	-	RII	12.5±3.4	Carter (1951a)
	♂	?BB	15.4±0.7	Owen (1953b)
	♀	?BB	16.3±0.7	,, ,,
	♀	CBB	18.3±4.3	Carter and Phillips (1954)
	♂	CBB	16.3±1.5	Carter et al. (1956)
	♀	CBB	20.0±5.1	,, ,, ,, ,,
	♂	CBB	11.9±5.0	Green and Mann (1961)
	-	CII	16.7±3.0	Lilly (1967)
a - Ra	♂	CBB	22.0±3.7	Carter and Phillips (1954)
	♀	CBB	21.0±5.4	,, ,, ,, ,,

Group Loci	Sex	Mating type	Crossover value	References
	♀	RBB	32.0±9.3	Carter and Phillips (1964)
	?	CBB	26.4±4.6	,, ,, ,, ,,
	?	RBB	25.0±4.4	,, ,, ,, ,,
	♂	CBB	22.6±2.1	Parsons (1958b)
	♂	RBB	24.1±2.1	,, ,,
	♀	CBB	20.5±2.0	,, ,,
	♀	RBB	20.6±2.0	,, ,,
	♂	CBB	17.5±3.1	Lane and Green (1960)
	♂	RBB	23.9±2.9	,, ,, ,, ,,
	♀	CBB	20.7±2.8	,, ,, ,, ,,
	♀	RBB	19.7±2.4	,, ,, ,, ,,
	♂	CBB	24.4±3.4	Green and Mann (1961)
	♂	RBB	28.5±2.5	,, ,, ,, ,,
a - ro	?	?	16.4±2.6	Falconer (1954)
a - Sd	?	?BB	40.3±3.0	Borger (1950)
	♂	CBB	45.5±4.5	Carter and Phillips (1954)
	♀	CBB	54.6±15.0	,, ,, ,, ,,
	♂	CBB	44.3±3.0	Wallace (1957a)
	♂	RBB	42.0±3.0	,, ,,
	♀	CBB	45.8±3.0	,, ,,
	♀	RBB	44.4±3.0	,, ,,
	♂	CBB	42.7±4.4	Lane and Dickie (1968)
a - stb	♂	CBB	40.9±3.0	,, ,, ,, ,,
	♀	CBI	34.8±5.4	,, ,, ,, ,,
a - Svp	?	-BB	9.2±3.6	Platz and Wolfe (1969)
a - un	♂	CBB	6.4±2.9	Carter (1947)
	♀	CBB	7.7±3.8	,, ,,
	♂	?BB	4.8±0.5	Fisher (1949)
	♀	?BB	4.7±0.5	,, ,,
	♂	?BB	4.6±0.4	Owen (1953b)
	♀	?BB	4.7±0.4	,, ,,
	♀	CBB	4.7±0.9	Runner (1959)
	♂	RBB	10.5±7.0	Snell and Bunker (1964)
	♀	RBB	5.5±5.6	,, ,, ,, ,,
	♂	CBB	2.4±1.7	Snell et al. (1967)
	♀	CBB	5.6±1.2	,, ,, ,, ,,
a - we	♂	CBB	10.5±1.1	Hertwig (1942)
	♀	CBB	16.7±1.6	,, ,,
	♀	RBB	18.5±1.8	,, ,,
	-	RII	15.7±4.0	,, ,,
	♂	?BB	9.0±0.6	Owen (1953b)
	♀	?BB	12.1±0.7	,, ,,
	♂	CBB	10.1±1.5	Parsons (1958b)
	♂	RBB	12.1±1.6	,, ,,
	♀	CBB	13.8±1.7	,, ,,

Group Loci	Sex	Mating type	Crossover value	References
	♀	RBB	13.9±1.8	Parsons (1958b)
	♂	RBB	21.1±3.0	Snell and Bunker (1964)
	♀	RBB	18.2±3.0	„ „ „ „
	♂	CBB	10.7±3.4	Snell *et al.* (1967)
	♀	CBB	17.1±2.0	„ „ „ „
bp - un	♀	RBB	5.0±0.8	Runner (1959)
Cs - Ra	?	RBB	45.0±4.5	Dickerman *et al.* (1968)
Cs - Sd	?	RBB	35.3±3.2	„ „ „ „
fi - pa	?	RII	19.0±3.6	Carter (1951a)
	♂	CBB	26.6±4.3	Bodner (1961a)
	♀	CBB	29.5±2.3	„ „
	♀	RBB	24.5±2.4	„ „
fi - Sd	♂	CBB	15.5±2.2	Wallace (1957a)
	♂	RBB	25.0±2.6	„ „
	♀	CBB	20.1±2.4	„ „
	♀	RBB	21.0±2.5	„ „
fi - stb	-	RII	<8.8	Lane and Dickie (1968)
H-3 - un	♂	-BB	15.8±8.4	Snell and Bunker (1964)
	♀	-BB	16.4±2.9	„ „ „ „
	♂	-BB	8.3±3.0	Snell *et al.* (1967)
	♀	-BB	11.3±2.4	„ „ „ „
H-6 - pa	-	-II	6.9±3.2	Lilly (1967)
H-13 - un	♀	-BB	5.6±1.6	Snell *et al.* (1967)
H-13 - we	♀	-BB	12.8±2.4	„ „ „ „
hy-1 - pa	-	RII	33.4±6.8	Clark (1936)
lst - Ra	?	CIB	47.3±5.9	Forsthoefel and Shenk (1970)
mg - Ra	♂	CBB	29.7±2.4	Lane and Green (1960)
	♀	RBB	25.9±2.0	„ „ „ „
pa - Ra	♀	CBB	40.0±6.5	Carter and Phillips (1954)
	♀	RBB	35.7±12.8	„ „ „ „
	♂	CBB	45.2±7.6	Green and Mann (1961)
pa - ro	-	-	*Circa* 0.6	Falconer (1954)
pa - Sd	?	?BB	45.5±3.0	Borger (1950)
pa - un	♂	?BB	6.7±0.8	Owen (1953b)
	♀	?BB	11.6±0.6	„ „
	♀	CBB	11.9±1.3	Falconer (1954)
pa - we	♂	?BB	2.3±0.3	Owen (1953b)
	♀	?BB	4.2±0.4	„ „
	♀	CBB	3.3±0.7	Falconer (1954)
Ra - Sd	♂	CBB	44.7±4.5	Carter and Phillips (1954)
	♀	RBB	45.5±15.0	„ „ „ „
	?	RBB	35.4±5.3	„ „ „ „
Ra - we	♂	CBB	34.2±2.4	Parsons (1958b)
	♂	RBB	32.4±2.3	„ „
	♀	CBB	32.1±2.3	„ „

Group	Loci	Sex	Mating type	Crossover value	References
		♀	RBB	33.3±2.3	Parsons (1958b)
	ro - Sd	?	?	37.5±3.8	Falconer (1954)
	ro - un	♂	CBB	10.2±1.7	,, ,,
		♀	CBB	6.2±1.6	,, ,,
	ro - we	♂	CBB	4.3±1.2	,, ,,
		♀	CBB	2.9±1.1	,, ,,
	Sd - stb	♂	CBB	11.3±1.8	Lane and Dickie (1968)
		♀	CBI	10.2±5.1	,, ,, ,, ,,
	un - we	♂	?BB	4.5±0.4	Owen (1953b)
		♀	?BB	7.4±0.5	,, ,,
		♂	CBB	6.6±1.4	Falconer (1954)
		♀	CBB	6.7±0.8	,, ,,
		♂	CBB	10.5±7.1	Snell and Bunker (1964)
		♀	CBB	12.7±2.6	,, ,, ,, ,,
		♂	CBB	8.3±3.0	Snell et al. (1967)
		♀	CBB	8.3±1.4	,, ,, ,, ,,
VI	bt - Ca	♂	RBB	10.7±1.8	Flanagan and Isaacson (1967)
		♀	RBB	9.9±3.3	,, ,, ,, ,,
	bt - N	♂	CBB	13.6±2.8	Murray and Snell (1945)
		♀	CBB	9.4±2.4	,, ,, ,, ,,
	Ca - N	♂	RBB	2.3±0.9	Cooper (1939)
		♂	CBB	4.7±1.9	Murray and Snell (1945)
		♀	RBB	4.4±2.2	MacNeil (1956)
		♂	RBB	0.9±0.4	Flanagan and Isaacson (1967)
		♀	RBB	1.4±0.5	,, ,, ,, ,,
	N - Sha	♂	CBB	0.73±0.42	,, ,, ,, ,,
		♂	RBB	1.1±0.3	,, ,, ,, ,,
		♀	CBB	1.7±0.7	,, ,, ,, ,,
		?	CBB	0.47±0.28	,, ,, ,, ,,
		?	RBB	1.2±0.5	,, ,, ,, ,,
VII	Al - Re	♂	RBB	30.4±3.9	Dickie (1955)
		♀	RBB	29.5±4.7	,, ,,
	Al - sh-2	♀	CBB	28.3±1.7	,, ,,
	Al - vt	♀	CBB	44.8±5.3	,, ,,
	Al - wa-2	♀	CBB	39.5±2.3	,, ,,
	Es-3 - Re	?	-BB	10.5±5.0	Roderick et al. (1970)
	nu - Re	♀	CBB	11.1±4.3	Flanagan (1966)
		?	CBI	19.4±4.6	,, ,,
	nu - Tr	♀	CBB	7.4±3.6	,, ,,
	oe - Re	?	CBB	15.2±1.8	Kelton and Rauch (1968)
		-	RBB	19.3±2.6	,, ,, ,. ,,
	oe - Sh-2	?	CBB	19.5±1.8	,, ,, ,, ,,
		-	RII	15.2±1.6	,, ,, ,, ,,

Group Loci	Sex	Mating type	Crossover value	References
oe - tn	-	RII	42.8±3.9	Kelton and Rauch (1968)
oe - wa-2	?	CBB	35.7±3.4	,, ,, ,, ,,
	-	RII	36.6±3.4	,, ,, ,, ,,
Re - sh-2	♂	CBB	18.4±6.3	Falconer (1947)
	♀	CBB	21.7±8.6	,, ,,
	♂	CBB	21.4±2.9	Carter and Phillips (1953)
	♂	RBB	17.2±2.6	,, ,, ,, ,,
	♀	CBB	22.8±5.6	Falconer and Sobey (1953)
	♀	CBB	29.1±3.8	Nasrat (1956)
	?	RBB	27.3±2.9	Kelton and Rauch (1968)
Re - shm	-	CII	15.0±7.0	Green (1968)
Re - ti	♂	CBB	12.8±5.4	Searle (1961)
	♂	RBB	21.9±2.3	,, ,,
	♀	CBB	16.1±3.0	,, ,,
	♀	RBB	22.2±3.0	,, ,,
Re - Tr	♀	CBB	20.9±2.9	Falconer and Sobey (1953)
	♀	RBB	23.7±3.6	,, ,, ,, ,,
	♀	CBB	18.5±5.4	Flanagan (1966)
Re - vt	♂	CBB	16.8±2.9	Mitchie (1952, 1955a)
	♂	RBB	19.3±3.4	,, ,, ,,
	♀	CBB	32.6±4.9	,, ,, ,,
	♀	RBB	23.4±4.0	,, ,, ,,
Re - wa-2	♀	CBB	40.1±6.3	Falconer (1947)
	♀	CBB	38.1±3.5	Carter (1951b)
	♀	RBB	46.5±3.7	,, ,,
	♂	CBB	44.1±3.5	Carter and Phillips (1953)
	♂	RBB	43.1±3.5	,, ,, ,, ,,
sh-2 - Tr	♀	CBB	2.9±2.0	Falconer and Sobey (1953)
sh-2 - vt	-	RII	1.7±1.7	Michie (1955b)
sh-2 - wa-2	♂	CBB	26.8±5.9	Snell and Law (1939)
	♀	CBB	25.0±2.8	,, ,, ,, ,,
	♂	CBB	30.1±3.0	Wright (1947b)
	♂	RBB	32.2±3.1	,, ,,
	?	CBB	31.1±2.3	Heston et al. (1952)
	?	CBB	29.4±3.6	Law (1952)
	♂	CBB	24.9±3.0	Carter and Phillips (1953)
	♂	RBB	31.2±3.3	,, ,, ,, ,,
	♀	CBB	23.0±1.1	,, ,, ,, ,,
	♀	RBB	26.7±3.1	,, ,, ,, ,,
	♀	CBB	26.7±5.0	Falconer and Sobey (1953)
	-	CII	37.4±3.2	Michie (1955b)
	♀	CIB	23.3±13.8	Carter et al. (1956)
	?	CBB	21.3±2.7	Kelton and Rauch (1968)
	-	CII	27.0±4.7	,, ,, ,, ,,

Group	Loci	Sex	Mating type	Crossover value	References
	ti - vt	♀	CBB	9.8±1.6	Searle (1961)
	ti - wa-2	♀	CBB	29.4±1.9	„ „
	Tr - wa-2	♀	CBB	33.3±5.4	Falconer and Sobey (1953)
	vt - wa-2	-	RII	28.0±3.5	Michie (1955b)
		♀	CBB	22.1±1.7	Searle (1961)
		?	?BB	25.4±1.0	Tatchell (1961)
VIII	*an - b*	♂	CBI	3.4±1.4	Hertwig (1942)
		-	CII	10.6±2.5	„ „
	asp - b	?	CBB	38.3±6.3	Schlesinger *et al.* (1966)
		?	CBI	38.7±6.1	Collins and Fuller (1968)
		-	CII	40.7±6.5	„ „ „ „
		?	CBB	44.1±1.7	Collins (1970)
		-	CII	40.6±2.8	„ „
	asp - Gpd-1	?	-BB	47.3±5.1	„ „
	asp - m	-	RII	41.5±3.3	„ „
	asp - Pt	?	CBB	43.7±2.5	„ „
	b - Lv	♀	-BB	4.8±2.1	Hutton (1969a), Hutton and Coleman (1969)
	b - m	♂	RBB	8.2±3.5	Woolley (1945)
		♀	RBB	7.1±1.1	„ „
		♂	CBB	6.7±1.9	Snell (1946)
		♀	CBB	8.0±6.3	„ „
		♀	CBB	7.1±1.1	Sirlin (1957)
		♂	CBB	5.4±2.3	Lane (1963)
		♀	CBB	8.9±1.5	„ „
		♂	CBB	7.2±2.5	Johnson (1969a)
		♀	CBB	9.3±1.9	„ „
		?	CBB	8.0±5.4	Hutton and Roderick (1970)
	b - Gpd-1	♀	-BB	31.7±4.6	Hutton (1969a), Hutton and Coleman (1969)
		?	-BB	31.4±4.6	Collins (1970), Collins and Hutton (1970)
		?	-BB	32.0±9.3	Hutton and Roderick (1970)
	b - Mup-a	-	-II	4.1±2.0	Hudson *et. al.* (1967)
		?	-BB	6.6±2.3	Finlayson *et al.* (1969)
	b - Ps	♂	CBB	6.7±2.4	Johnson (1969b)
		♀	CBB	9.0±1.8	„ „
	b - Pt	♂	CBB	3.5±1.7	Hollander and Strong (1951)
		♂	RBB	5.5±1.4	„ „ „ „
		♀	CBB	10.0±9.5	„ „ „ „
		♀	RBB	5.2±1.5	„ „ „ „
		♀	CBB	3.7±1.5	Yoon (1961)
		♂	CBB	3.1±1.8	„ „
		♂	RBB	7.8±1.7	„ „

Group	Loci	Sex	Mating type	Crossover value	References
		♀	CBB	3.7±0.6	Yoon (1961)
		♀	RBB	7.6±1.1	,, ,,
		?	CBB	4.1±1.9	Findlayson et al. (1969)
		?	CBB	5.6±1.2	Collins (1970), Collins and Hutton (1970)
	b - vc	♀	CBB	2.5±2.5	Sirlin (1956)
		♀	CBI	6.6±1.9	,, ,,
		-	CII	5.4±3.1	,, ,,
		♀	CBB	10.8±2.2	Sirlin (1957)
	b - wi	♂	CBB	4.3±1.6	Lane (1963)
		♀	CBB	3.6±1.0	,, ,,
	b - wd	♀	CBB	32.5±3.7	Yoon (1961)
		-	CII	29.4±3.5	,, ,,
		-	RII	38.2±6.4	,, ,,
	Gpd-1 - Lv	♀	-BB	36.5±4.7	Hutton and Coleman (1969)
	Gpd-1 - m	?	-BB	24.0±8.5	Hutton and Roderick (1970)
	Gpd-1 - Pt	?	-BB	25.5±4.3	Collins (1970), Collins and Hutton (1970)
	m - Ps	♂	CBB	1.0±0.8	Johnson (1969b)
		♀	CBB	4.0±0.4	,, ,,
	m - Pt	♂	CBB	2.2±1.5	Lane (1963)
		♀	CBB	3.4±0.6	,, ,,
	m - vc	♀	CBB	18.0±2.7	Sirlin (1957)
	Mup-a - Pt	?	-BB	10.7±2.8	Finlayson et al. (1969)
	Pt - wd	♀	CBB	36.2±3.8	Yoon (1961)
	Pt - wi	♂	CBB	4.8±1.6	Lane (1963)
		♂	RBB	8.2±2.0	,, ,,
		♀	CBB	10.3±1.1	,, ,,
		♀	RBB	16.2±3.1	,, ,,
IX	Fu - H-2	?	-BB	5.3±0.3	Snell (1952)
		♂	-BB	4.1±1.2	Allen (1955a)
		♀	-BB	6.9±3.0	,, ,,
	Fu - Ki	?	RBB	< 3.8	Dunn and Gluecksohn-Waelsch (1954)
	Fu - T	♂	RBB	4.3±1.1	Dunn and Caspari (1942, 1945)
	H-2 - Ki	♂	- BB	10.2±2.2	Allen (1955b)
		?	- BB	5.0±2.2	Snell (1952)
	H-2 - T	♂	-BB	4.1±2.3	,, ,,
		♀	-BB	12.5±11.7	,, ,,
		♂	-BB	8.3±1.0	Allen (1955b)
		♀	-BB	15.4±1.6	,, ,,
		♂	-BB	14.7±6.1	Snell et al. (1967)
		♀	-BB	26.3±10.1	,, ,, ,,
		♂	-BB	15.2±3.4	Shreffler (1964b)

Group	Loci	Sex	Mating type	Crossover value	References
		♀	-BB	17.3±3.4	Shreffler (1964b)
		♂	-BB	7.9±1.3	Pizarro and Dunn (1970)
		♀	-BB	15.2±4.0	,, ,, ,, ,,
	H-2 - f	♂	-BB	2.8±0.8	,, ,, ,, ,,
		♀	-BB	1.9±1.5	. ,, ,, ,,
	H-2 - Tla	?	-BB	1.5±0.9	Boyse *et al.* (1964)
	Ki - T	♂	RBB	2.5±6.6	Dunn and Caspari (1942, 1945)
		♀	RBB	5.9±1.5	,, ,, ,, ,,
		♂	CBB	4.7±1.5	Dunn and Caspari (1945)
		?	RBB	5.3±3.0	Dunn and Gluecksohn-Waelsch (1953b)
	Ki - tf	♂	CBB	0.4±0.3	Dunn *et al.* (1962)
		♀	CBB	1.2±4.6	,, ,, ,,
	T - tf	♂	CBB	9.2±2.4	Lyon (1959), Lyon and Phillips (1959)
		♀	RBB	8.4±1.7	Lyon (1959), Lyon and Phillips (1959)
		♀	CBB	5.8±2.8	Lyon (1959)
		♂	RBB	2.9±1.4	Lyon and Phillips (1959)
		♀	RBB	11.0±2.0	,, ,, ,, ,,
		♂	RBB	4.4±1.6	Dunn *et al.* (1962)
		♀	RBB	10.4±2.9	,, ,, ,,
		♂	CBB	7.5±1.4	Lyon and Meredith (1964a)
		♂	RBB	2.2±1.3	,, ,, ,, ,,
		♀	CBB	9.4±1.5	,, ,, ,, ,,
		♀	RBB	8.6±2.6	,, ,, ,, ,,
		♂	CBB	8.7±1.6	Searle (1966)
		♂	RBB	4.4±1.8	,, ,,
		♀	CBB	9.6±1.8	,, ,,
		♀	RBB	20.0±3.3	,, ,,
		♂	RBB	4.7±9.7	Dunn and Bennett (1968)
		♂	CBB	1.7±1.2	Pizarro and Dunn (1970)
		♂	RBB	4.0±1.1	,, ,, ,, ,,
		♀	RBB	12.3±3.7	,, ,, ,, ,,
X	*gr - v*	♀	CBB	12.9±2.8	Bloom and Falconer (1966)
		-	RII	25.4±7.0	,, ,, ,, ,,
	ii - v	♂	RBB	10.2±4.3	Snell (1945)
		♀	RBB	40.0±2.4	,, ,,
		-	RII	25.2±8.3	,, ,,
XI	*dw - Mi*	?	CIB	57.0±5.3	Kroon and Buis (1970)
	dw - ob	-	RII	33.5±5.2	,, ,, ,, ,,
	Lc - Mi	♂	CBB	10.7±1.2	Phillips (1960)

Group	Loci	Sex	Mating type	Crossover value	References
		♂	RBB	20.0±5.7	Phillips (1960)
		♀	CBB	11.6±1.7	,, ,,
		♀	RBB	11.4±5.4	,, ,,
		?	CBB	11.1±1.1	Searle (1964)
	Lc - px	?	CBI	5.9±1.6	,, ,,
	Lc - wa-1	♂	RBB	7.8±1.9	Phillips (1960)
		♀	RBB	8.5±2.0	,, ,,
		?		5.9±1.2	Searle (1964)
	Ldr-1 - Mi	♀	-BB	28.7±4.4	Hutton and Roderick (1970)
	Mi - px	?	CBI	5.1±1.5	Searle (1964)
	Mi - wa-1	♂	CBB	10.3±3.8	Bunker and Snell (1948)
		♀	CBB	8.7±2.0	,, ,, ,, ,,
		♂	CBB	4.2±1.5	Phillips (1960)
		♀	CBB	3.2±1.4	,, ,,
		?	RBB	4.2±1.5	Searle (1964)
	eo - wa-1	♂	CBB	10.2±2.0	Bennett and Gresham (1956)
		♀	CBB	4.9±1.7	,, ,, ,, ,,
XII	bm - ep	-	RII	8.6±3.5	Lane and Dickie (1968)
	Dc - ep	?	CBB	34.7±5.9	Deol and Lane (1966)
	ep - ru	♂	CBB	2.5±1.0	Lane and Green (1967)
		♂	RBB	2.6±0.7	,, ,, ,, ,,
		♀	CBB	1.2±0.4	,, ,, ,, ,,
		♀	RBB	1.2±0.8	,, ,, ,, ,,
		-	RII	3.0±2.1	,, ,, ,, ,,
XIII	Dh - dr	♀	CBB	26.3±3.9	Lyon (1961a)
	Dh - fz	♂	CBB	41.5±1.1	Searle (1964)
		♀	CBB	43.9±2.8	,, ,,
	Dh - ln	♂	CBB	1.4±1.4	Lyon (1961a)
		♂	CBB	3.8±4.6	Searle (1964)
		♀	CBB	2.2±0.9	,, ,,
	dr - fz	-	RII	47.1±4.2	Lyon (1961a)
	dr - ln	♀	RBB	28.2±3.2	,, ,,
		-	RII	39.7±4.2	,, ,,
	fz - ln	♂	CBB	36.1±2.1	Snell et al. (1954)
		♀	CBB	44.0±1.1	,, ,, ,, ,,
		♂	CBB	40.6±2.1	Carter et al. (1956)
		♀	CBB	41.9±2.1	,, ,, ,, ,,
		♂	CBB	36.4±1.4	Parsons (1958a)
		♂	RBB	34.6±1.3	,, ,,
		♀	CBB	43.8±1.4	,, ,,
		♀	RBB	42.0±1.3	,, ,,
		♂	?BB	39.1±4.1	Dickie and Woolley (1950)
		♀	?BB	49.1±4.0	,, ,, ,, ,,

Group	Loci	Sex	Mating type	Crossover value	References
		♂	CBB	33.2±2.5	Dickie (1964)
		♀	CBB	42.4±2.8	,, ,,
		♂	CBB	37.8±0.9	Searle (1964)
		♀	CBB	40.8±2.3	,, ,,
	fz - Lp	♂	CBB	54.8±5.2	Snell et al. (1954)
		♀	CBB	50.0±9.8	,, ,, ,, ,,
	fz - py	♂	CBB	48.6±1.7	Parsons (1958a)
		♂	RBB	50.1±1.3	,, ,,
		♀	CBB	50.0±1.3	,, ,,
		♀	RBB	48.6±1.4	,, ,,
	fz - Sp	♂	CBB	31.6±2.6	Snell et al. (1954)
		♀	CBB	39.8±3.2	,, ,, ,, ,,
		♂	CBB	33.3±1.3	Parsons (1958a)
		♂	RBB	30.5±1.3	,, ,,
		♀	CBB	41.6±1.1	,, ,,
		♀	RBB	40.3±1.6	,, ,,
		?	?BB	41.3±5.2	Dickie and Woolley (1950)
		♂	CBB	31.3±2.4	Dickie (1964)
		♀	CBB	38.6±2.8	,, ,,
	Id-1 - ln	♀	-BB	16.7±5.8	Hutton and Roderick (1970)
	Id-1 - Sp	♀	-BB	11.1±5.4	,, ,, ,, ,,
	ln - Lp	♂	CBB	31.3±4.7	Snell et al. (1954)
		♀	CBB	36.8±9.4	,, ,, ,, ,,
	ln - py	♂	CBB	23.5±2.5	Fisher (1953)
		♂	RBB	22.1±3.4	,, ,,
		♀	CBB	36.7±2.6	,, ,,
		♀	RBB	43.2±2.4	,, ,,
		♂	CBB	23.0±1.4	Parsons (1958a)
		♂	RBB	25.2±1.7	,, ,,
		♀	CBB	37.0±1.3	,, ,,
		♀	RBB	39.4±1.6	,, ,,
	ln - Sp	♂	CBB	5.8±1.3	Snell et al. (1954)
		♀	CBB	8.5±2.2	,, ,, ,, ,,
		♂	CBB	5.3±0.6	Parsons (1958a)
		♂	RBB	3.9±0.6	,, ,,
		♀	CBB	4.7±0.6	,, ,,
		♀	RBB	5.9±0.6	,, ,,
		♂	CBB	4.1±1.0	Dickie (1964)
		♀	CBB	5.1±1.3	,, ,,
	py - Sp	♂	CBB	30.1±2.0	Parsons (1958a)
		♀	RBB	24.6±1.3	,, ,,
		♂	CBB	40.5±1.2	,, ,,
		♀	RBB	41.2±1.6	,, ,,
XIV	bg - sa	♂	CBB	3.4±1.5	Lyon and Meredith (1969)

Group Loci	Sex	Mating type	Crossover value	References
	♀	CBB	8.4±3.6	Lyon and Meredith (1969)
bg - Xt	♂	CBB	0.68±0.63	,, ,, ,, ,,
	♀	CBB	0.44±0.43	,, ,, ,, ,,
ch - cr	?	RBB	7.7±5.2	Phillips (1956)
ch - f	?	CBB	27.3±13.4	,, ,,
	?	RBB	13.7±3.0	,, ,,
cr - f	♀	RBB	33.1±1.4	King (1956)
	♀	RBB	34.2±3.0	,, ,,
	?	CBB	28.3±6.6	Phillips (1956)
	♂	CBB	14.6±2.1	Lyon et al. (1967)
	♀	CBB	17.7±3.7	,, ,, ,, ,,
cr - pe	♂	RBB	40.7±9.5	,, ,, ,, ,,
	♀	RBB	45.4±3.0	,, ,, ,, ,,
	♂	RBB	<.12.8	,, ,, ,, ,,
	♀	CBB	1.8±1.1	,, ,, ,, ,,
	♀	RBB	1.9±0.8	,, ,, ,, ,,
f - pe	-	CII	16.7±5.0	Lyon and Meredith (1969)
f - Xt	♂	CBB	17.1±2.0	Lyon et al. (1967)
	♂	RBB	25.4±3.9	,, ,, ,, ,,
	♀	CBB	22.9±3.0	,, ,, ,, ,,
	♀	RBB	20.0±8.9	,, ,, ,, ,,
mu - pe	♂	CBB	19.4±6.6	,, ,, ,, ,,
	♂	RBB	24.5±3.4	,, ,, ,, ,,
	♀	CBB	27.7±3.8	,, ,, ,, ,,
	♀	RBB	32.4±5.1	,, ,, ,, ,,
	-	CCI	21.9±4.2	,, ,, ,, ,,
	-	RII	39.2±6.1	,, ,, ,, ,,
cr - Xt	♂	CBB	1.9±1.9	Johnson (1959a)
	♂	RBB	1.4±1.4	,, ,,
	♀	CBB	<,5.2	,, ,,
	♀	RBB	<.14.8	,, ,,
	♂	CBB	0.68±0.48	Lyon et al. (1967)
mu - Xt	♂	CBB	17.1±4.1	Lyon and Meredith (1969)
	♂	RBB	5.4±2.2	,, ,, ,, ,,
	♀	CBB	18.6±2.6	,, ,, ,, ,,
pe - Xt	♂	CBB	34.4±3.5	Lyon et al. (1967)
	♂	RBB	25.2±3.4	,, ,, ,, ,,
	♀	CBB	42.3±2.4	,, ,, ,, ,,
	♀	RBB	44.9±5.6	,, ,, ,, ,,
	♂	CBB	30.6±3.8	Lyon and Meredith (1969)
	♂	RBB	36.5±6.7	,, ,, ,, ,,
	♀	CBB	38.0±4.2	,, ,, ,, ,,
	♀	RBB	35.7±5.2	,, ,, ,, ,,
	-	CII	32.6±7.1	,, ,, ,, ,,
sa - Xt	♂	CBB	3.7±1.3	,, ,, ,, ,,

Group	Loci	Sex	Mating type	Crossover value	References
		♀	CBB	8.0±1.8	Lyon and Meredith (1969)
XVI	de - Va	?	CBB	30.8±4.5	Currey (1959)
		?	RBB	24.1±4.8	,,　　　,,
XVII	g - lx	♂	RBB	45.8±8.2	Sidman and Green (1965)
	g - rd	?	RBB	5.1±2.9	Paigen and Noell (1961)
		♂	CBB	15.8±2.3	Sidman and Green (1965)
	g - W	♂	CBB	27.1±2.8	Paigen and Noell (1961)
	le - lx	♂	RBB	29.2±4.0	Lane (1967)
		♀	RBB	32.8±7.8	,,　　　,,
	le - pi	♀	CBB	20.2±1.4	,,　　　,,
		♀	RBB	21.7±3.2	,,　　　,,
		-	CII	15.3±2.9	,,　　　,,
		-	RII	21.0±6.7	,,　　　,,
	le - W	♂	CBB	13.0±1.8	,,　　　,,
		♀	CBB	19.5±1.3	,,　　　,,
		♂	CBB	7.3±1.8	Lane and Green (1967)
		♂	RBB	6.1±1.7	,,　　,,　　,,　　,,
		♀	CBB	18.4±3.2	,,　　,,　　,,　　,,
		♀	RBB	12.6±2.5	,,　　,,　　,,　　,,
	lx - pi	♂	RBB	20.3±4.9	Lane (1967)
		♀	RBB	10.1±3.1	,,　　　,,
	lx - rd	♂	RBB	47.5±7.0	Sidman and Green (1965)
	lx - rl	♀	CBB	37.7±2.1	Falconer (1952a)
		♀	RBB	7.2±1.3	,,　　　　,,
		-	CII	18.1±4.4	,,　　　　,,
		-	RII	10.8±6.2	,,　　　　,,
	lx - W	♂	CBB	11.5±3.6	Carter (1949, 1951c)
		♂	RBB	20.2±2.8	,,　　,,　　,,
		♂	CIB	28.0±8.6	,,　　,,　　,,
		♂	RIB	21.0±6.7	,,　　,,　　,,
		♀	CBB	14.6±3.4	,,　　,,　　,,
		♀	RBB	19.7±3.3	,,　　,,　　,,
		♀	RIB	19.0±7.5	,,　　,,　　,,
		♂	RBB	13.8±6.4	Sidman and Green (1965)
		♂	RBB	18.6±2.8	Lane (1967)
		♀	RBB	13.0±2.9	,,　　　,,
	rl - W	♂	CBB	14.8±14.7	Falconer (1952a)
		♂	CBB	37.9±7.8	,,　　　　,,
		-	CII	25.5±4.7	,,　　　　,,
		-	RII	35.3±7.3	,,　　　　,,
	Ph - Rw	♂	RBB	<1.8	Searle and Truslove (1970)
		♀	CBB	<6.8	,,　　,,　　,,　　,,
	pi - W	♂	CBB	8.8±2.7	Dickie and Woolley (1946)

Group	Loci	Sex	Mating type	Crossover value	References
		♀	CBB	5.3±2.3	Dickie and Woolley (1946)
		♂	CBB	0.6±0.6	Lane (1967)
		♀	CBB	3.0±0.6	„ „
	Rw - W	♂	RBB	<4.7	Searle and Truslove (1970)
		♀	RBB	<1.5	„ „ „ „
	Ph - W	♂	RBB	1.2±1.2	Gruneberg and Truslove (1960)
		♀	RBB	<0.7	„ „ „ „
	Pgm-1 - W	♀	BB	4.4±3.0	Hutton and Roderick (1970)
		?	BB	1.9±1.8	„ „ „ „
	rd - W	♂	CBB	12.2±2.1	Sidman and Green (1965)
		♀	CBB	17.6±9.2	„ „ „ „
		?	CBB	15.2±3.1	„ „ „ „
XVIII	Es 1 - Es 2	?	-BB	11.1±2.5	Popp (1967)
		-	-II	7.7±2.6	Petras and Biddle (1967)
		♂	-BB	13.0±4.6	Ruddle et al. (1969)
		♀	-BB	10.5±2.2	„ „ „ „
	Es-1 - Es 5	-	-II	9.5±2.7	Petras and Biddle (1967)
	Es-1 - la	-	-II	3.8±2.0	Yoon (1969)
	Es-1 - Os	♂	-BB	0.7±0.7	Ruddle and Roderick (1965b)
		♀	-BB	2.0±1.0	Popp (1965)
	Es-1 - tg	♀	-BB	5.8±2.1	Tsuji and Meier (1969)
	Es-2 - Es 5	-	RII	1.7±1.3	Petras and Biddle (1967)
	Es-2 -Es 6	-	RII	Linked	Petras and Sinclair (1969)
	Hk - Os	♂	RBB	6.4±2.8	Green et al. (1963)
		♀	CBB	13.7±3.6	„ „ „ „
		♀	RBB	12.2±4.7	„ „ „ „
	Os - tg	-	RII	<13.0	Green and Sidman (1962), Green et al. (1963)
XX	Bn - jp	♀	CBB	34.0±4.5	Phillips (1954)
		♀	RBB	26.1±6.2	„ „
	Bn - Mo	♀	CBB	9.1±2.7	Falconer (1954)
		♀	RBB	11.4±2.2	„ „
		♀	RBB	9.5±2.8	Phillips (1961)
		♀	CBB	13.6±1.6	Auerbach et al. (1962)
	Bn - Str	♀	RBB	4.3±2.9	Phillips (1963)
		♀	RBB	5.5±2.6	Lyon (1966a)
	Bn - Ta	♀	CBB	8.2±2.0	Falconer (1954)
		♀	RBB	11.0±2.5	„ „
		♀	CBB	21.4±10.0	Phillips (1954)
		♀	RBB	14.7±3.4	„ „
		♀	CBB	33.0±5.0	Phillips (1961)
		♀	RBB	10.7±1.4	Auerbach et al. (1962)
		♀	RBB	13.2±5.2	Lyon (1966a)
	jp - spf	♀	RBB	50.0±7.0	Cattanach (1966)

Group Loci	Sex	Mating type	Crossover value	References
jp - *Ta*	♀	RBB	20.8 ± 2.8	Phillips (1954)
Mo - *sla*	♀	RBB	2.9 ± 0.9	Falconer and Isaacson (1962)
Mo - *Str*	♀	CBB	7.9 ± 2.3	Lyon (1966a)
"	♀	RBB	10.5 ± 2.3	„ „
Mo - *Ta*	♀	RBB	3.7 ± 1.0	Falconer (1953)
	♀	RBB	6.2 ± 1.5	Falconer (1954)
	♀	RBB	5.9 ± 5.7	Lyon (1960a)
	♀	CBB	23.8 ± 9.3	Phillips (1961)
	♀	RBB	4.5 ± 0.7	„ „
	♀	CBB	2.9 ± 0.9	Falconer and Isaacson (1962)
	♀	RBB	2.9 ± 0.8	Auerbach *et al.* (1962)
	♀	CBB	2.9 ± 1.4	Lyon (1966a)
	♀	RBB	5.9 ± 1.2	„ • „
	♀	CBB	5.1 ± 2.5	Lyon and Hawkes (1970)
Mo - *Tfm*	♀	CBB	6.4 ± 2.8	„ „ „ „
sla - *Ta*	♀	CBB	8.7 ± 2.1	Falconer and Isaacson (1962)
	♀	RBB	2.0 ± 0.7	„ „ „ „
Str - *Ta*	♀	CBB	4.3 ± 0.8	Lyon (1963, 1966a)
	♀	RBB	8.7 ± 1.5	„ „ „
Tfm - *Ta*	♀	CBB	1.3 ± 1.3	Lyon and Hawkes (1970)

SEX DIFFERENCES IN CROSSINGOVER

The collection of data on the frequency of crossingover between the sexes has not been pursued on a systematic basis but, even so, information has become available for 137 chromosomal intercepts. The full results are set forth in Table 11.4. In four cases, the amount of crossingover has been exactly the same in both sexes. If these are ignored, 51 intercepts show a higher frequency in the male while 82 showed a higher frequency in the female. The observed differences attained significance for 36 intercepts, of which the male gametogenesis displayed the higher frequency in nine cases and the female in 27 cases. Presumably, most of the insignificant differences would have been more decisive, had the sample sizes been larger.

The sex difference in the mean frequency of chiasmata per bivalent as revealed by Tables 11.10 and 11.12 implies that a difference will exist in genetic crossingover. Furthermore, the difference would be in the direction of a greater frequency for the female. The above summary is in full accord with expectation. This is a general conclusion, of course, and one which may not hold for every chromosome, nor for all regions of a chromosome. None-the-less, data on 17 chromosomes (or regions thereof) indicate a higher rate of crossingover for the female. Twelve of the chromosomes display differences large enough to be statistically significant and the female accounts for eight of these.

The presence or otherwise of a sex difference for a particular chromosome intercept is of interest in itself but Carter (1954) has shown that the information can be of broader significance. In an intriguing analysis of the chiasma data of Slizynski (1955a), Carter demonstrated that the difference in mean chiasma frequency between the sexes is likely to have repercussions on the observed crossover value for the sexes in a not necessarily uniform manner. The difference in mean chiasma frequency probably results from lessened chiasma interference in the female and this implies that the frequency distribution of successive chiasmata along the chromosome will differ. Thus, since the crossover value per intercept is a function of this, successive intercepts along the chromosome could coincide with differences in chiasma densities between the sexes. A minimum in the female coinciding with a maximum in the male, for instance, could lead to a reversal of the general tendency for crossingover to be greater in the former.

Carter's analysis culminated in the following speculative conclusions for a chromosome with a terminal centromere (effectively, as for all or most of the mouse chromosomes). Near the centromere, the female should have the higher frequency of crossingover; until about 1.1 to 1.4 chiasma interference scale units (30 and 50 crossover units), whence the male should have the higher frequency. Beyond this interval, the female will again show the higher frequency, but apparently not to the same extent as the region proximal to the centromere. The interval in which the male could show the higher frequency may vary in length (and perhaps position) per chromosome depending on the strength of chiasma interference. The main implication is that only one such region should occur for each chromosome.

At this time, only linkage group VI provides positive evidence of a region giving rise to a higher rate of male crossingover.

Seven intercepts from *bt* to *N* (four of them significantly) display the pattern. The two extreme intercepts, *N* - *Ve* and *bt* - *aw*, show an identical and an insignificant higher frequency of female crossingover, respectively. As other genes are added, it will be interesting to observe if the *bt* and *N* loci constitute boundaries for the region of higher crossingover in the male.

Four other groups have one intercept each in which the male shows a significantly greater amount of crossingover. These are: III *Fks* - *s*, VII *sh-2* - *wa-2*, XII *ep* - *ru* and XVII *pi* - *W*. In no instance is there indisputable supporting evidence for a region of higher male crossingover. For example, the intercept *sh-2* - *wa-2* has been mentioned as a possibility but this intercept is on the end of the map and meagre evidence is available for adjacent intercepts. The nearest intercept shows only a slightly higher rate for the male while the next two show significantly higher rates for the female. *Fks* - *s* has data on two adjacent intercepts but both show the female possessing the higher rate. Intercepts *ep* - *ru* and *pi* - *W* are barely significant and, without supporting data, cannot be taken seriously at this time. Very little can be made of the above, for negative data is often unhelpful at the best of times. In the present case, it must be remembered that no systematic experimental study of sex differences in crossingover has been made; the rearing of adequate samples over sequences of intercepts, followed by analysis for direction and magnitude of such differences as may emerge.

The remarks of the preceding paragraphs are critical and are intended to draw attention to the general lack of data on sex differences. Only too often, it is possible to read that separate data for each sex have been examined, found to be insignificantly different, and subsequently pooled in the final report as if the lack of difference is of no import. In a formal, yet narrow sense, this is true but Carter's analysis should make it apparent that all information on sex differences is potentially useful. Even if the difference is insignificant, combination of individual data could build up to a significant result. If Carter's analysis is substantially correct, there are only two very short regions in the chromosome in which it may be held that a sex difference is non-existent. These will border the region of higher crossingover in the male and deserve to be isolated as part of the analysis of sex differences.

TABLE 11.4

Sex differences in crossover values for chromosome intercepts of the mouse.

Group	Loci	Sex	Crossover value	References
I	c - fr	♂	18.0±4.9	Falconer and Snell (1952)
		♀	16.8±3.1	„ „ „ „
		Diff.	1.2±5.8	
	c - H-1	♂	5.4±3.7	Snell (1958), Snell and Stevens (1961)
		♀	8.4±2.1	Snell (1958), Snell and Stevens (1961)
		Diff.	3.0±4.3	
	c - Hbb	♂	2.9±0.7	Popp and Amand (1960, 1964), Hutton et al. (1962), Wolfe et al. (1963)
		♀	6.6±6.9	Popp and Amand (1960, 1964), Hutton et al. (1962), Wolfe et al. (1963), Hutton and Roderick (1970)
		Diff.*	3.7±0.1	
	c - Nil	♂	3.7±1.4	Wallace and Herbertson (1969), Wallace (1970; appendix)
		♀	8.6±1.8	
		Diff.*	4.9±2.3	
	c - ol	♂	14.1±3.2	Hertwig (1942)
		♀	10.3±3.3	„ „
		Diff.	3.8±4.6	
	c - p	♂	12.3±0.5	Dunn (1920a), Castle and Wachter (1924), Detlefsen and Clemente (1924), Feldman (1924), Detlefsen (1925), Gruneberg (1935, 1936b), Falconer and Snell (1952), Popp and Amand (1960), Amand (1970; appendix)
		♀	16.8±4.2	Dunn (1920a), Castle and Wachter (1924), Detlefsen and Clemente (1924), Feldman (1924), Detlefsen (1925), Gruneberg (1935, 1936b), Falconer and Snell (1952), Popp and Amand (1960), Snell and Stevens (1961,) Amand (1970: appendix)

Group Loci	Sex	Crossover value	References
	Diff.*	4.6±0.7	
c - pu	♂	30.4±1.8	Amand (1970; appendix)
	♀	32.3±1.8	,, ,, ,,
	Diff.	1.9±2.6	
c - sh-1	♂	2.6±0.5	Gates (1928b, 1929, 1931), Gruneberg (1935, 1936b), Falconer and Snell (1952)
	♀	4.1±2.0	Gates (1928b, 1929, 1931), Gruneberg (1935, 1936b), Falconer and Snell (1952)
	Diff.*	1.5±0.5	
fr - p	♂	29.3±4.4	Falconer and Snell (1952), Eicher (1967)
	♀	34.9±3.9	Falconer and Snell (1952)
	Diff.	5.6±5.9	
fr - sh-1	♂	16.7±3.6	Falconer and Snell (1952), Eicher (1967)
	♀	15.4±3.0	Falconer and Snell (1952)
	Diff.	1.2±4.7	
Hbb - p	♂	15.6±1.4	Popp and Amand (1960), Popp (1962a)
	♀	20.3±1.4	Popp and Amand (1960), Popp (1962a)
	Diff.*	4.7±2.0	
Hbb - sh-1	♂	2.0±0.5	Popp (1962a)
	♀	1.4±0.4	,, ,,
	Diff.	0.6±0.6	
Nil - p	♂	3.2±1.3	Wallace and Herbertson (1969), Wallace (1970; appendix)
	♀	2.5±1.3	Wallace and Herbertson (1969), Wallace (1970; appendix)
	Diff.	0.7±1.8	
p - pu	♂	16.3±1.4	Amand (1970; appendix)
	♀	16.0±1.4	,, ,, ,,
	Diff.	0.3±2.0	
p - ru-2	♂	2.5±0.7	Eicher (1970), Lilly (1970; appendix)
	♀	3.2±1.2	Eicher (1970)
	Diff.	0.7±1.5	
p - sh-1	♂	13.9±2.1	Snell (1946), Falconer and Snell (1952), Popp (1962a), Eicher (1967)
	♀	15.2±1.4	Snell (1964), Falconer and Snell (1952), Popp (1962a)
	Diff.	1.3±2.5	

Group	Loci	Sex	Crossover value	References
II	cw - se	♂	36.3±3.3	Falconer and Isaacson (1966), Lyon et al. (1968)
		♀	37.8±2.6	Falconer and Isaacson (1966), Lyon et al. (1968)
		Diff.	1.5±4.2	
	cw - tk	♂	42.5±3.4	Falconer and Isaacson (1966), Lyon et al. (1968)
		♀	44.1±2.7	Falconer and Isaacson (1966), Lyon et al. (1968)
		Diff.	2.7±4.3	
	d - se	♂	0.20±0.09	Castle et al. (1936), Goodwin and Vincent (1955)
		♀	0.11±0.09	Snell (1931), Castle et al. (1936), Goodwin and Vincent (1955)
		Diff.	0.09±0.11	
	d - tk	♂	3.4±0.8	Green and Lane (1967)
		♀	3.2±1.0	„　　„　　„　　„
		Diff.	0.2±1.3	
	d - Trf	♂	8.5±2.6	Shreffler (1963)
		♀	16.9±2.6	„　　„
		Diff.*	8.3±3.7	
	dse - lu	♂	13.6±5.1	Green (1961)
		♀	18.5±4.0	„　　„
		Diff.	4.9±6.5	
	du - se	♂	15.0±7.0	„　　„
		♀	21.6±4.1	„　　„
		Diff.	6.3±8.1	
	lu - Trf	♂	20.7±5.2	Shreffler (1963)
		♀	23.8±5.3	„　　„
		Diff.	3.1±7.4	
	se - sv	♂	1.8±1.7	Deol and Green (1966)
		♀	2.7±0.7	„　　„　　„　　„
		Diff.	0.9±1.8	
	se - tk	♂	5.2±1.5	Falconer and Isaacson (1966), Lyon et al. (1968)
		♀	5.6±1.2	Falconer and Isaacson (1966), Lyon et al. (1968)
		Diff.	0.4±1.9	
III	Fkl - s	♂	23.3±3.9	Lyon (1970; appendix)
		♀	9.1±3.1	„　　„　　„
		Diff.*	14.2±5.1	
	hr - s	♂	2.3±7.7	Snell (1931)
		♀	10.3±1.9	Snell (1931), Kidwell et al. (1966)
		Diff.	8.8±7.8	

Group	Loci	Sex	Crossover value	References
	pn - s	♂	24.4±4.7	Kidwell *et al.* (1961)
		♀	27.0±1.0	Kidwell *et al.* (1961, 1966)
		Diff.	2.6±6.7	
IV	av - si	♂	35.0±3.8	Schaible (1970; appendix)
		♀	32.1±3.1	„ „ „
		Diff.	2.9±4.8	
	gl - Sl	♂	17.7±13.2	Lane (1970; appendix)
		♀	31.1±4.8	„ „ „
		Diff.	13.4±15.9	
V	a - As	♂	0.51±0.22	Phillips (1966)
		♀	0.40±0.19	„ „
		Diff.	0.11±0.29	
	a - fi	♂	26.9±1.8	Wallace (1957a)
		♀	35.8±2.1	„ „
		Diff.*	8.9±2.8	
	a - H-3	♂	12.6±3.2	Snell and Bunker (1964), Snell *et al.* (1967)
		♀	18.4±2.1	Snell and Bunker (1964), Snell *et al.* (1967)
		Diff.	5.7±3.8	
	a - lst	♂	19.1±4.4	Forsthoefel and Shenk (1970)
		♀	28.0±5.9	„ „ „ „
		Diff.	8.9±7.4	
	a - mg	♂	9.8±1.1	Lane and Green (1960)
		♀	13.2±1.0	„ „ „ „
		Diff.*	3.3±1.5	
	a - pa	♂	15.8±0.6	Roberts and Quisenberg (1935), Owen (1953b), Green and Mann (1961)
		♀	17.1±0.6	Roberts and Quisenberg (1935), Owen (1953b), Carter and Phillips (1954)
		Diff.	1.4±0.9	
	a - Ra	♂	23.6±0.1	Carter and Phillips (1954), Parsons (1958b), Lane and Green (1960), Green and Lane (1961)
		♀	19.7±1.1	Carter and Phillips (1954), Parsons (1958b), Lane and Green (1960)
		Diff.*	3.9±1.5	

Group Loci	Sex	Crossover value	References
a - Sd	♂	43.5±1.8	Carter and Phillips (1954), Wallace (1957a), Lane and Dickie (1968)
	♀	45.3±2.0	Carter and Phillips (1954), Wallace (1957a)
	Diff.	1.8±2.7	
a - stb	♂	40.9±3.0	Lane and Dickie (1968)
	♀	34.8±5.4	,, ,, ,, ,,
	Diff.	6.1±6.1	
a - Sut	♂	6.3±1.7	Egorov and Blandova (1970; appendix)
	♀	3.1±1.4	Egorov and Blandova (1970; appendix)
	Diff.	3.2±2.2	
a - un	♂	4.6±0.3	Carter (1947), Fisher (1949), Owen (1953b), Snell and Bunker (1964), Snell et al. (1967)
	♀	4.8±0.2	Carter (1947), Fisher (1949), Owen (1953b), Snell and Bunker (1964), Snell et al. (1967)
	Diff.	0.2±0.4	
a - we	♂	9.9±0.5	Hertwig (1942), Owen (1953b), Parsons (1958b), Snell and Bunker (1964), Snell et al. (1967)
	♀	13.8±0.5	Hertwig (1942), Owen (1953b), Parsons (1958b), Snell and Bunker (1964), Snell et al. (1967)
	Diff.*	3.9±0.7	
fi - pa	♂	26.6±4.3	Bodner (1961)
	♀	27.1±1.7	,, ,,
	Diff.	0.5±6.8	
fi - Sd	♂	19.6±1.6	Wallace (1957a), Carter et al. (1956)
	♀	19.7±1.6	Wallace (1957a), Carter et al. (1956)
	Diff.	0.1±2.3	
H-3 - un	♂	9.2±1.2	Snell and Bunker (1964), Snell et al. (1967)
	♀	13.4±1.8	Snell and Bunker (1964), Snell et ql. (1967)
	Diff.*	4.2±1.1	

Group	Loci	Sex	Crossover value	References
	H-3 - we	♂	1.0±1.0	Snell and Bunker (1964), Snell et al. (1967)
		♀	2.3±1.0	Snell and Bunker (1964), Snell et al. (1967)
		Diff.	1.3±1.0	
	mg - Ra	♂	29.7±2.4	Lane and Green (1960)
		♀	25.9±2.0	„ „ „ „
		Diff.	3.8±3.1	
	pa - Ra	♂	45.2±7.6	Green and Mann (1961)
		♀	39.4±5.8	Carter and Phillips (1954)
		Diff.	5.8±9.5	
	pa - un	♂	6.7±0.8	Owen (1953b)
		♀	11.6±0.6	Owen (1953b), Falconer (1954)
		Diff.*	4.9±0.8	
	pa - we	♂	2.3±0.3	Owen (1953b)
		♀	4.0±0.4	Owen (1953b), Falconer (1954)
		Diff.*	1.7±0.5	
	Ra - Sd	♂	44.7±4.5	Carter and Phillips (1954)
		♀	45.5±15.0	„ „ „ „
		Diff.	0.8±15.7	
	Ra - we	♂	33.2±1.7	Parsons (1958b)
		♀	32.7±1.6	„ „
		Diff.	0.5±2.3	
	ro - un	♂	10.2±1.7	Falconer (1954)
		♀	6.2±1.5	„ „
		Diff.	4.1±2.3	
	ro - we	♂	4.3±1.2	„ „
		♀	2.9±1.1	„ „
		Diff.	1.4±1.6	
	Sd - stb'	♂	11.3±1.8	Lane and Dickie (1968)
		♀	10.2±5.1	„ „ „ „
		Diff.	1.1±5.5	
	un - we	♂	4.7±0.4	Owen (1953b), Falconer (1954), Snell and Bunker (1964), Snell et al. (1967)
		♀	7.4±0.4	Owen (1953b), Falconer (1954), Snell and Bunker (1964), Snell et al. (1967)
		Diff.*	2.7±0.6	
VI	bt - Ca	♂	11.7±0.5	Flanagan and Isaacson (1967), Mallyon and Wallace (1970; appendix), Amand (1970; appendix)

Group Loci	Sex	Crossover value	References
	♀	4.8±0.4	Flanagan and Isaacson (1967), Mallyon and Wallace (1970; appendix), Amand (1970; appendix)
	Diff.*	6.9±0.6	
bt - hl	♂	8.5±0.7	Hollander (1959, 1970; appendix)
	♀	7.6±0.5	,, ,, ,, ,,
	Diff.	0.9±0.9	
bt - Ht	♂	7.8±0.7	W.St. Amand (1970; appendix)
	♀	5.4±0.6	,, ,, ,, ,,
	Diff.	1.6±0.9	
bt - N	♂	13.1±0.6	Murray and Snell (1945), Mallyon and Wallace (1970; appendix)
	♀	4.5±0.4	Murray and Snell (1945), Mallyon and Wallace (1970; appendix)
	Diff.*	8.6±0.8	
bt - uw	♂	38.3±2.8	M. M. Dickie (1970; appendix)
	♀	43.7±2.6	,, ,, ,, ,, ,,
	Diff.	5.4±3.8	
bt - Ve	♂	9.8±1.2	Stieler (1970; appendix)
	♀	7.3±1.5	,, ,, ,,
	Diff.	2.5±2.0	
Ca - hl	♂	4.7±0.5	Hollander (1959, 1970; appendix)
	♀	1.4±0.2	,, ,, ,, ,,
	Diff.*	3.3±0.5	
Ca - Ht	♂	4.9±0.6	W.St. Amand (1970; appendix)
	♀	1.7±0.3	,, ,, ,, ,,
	Diff.*	3.2±0.7	
Ca - N	♂	1.3±1.6	Cooper (1939), Murray and Snell (1945), Hollander (1959, 1970; appendix), Flanagan and Isaacson (1967), Mallyon and Wallace (1970; appendix)
	♀	0.38±0.10	Cooper (1939, Murray and Snell (1945), McNeil (1956), Hollander (1959, 1970; appendix), Flanagan and Isaacson (1967), Mallyon and Wallace (1970; appendix)
	Diff.*	0.93±0.18	
Ca - uw	♂	48.3±4.7	M. M. Dickie (1970; appendix)
	♀	40.5±4.1	,, ,, ,, ,, ,,

Group	Loci	Sex	Crossover value	References
		Diff.	7.7±6.1	
	N - Sha	♂	1.0±0.2	Flanagan and Isaacson (1967)
		♀	1.4±0.5	„ „ „ „
		Diff.	0.4±0.6	
	N - Ve	♂	1.8±0.7	Stieler (1970; appendix)
		♀	1.8±5.7	„ „ „
		Diff.	0±5.8	
VII	*Al - Re*	♂	30.4±3.9	Dickie (1955)
		♀	29.5±4.7	„ „
		Diff.	0.9±6.1	
	Re - sh-2	♂	19.1±1.8	Falconer (1947), Carter and Phillips (1953)
		♀	27.8±2.6	Falconer (1947), Falconer and Sobey (1953), Nasrat (1956)
		Diff.*	8.7±3.1	
	Re - ti	♂	20.5±2.1	Searle (1961)
		♀	19.2±2.1	„ „
		Diff.	1.3±3.0	
	Re - tn	♂	15.7±6.2	Lane (1970; appendix)
		♀	22.9±2.7	„ „ „
		Diff.*	7.2±2.1	
	Re - vb	♂	17.6±7.6	„ „ „
		♀	18.0±2.7	„ „ „
		Diff.	0.4±8.1	
	Re - vt	♂	16.6±2.2	Michie (1952, 1955a)
		♀	27.3±3.1	„ „ „
		Diff.*	10.7±3.8	
	Re - wa-2	♂	43.6±2.5	Carter and Phillips (1953)
		♀	42.0±2.4	Falconer (1947), Carter (1951b)
		Diff.	1.6±3.4	
	sh-2 - wa-2	♂	29.4±1.4	Snell and Law (1939), Wright (1947b), Fisher *et al.* (1947), Carter and Phillips (1953)
		♀	23.6±0.9	Snell and Law (1939), Carter and Phillips (1953), Falconer and Sobey (1953), Carter *et al.* (1956)
		Diff.*	5.8±1.7	
VIII	*b - m*	♂	6.6±1.2	Woolley (1945), Snell (1946), Lane (1963), Johnson (1969b)
		♀	7.7±0.6	Wooley (1945), Snell (1946), Sirlin (1957), Lane (1963), Johnson (1969b)

Group	Loci	Sex	Crossover value	References
		Diff.	1.1±1.4	
	b - Ps	♂	6.7±2.4	Johnson (1969b)
		♀	9.0±1.9	„ „
		Diff.	2.3±3.1	
	b - Pt	♂	5.1±0.8	Hollander and Strong (1951), Lane (1963), Meredith (1970; appendix)
		♀	4.5±0.6	Hollander and Strong (1951), Yoon (1961), Lane (1963)
		Diff.	0.6±0.9	
	b - sno	♂	15.0±1.2	Hollander (1970; appendix)
		♀	18.3±1.2	„ „ „
		Diff.	3.3±1.7	
	b - wi	♂	4.3±1.6	Lane (1963)
		♀	3.6±1.0	„ „
		Diff.	0.7±1.9	
	m - Ps	♂	1.0±0.8	Johnson (1969b)
		♀	4.0±0.2	„ „
		Diff.	3.0±1.0	
	m - Pt	♂	2.2±1.5	Lane (1963)
		♀	3.4±0.6	„ „
		Diff.	1.3±1.6	
	Pt - wi	♂	6.1±1.3	„ „
		♀	11.0±1.1	
		Diff.*	4.9±1.8	
IX	Fu - H-2	♂	4.1±1.2	Allen (1955a)
		♀	6.9±3.0	„ „
		Diff.	2.8±3.2	
	H-2 - T	♂	9.1±0.7	Snell (1952), Allen (1955b), Shreffler (1964b), Snell et al. (1967), Pizarro and Dunn (1970), Green and Stimpfling, (1970; appendix)
		♀	15.9±1.2	Snell (1952), Allen (1955b), Shreffler (1964b), Snell et al. (1967), Pizarro and Dunn (1970), Green and Stimpfling, (1970; appendix)
		Diff.*	6.7±1.4	
	H-2 - tf	♂	3.2±0.7	Pizarro and Dunn (1970), Green and Stimpfling (1970; appendix)
		♀	4.6±1.1	Pizarro and Dunn (1970), Green and Stimpfling (1970; appendix)

Group	Loci	Sex	Crossover value	References
		Diff.	1.4±1.3	
	Ki - T	♂	2.8±0.7	Dunn and Caspari (1942, 1945)
		♀	5.9±1.5	,, ,, ,, ,, ,,
		Diff.	3.1±1.6	
	Ki - tf	♂	0.4±0.3	Dunn et al. (1962)
		♀	1.2±0.5	,, ,, ,, ,,
		Diff.	0.8±0.5	
	T - tf	♂	4.3±0.4	Lyon (1956), Lyon and Phillips (1959), Dunn et al. (1962), Lyon and Meredith (1964a), Searle (1966), Dunn and Bennett (1968; 1970; appendix), Green and Stimpfling (1970; appendix)
		♀	9.7±0.7	Lyon (1956), Lyon and Phillips (1959), Dunn et al. (1962), Lyon and Meredith (1964a), Searle (1966), Dunn and Bennett (1968; 1970; appendix), Green and Stimpfling (1970; appendix)
		Diff.*	5.3±0.8	
X	ji - v	♂	10.2±4.3	Snell (1945)
		♀	40.0±15.5	,, ,,
		Diff.	29.8±16.1	
XI	Lc - Mi	♂	10.7±1.2	Phillips (1960)
		♀	11.6±1.7	,, ,,
		Diff.	0.9±2.1	
	Lc - wa-1	♂	7.8±1.9	Phillips (1960)
		♀	8.5±2.0	,, ,,
		Diff.	0.7±2.8	
	Mi - ob	♂	25.2±5.6	Lane (1970; appendix)
		♀	25.5±6.9	,, ,, ,,
		Diff.	0.3±8.9	
	Mi - wa-1	♂	5.0±1.4	Bunker and Snell (1948), Phillips (1960)
		♀	5.0±1.2	Bunker and Snell (1948), Phillips (1960)
		Diff.	0±2.2	
	Mi - tc	♂	5.0±1.1	Lane (1970; appendix)
		♀	6.9±1.2	,, ,, ,,
		Diff.	1.9±1.6	

Group	Loci	Sex	Crossover value	References
XII	ep - ru	♂	2.6±0.6	Green and Lane (1961)
		♀	1.2±0.4	„ „ „ „
		Diff.*	1.4±0.7	
XIII	Dh - fz	♂	41.5±1.1	Searle (1964)
		♀	43.9±2.8	„ „
		Diff.	2.4±3.0	
	Dh - ln	♂	3.6±0.5	Lyon (1961a), Searle (1964)
		♀	2.2±0.9	Searle (1964)
		Diff.	1.4±1.0	
	fz - ln	♂	35.5±0.7	Dickie and Woolley (1950), Snell et al. (1954), Carter et al. (1956), Parsons (1958), Dickie (1964), Searle (1964)
		♀	43.1±0.8	Dickie and Woolley (1950), Snell et al. (1954), Carter et al. (1956), Parsons (1958), Dickie (1964), Searle (1964)
		Diff.*	7.6±1.0	
	fz - Lp	♂	54.8±5.2	Snell et al. (1954)
		♀	50.0±5.2	„ „ „ „
		Diff.	4.8±7.4	
	fz - py	♂	49.4±1.1	Parsons (1958a)
		♀	49.3±0.9	„ „
		Diff.	0.1±1.4	
	fz - Sp	♂	31.9±0.8	Dickie and Woolley (1950), Snell et al. (1954), Parsons (1958a), Dickie (1964)
		♀	40.9±0.8	Dickie and Woolley (1950), Snell et al. (1954), Parsons (1958a), Dickie (1964)
		Diff.*	9.0±1.2	
	ln - Lp	♂	31.3±4.7	Snell et al. (1954)
		♀	36.8±9.4	„ „ „ „
		Diff.	5.5±10.5	
	ln - py	♂	24.5±0.9	Fisher (1953), Parsons (1958a)
		♀	38.5±0.9	„ „ „ „
		Diff.*	14.0±1.3	
	ln - Sp	♂	4.6±0.4	Snell et al. (1954), Parsons (1958a), Dickie (1964)
		♀	5.4±0.4	Snell et al. (1954), Parsons (1958a), Dickie (1964)
		Diff.	0.8±0.5	
	ln - th	♂	1.3±0.7	Larsen (1965; appendix)
		♀	2.5±0.7	„ „ „

Group	Loci	Sex	Crossover value	References
		Diff.	1.2±0.9	
	$py.- Sp$	♂	27.6±1.1	Parsons (1958a)
		♀	40.4±1.0	,, ,,
		Diff.*	12.8±1.4	
	$Sp - th$	♂	4.6±1.2	Larsen (1965; appendix)
		♀	6.5±1.1	,, ,, ,,
		Diff.	1.9±1.6	
XIV	$bg - sa$	♂	5.8±0.9	Lyon and Meredith (1969), Amand (1970; appendix)
		♀	9.1±1.0	Lyon and Meredith (1969), Amand (1970; appendix)
		Diff.*	3.3±1.3	
	$bg - Xt$	♂	0.65±6.3	Lyon and Meredith (1969), Amand (1970; appendix)
		♀	0.44±4.3	Lyon and Meredith (1969), Amand (1970; appendix)
		Diff.	0.24±0.76	
	$cr - f$	♂	14.6±2.1	Lyon et al. (1967)
		♀	17.7±3.7	,, ,, ,, ,,
		Diff.	3.1±4.2	
	$cr - pe$	♂	40.7±9.5	,, ,, ,, ,,
		♀	45.4±3.0	,, ,, ,, ,,
		Diff.	4.6±9.9	
	$cr - Xt$	♂	0.8±0.4	Johnson (1969a), Lyon et al. (1967)
		♀	1.5±0.6	Johnson (1969a), Lyon et al. (1967)
		Diff.	0.7±0.7	
	$f - Xt$	♂	18.8±1.8	Lyon et al. (1967)
		♀	22.7±2.4	,, ,, ,, ,,
		Diff.	3.9±3.0	
	$mu - pe$	♂	23.5±3.0	Lyon and Meredith (1969)
		♀	29.3±3.1	,, ,, ,, ,,
		Diff.	5.8±4.4	
	$mu - Xt$	♂	8.0±1.9	,, ,, ,, ,,
		♀	18.6±2.6	,, ,, ,, ,,
		Diff.*	10.1±3.2	
	$pe - Xt$	♂	30.6±2.0	Lyon et al. (1967), Lyon and Meredith (1969)
		♀	40.9±1.8	Lyon et al. (1967), Lyon and Meredith (1969)
		Diff.*	10.3±2.7	
	$sa - Xt$	♂	3.7±1.3	Lyon and Meredith (1969)
		♀	8.0±1.8	,, ,, ,, ,,

Group Loci	Sex	Crossover value	References
	Diff.	4.3±2.2	
XVI *ft - ma*	♂	4.8±1.0	Lane (1970: appendix)
	♀	3.9±1.2	,, ,, ,,
	Diff.	0.9±1.5	
fr - Va	♂	22.1±3.6	,, ,, ,,
	♀	28.5±3.4	,, ,, ,,
	Diff.	6.4±4.9	
ma - Va	♂	28.6±3.8	,, ,, ,,
	♀	22.6±3.7	,, ,, ,,
	Diff.	6.0±5.0	
XVII *go - Hm*	♂	28.4±3.7	Hollander (1970; appendix)
	♀	31.2±3.4	,, ,, ,,
	Diff.	2.8±5.0	
Hm - W	♂	20.4±3.0	Green (1970; appendix)
	♀	20.7±1.4	Green (1970; appendix), Green and Woodworth (1970; appendix)
	Diff.	0.3±3.3	
le - lx	♂	29.2±4.0	Lane (1970)
	♀	32.8±7.8	,, ,,
	Diff.	3.6±8.8	
le - W	♂	9.3±1.0	Lane (1967), Lane and Green (1967)
	♀	17.3±1.0	Lane (1967), Lane and Green (1967)
	Diff.*	8.0±1.4	
lx - pi	♂	20.3±4.9	Lane (1967)
	♀	10.1±3.1	,, ,,
	Diff.	10.3±5.8	
lx - W	♂	17.9±1.6	Carter (1951c), Sidman and Green (1965), Lane (1967)
	♀	16.0±1.6	Carter (1951c), Lane (1967)
	Diff.	1.9±2.3	
pi - W	♂	8.8±2.7	Dickie and Woolley (1946)
	♀	3.1±0.6	Lane (1967)
	Diff.*	5.7±2.7	
rd - W	♂	12.4±1.7	Sidman and Green (1965)
	♀	10.5±3.2	,, ,, ,, ,,
	Diff.	1.9±3.6	
XVIII *e - Os*	♂	28±3.0	Hollander (1970; appendix)
	♀	28±7.0	,, ,, ,,
	Diff.	0±7.3	

Group Loci	Sex	Crossover value	References
e - Q	♂	30±3.0	Hollander (1970; appendix)
	♀	30±7.0	„ „ „
	Diff.	0±7.4	
Es-1 - Es-2	♂	13.0±4.6	Ruddle et al. (1969)
	♀	10.5±2.2	„ „ „ „
	Diff.	2.5±5.1	
Es-1 - Os	♂	0.7±0.7	Ruddle and Roderick (1965)
	♀	2.0±1.0	Popp (1965)
	Diff.	1.3±1.2	
Hk - Os	♂	6.4±2.8	Green et al. (1963)
	♀	13.2±2.8	„ „ „ „
	Diff.	6.7±4.0	
Os - Q	♂	5.8±1.2	Hollander (1970; appendix)
	♀	2.8±1.9	„ „ „
	Diff.	3.0±2.3	

* Significant difference

MULTIPOINT CROSSES

There have been approximately 55 published cases of multipoint crosses. The overwhelming majority are three-point and the few exceptions are four-point. In most instances, these have been conducted with the objective of pin-pointing the order of loci rather more directly and accurately than a succession of two-point crosses. The number of experiments which have been mainly designed for other purposes are few (Owen, 1953; Wallace, 1957a; Parsons, 1958a, 1958b). In this section, the multi-point crosses are examined for degree of interference.• To this end, both the coincidence and Kosambi coefficients have been computed and compared with each other. It is rather unfortunate that the majority of the multi-point crosses had to be rejected for various reasons. Several involved problems of misclassification of phenotypes but the main reason was insufficient observations in the double crossover class. Initially, it was proposed to reject all data which did not contain at least five double crossovers but this was later amended to four. This admitted a larger source of error, but it extended the range of cases. In brief, this sort of analysis is relatively new and is explorative. The paucity of entries in the

TABLE 11.5 Coincidence and Kosambi coefficients for three-point crosses

Linkage group	Triad	Length (CO units)	Sex	Coincidence	Kosambi Coefficient	References
I	c - p - Sh-1	19	♀	0.286±0.123*	0.743±0.340	Gruneberg (1935, 1936b)
II	cw - se - tk	40	♀	0.554±0.254	0.641±0.294	Falconer and Isaacson (1966)
III	pn - hr - s	30	♀	0.336±0.161*	0.690±0.326	Kidwell et al. (1966)
V	Sd - stb - a	51	♂	1.013±0.108	1.185±0.469	Lane and Dickie (1968)
	Sd - fi - a	52	♂	0.614±0.110*	0.711±0.154	Wallace (1957a)
			♂♀	0.766±0.110*	0.849±0.148	,, ,,
	Sd - a - Ra	66	♂	1.139±0.187	1.273±0.326	Carter and Phillips (1954)
	we - a - Ra	34	♂	0.235±0.107*	0.354±0.157*	Parsons (1958b)
			♂♀	0.299±0.099*	0.456±0.172*	,, ,,
VII	wa-2 - vt - ti	34	♀	0.418±0.174*	0.711±0.314	Searle (1961)
VIII	wi - b - Pt	8	♀♀	1.191±0.569	5.142±2.801	Lane (1963)
	b - Pt - m	11	♀♀	0.337±0.148*	2.249±1.075	,, ,,
XIII	fz - Sp - ln	41	♂♀	0.675±0.222	0.987±0.346	Snell et al. (1954), Dickie (1964)
	fz - Sp - ln	41	♂♀	0.480±0.173*	0.554±0.212*	,, ,, ,, ,,
			♂♀	0.562±0.111*	0.782±0.156	Parsons (1958a)
	fz - ln - py	71	♂	0.576±0.085*	0.694±0.105*	,, ,,
			♂	0.604±0.042*	0.611±0.051*	,, ,,
			♀	0.968±0.028	0.982±0.046	,, ,,

	fz - Sp - py	71	♂	0.613±0.041*	0.620±0.051*	Parsons (1958a)
			♀	0.962±0.028	0.976±0.051	,,
	fz - ln - Lp	72	♂	0.713±0.122*	0.665±0.665	Snell et al. (1954)
			♀	0.927±0.237	0.927±0.374*	,, ,, ,, ,,
	Sp - ln - py	37	♂	0.410±0.114*	0.743±0.211	Parsons (1958a)
			♀	0.697±0.093*	0.862±0.124	,,
XIV	Xt - mu - pe	39	♀	0.776±0.205	1.045±0.111	Lyon and Meredith (1969)
V	Sd - pa - a	61	**	0.861±0.160	0.946±0.223	Borger (1950)
	mg - a - Ra	34	**	0.110±0.074*	0.359±0.160*	Lane and Green (1960)

Note.   * Significant difference   ** Both sexes pooled

double crossover class is due in some instances to the closeness of linkage but, in the majority, it seemingly arises from the fact that the researcher was pre-occupied with other matters and terminated the experiment before adequate double crossover animals have been obtained.

The results of the analysis for interference displayed by three-point crosses is shown by Table 11.5. The data are inclined to be fragmentary and it may not be wise to draw firm conclusions at this stage. However, a few items may be mentioned. The C and K values are positively and significantly correlated ($r = 0.55$; $t = 5.3$, df. = 23). K is inversely correlated with intercept length but not significantly ($r = -0.3125$; $t = 1.6$, df. = 23). This latter relation is not to be expected if Owen's deductions are correct. Out of the 27 entries, 17 of the C values and eight of the K values differ significantly from unity. It is particularly interesting that not one of the significant values is greater than unity. This could indicate a higher degree of interference either in general or because the chromosome regions represented by these intercepts are remote from the centromere; indeed, beyond the median region of the chromosome arm, if Owen's hypothesis is accepted.

The only linkage group for which the centromere has been accurately positioned is II, the centromere being about two crossover units from cw. This locus has featured in a three-point cross and the triad of genes has given low (but insignificant) values for C and K. The sample size is small and the standard errors are large; hence the realization of a low (rather than a high) K value may not represent a great discrepancy between observation and expectation. Although the data currently available should be regarded as preliminary, these support the suggestion of Parsons (1958b) that interference may be more intense for the mouse than, for example, in *Drosophila melanogaster*.

The table creates the impression that female gametogenesis is giving higher K values than the male. Out of the eight crosses for which data are available for each sex, six have larger values for the female. The means are 0.705 ± 0.060 and 0.793 ± 0.053 for the male and female, respectively, but the difference is insignificant. A real difference in K could occur and certainly cannot be ruled out because of the observed difference in chiasma frequency and distribution (cf. the comments of Carter, 1954) and the sex difference of crossover values. The usual qualifications apply, of course, that a sex difference may not exist for all segments of a chromosome or for all chromosomes. Parsons (1958a, b) has in fact remarked that each sex in the mouse may have to be analysed in terms of a different interference function.

The last two entries of the table give data which has been pooled for the sexes. Borgers' (1950) observations are too extensive to be ignored despite the act of pooling. Green and Lane (1960) presented their data separately for each sex but these do not give adequate double crossovers if examined alone. When pooled, the data yield the significant results of the table. Owen (1953b) gives data for a four-point cross and calculates the following K values (using his approximate formulae for low frequencies of double crossovers): male, $K = 0.162 \pm 0.164$ and female, $K = 0.966 \pm 0.683$. However, the data as a whole contain only three double crossovers and this is an insufficient number for quantitative purposes.

MAPPING FUNCTIONS

The mouse has featured particularly in the development of mapping functions, initially because of the discovery by Wright (1947b) of crossover values significantly in excess of 50 per cent. Wright's data are interesting yet peculiar. Two genes, only moderately linked (31 per cent), individually showed large and similar crossover values (56 per cent) with sex. This fact is mutually inconsistent and at best is indicative of a curious situation. However, subsequent observations have failed to confirm the partial sex-linkage (Carter and Phillips, 1953; Tatchell, 1961a) and the data remain somewhat of an oddity (Michie 1955c).

However, Wright's observations provided the stimulus for the development of a mapping function capable of explaining her data (Fisher, Lyon, and Owen 1947). Subsequently, Owen (1949, 1950) expanded the theory of these functions and has contributed a useful perspective view. The interesting aspect of this work is that functions are now available which are capable of allowing for crossover values in excess of 50 per cent;

something which the older functions did not do (Haldane 1919, Kosambi, 1944). However, it has yet to be adequately demonstrated that crossover values in excess of 50 per cent are common or even exist in mammalian genetics. There is a hint in the distribution of observed crossover values shown by Fig 11.17 that these are a possibility but statistical confirmation is required.

The amount of data which could indicate whether or not crossover values exceeding 50 per cent do exist for the mouse is meagre but has been examined by Carter (1954). The evidence is negative. Making use of chiasma data collected by Slizynski (1949, 1955) for the male, Carter has published a table for converting the crossover value to chromosome map distance and *vice versa*. The table conveniently summarized a monotonic relationship between the two variables which results in p increasing steadily to a maximum of 50 per cent. The relationship is approximated fairly closely by the fourth power curve proposed by Carter and Falconer (1951), particularly over the important range of crossover values from zero to 40 per cent.

Carter and Falconer's (1951) proposed mapping function is of interest because of the remark that linkage data of the mouse has been found to give a good fit to the relationship,

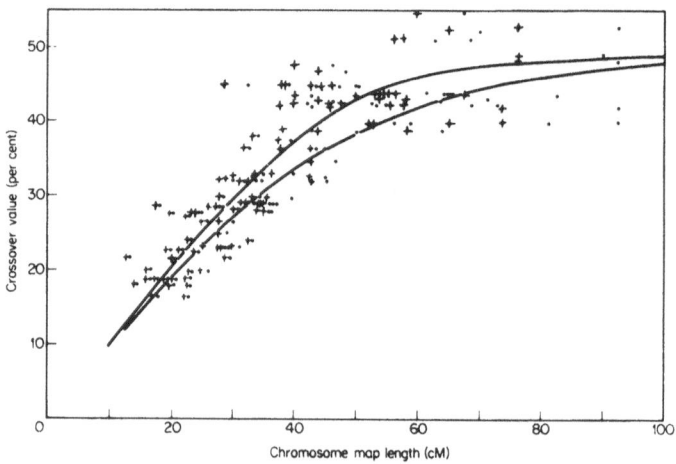

Fig. 11.17. Comparison of the predictions of the Carter-Falconer and Kosambi mapping functions (upper and lower curves, respectively).

particularly where large crossover values are concerned. Up to now, the commonly employed mapping function has been Kosambi's (e.g., see Wallace, 1957a), and the function has in fact given reasonable results. A comparison of the Carter-Falconer and Kosambi functions has been made to ascertain if one might be superior to the other. The results are shown in Fig. 11.17. The procedure has been to take all of the possible triads of linked loci per linkage group and plot the observed crossover value for the two outside loci against the chromosome map length as predicted by the sum of the two intervening intercepts. Two plots have been made, one for the Kosambi function (points) and one for the Carter-Falconer (crosses). The two curves show the theoretical values for the crossover value for chromosome map length. Both curves are practically identical for loci less than 10 cM apart and almost identical for loci up to 20 cM apart. For this reason, all triads of less than 20 cM have been ignored.

Both relationships may be held to be representative but simple χ^2 tests reveal that the Carter-Falconer provides a better fit than the Kosambi. It may be noted that the individual entries (points and crosses) occur as obvious pairs for triads of 35 to 40 cM or less. Beyond this point, the intervening intercepts become longer and the differences between the two predictions and the observed crossover values become greater. It is only by consideration of the differences for large intercepts that a decision will be possible between the validity of the various mapping functions. Inspection of the figure shows that the Carter-Falconer function (upper curve) bisects the scatter of entries rather evenly whereas the Kosambi function (lower curve) gives mean values which are consistently too low. The Carter-Falconer and Kosambi functions had been described as 'empirical' because their worth rests upon the goodness of fit to observation and the availability of tanh and tan tables by which numerical values can be readily obtained. There is no reason why the use of these functions should not continue until the appropriateness of the more sophisticated conceived functions have been experimentally verified.

The Kosambi function yields a simple and direct addition formula for the computation of crossover values for a triad of loci:

TABLE 11.6. A conversion table for the Carter-Falconer relationship between crossover percentage (p) and chromosome map length (x). p is shown by the outside row and column while x is shown by the main columns.

p	0.0	0.1	0.2	0.3	0.4	0.5	0.6	0.7	0.8	0.9
0	0	0.1	0.2	0.3	0.4	0.5	0.6	0.7	0.8	0.9
1	1.0	1.1	1.2	1.3	1.4	1.5	1.6	1.7	1.8	1.9
2	2.0	2.1	2.2	2.3	2.4	2.5	2.6	2.7	2.8	2.9
3	3.0	3.1	3.2	3.3	3.4	3.5	3.6	3.7	3.8	3.9
4	4.0	4.1	4.2	4.3	4.4	4.5	4.6	4.7	4.8	4.9
5	5.0	5.1	5.2	5.3	5.4	5.5	5.6	5.7	5.8	5.9
6	6.0	6.1	6.2	6.3	6.4	6.5	6.6	6.7	6.8	6.9
7	7.0	7.1	7.2	7.3	7.4	7.5	7.6	7.7	7.8	7.9
8	8.0	8.1	8.2	8.3	8.4	8.5	8.6	8.7	8.8	8.9
9	9.0	9.1	9.2	9.3	9.4	9.5	9.6	9.7	9.8	9.9
10	10.0	10.1	10.2	10.3	10.4	10.5	10.6	10.7	10.8	10.9
11	11.0	11.1	11.2	11.3	11.4	11.5	11.6	11.7	11.8	11.9
12	12.0	12.1	12.2	12.3	12.4	12.5	12.6	12.7	12.8	12.9
13	13.0	13.1	13.2	13.3	13.4	13.5	13.6	13.7	13.8	13.9
14	14.0	14.1	14.2	14.3	14.4	14.5	14.6	14.7	14.8	14.9
15	15.0	15.1	15.2	15.3	15.4	15.5	15.6	15.7	15.8	15.9
16	16.0	16.1	16.2	16.3	16.4	16.5	16.7	16.8	16.9	17.0
17	17.1	17.2	17.3	17.4	17.5	17.6	17.7	17.8	17.9	18.0
18	18.1	18.2	18.3	18.4	18.5	18.6	18.7	18.8	18.9	19.0
19	19.1	19.2	19.3	19.4	19.5	19.6	19.7	19.8	19.9	20.0
20	20.1	20.2	20.3	20.4	20.5	20.6	20.8	20.9	21.0	21.1
21	21.2	21.3	21.4	21.5	21.6	21.7	21.8	21.9	22.0	22.1
22	22.2	22.3	22.4	22.5	22.6	22.7	22.8	23.0	23.1	23.2
23	23.3	23.4	23.5	23.6	23.7	23.8	23.9	24.0	24.1	24.2

24	24.3	24.4	24.5	23.6	23.7	24.8	24.9	25.0	25.1	25.2
25	25.3	25.4	25.6	25.7	25.8	25.9	26.0	26.1	26.2	26.3
26	26.4	26.5	26.7	26.8	26.9	27.0	27.1	27.2	27.3	27.4
27	27.5	27.6	27.7	27.8	27.9	28.0	28.2	28.3	28.4	28.5
28	28.6	28.7	28.8	28.9	29.0	29.1	29.3	29.4	29.5	29.6
29	29.7	29.8	29.9	30.0	30.1	30.3	30.4	30.5	30.6	30.7
30	30.8	31.0	31.1	31.2	31.3	31.4	31.6	31.7	31.8	31.9
31	32.0	32.1	32.2	32.4	32.5	32.6	32.8	32.9	33.0	33.1
32	33.2	33.3	33.5	33.6	33.7	33.8	33.9	34.0	34.2	34.3
33	34.4	34.5	34.6	34.8	34.9	35.0	35.1	35.3	35.4	35.5
34	35.6	35.8	35.9	36.0	36.2	36.3	36.4	36.6	36.7	36.8
35	37.0	37.1	37.2	37.4	37.5	37.6	37.8	37.9	38.0	38.2
36	38.3	38.4	38.6	38.7	38.9	39.0	39.1	39.2	39.4	39.5
37	39.7	39.8	39.9	40.0	40.2	40.4	40.6	40.7	40.9	41.0
38	41.2	41.4	41.6	41.7	41.8	42.0	42.1	42.3	42.4	42.6
39	42.7	42.8	43.0	43.2	43.3	43.4	43.6	43.9	44.1	44.3
40	44.5	44.6	44.7	44.9	45.0	45.2	45.3	45.5	45.6	45.9
41	46.0	46.2	46.4	46.6	46.8	47.0	47.2	47.4	47.6	47.8
42	47.9	48.1	48.4	48.7	48.9	49.0	49.2	49.4	49.6	49.8
43	50.1	50.3	50.6	50.8	51.0	51.3	51.5	51.7	51.9	52.2
44	52.5	52.8	53.0	53.3	53.5	53.7	54.0	54.3	54.6	54.9
45	55.2	55.4	55.9	56.1	56.4	56.6	57.0	57.3	57.7	58.0
46	58.4	58.7	59.1	59.5	59.8	60.2	60.6	61.0	60.4	61.9
47	62.3	62.8	63.3	63.9	64.3	64.8	65.3	65.8	66.6	67.1
48	68.8	68.5	69.2	70.0	70.8	71.5	72.3	73.4	74.5	75.6
49	76.8	78.1	79.7	81.4	83.4	84.5	88.5	92.0	97.1	104.8

$$p_{12} = \frac{p_1 + p_2}{1 + 4p_1 p_2},$$

where p_1, p_2, and p_{12} are the crossover values for the intercepts a - b, b - c, and a - c. However, the reliability of any addition formulae cannot be greater than the reliability of the estimates of its p_1 and p_2 constituents and this imposes a practical limitation. This aspect is shown in part by the greater scatter of entries in the figure for long intercepts. The Carter-Falconer function does not have a simple direct addition formula but this disadvantage can be overcome by the provision of a conversion table. Table 11.6 presents such a tabulation and should be sufficiently accurate to match most experimental data. The table is least useful for high crossover values or long chromosome intercepts unless the data are of a high order of accuracy.

Parsons (1958a, b) has attempted to construct chromosome maps for three loci of linkage group V and four loci of group XIII. The maps are constructed using the segmental functions tabulated by Fisher and Yates (1953) which are based on a mapping function of Owen (1949). The agreement between expectation and observation is merely moderately good. Owen's function predicts that crossover values in excess of 50 per cent should occur and, in one instance (group XIII), this leads to a significant discrepancy. The function predicted a crossover value of 52.5 per cent for the terminal loci, whereas the observed crossover value was 49.3 per cent. It is briefly noted that some modification of the $\frac{1}{4}\chi_4^2$ interference function may be necessary to accommodate the various linkage groups/chromosomes of the mouse.

AFFINITY

Quasi-linkage association, particularly those of a sporadic nature, have been observed in a number of species and the mouse has provided evidence for one such association which has been defined as affinity (Michie, 1953; Wallace, 1953). The concept has been subsequently developed and placed on a rational experimental basis (Michie, 1955d; Wallace, 1957b, 1958b, 1959, 1961). Aside from the general significance of the phenomenon, the mouse investigations have yielded interesting quantitative data.

The mouse occupies rather a special position for the detection of affinity. Two avenues of approach have produced significant results. Several crosses have been undertaken between laboratory mice and animals of sub-species, or near sub-species status, which have been long separated from the main stream of *Mus*. These latter animals could have evolved distinct and integrated karyotypes, one aspect of which is a measure of affinity between the chromosomes. Should the laboratory mice be of a multiple mutant stock, affinity could manifest in the F_2 or backcross by association of the tagged chromosomes. In this respect, the affinity effect inherent in the observations of Gates (1926), Little (1927), and Green (1931) has been brought out by Michie (1955d).

The existence of numerous inbred strains of laboratory mice could imply that these will be homozygous for affined chromosomes (up to 20, obviously). Crosses between strains differing in many genes could result in the expression of associations. The existence of inbred strains means that, once detected, the affinity effect should be reproducible and open to experimental attack. This was the successful approach of Wallace (1958b, 1961). Much will depend upon the proximity of mutant genes in relation to their affined chromosomes, apart from the preferential staying together of the chromosomes. Detection in this respect will be a hit and miss affair, as well as being manifested only between certain crosses (itself, a sign of affinity).

The formal affinity hypothesis postulates that certain centromeres will exhibit preferential segregation at meiosis, tending to travel in unison towards the same pole. Should two or more genes on different chromosomes be more or less contiguous with the centromeres, these will exhibit quasi-linkage because of the preferential centromeric segregation. Thus – if the underlying premise is granted – it should be possible to deduce from the behaviour of the genes that these are in the vicinity of the centromere. It follows that, given affinity between members of two or more linked groups of genes, the position of the centromere can be assessed relative to the genes of each group. Affinity, therefore, emerges as a

TABLE 11.7 Mapping of centromere position by affinity

Linkage	Sex	Position of centromere (C)		
II	♀	C $\dfrac{}{0.8}$		*dse*
V	♂	*fi* $\dfrac{}{10.7}$	C $\dfrac{}{16.9}$	*Sd*
	♀	*fi* $\dfrac{}{8}$	C $\dfrac{}{15}$	*Sd*
VI	♀	C $\dfrac{}{17.2}$		*Ca*
XVII	♀	C $\dfrac{}{17.5}$		*W*

prima facie method of locating the position of centromeres; or, failing, the relative position of points of affinity.

The positioning of the centromere within a linkage group requires, as a minimum, that affinity be expressed between a gene on one chromosome and two genes on another. The strength of affinity and the crossover value for the two genes provide the basic data for the centromere position. Thus far, only Wallace (1958b, 1961) has collected adequate material from which worthwhile calculations could be made. The genes *Ca, fi, Sd, se,* and *W* of linkage groups II, III, V, and VI were involved in her experiments in a remarkably interwoven yet consistent pattern of relationships. Wallace's conclusions are shown by Table 11.7. Three estimates of the 'separation value' for the centromeres of groups III, VI, and V gave the almost identical value of 38 per cent in each case.

There appears to be a lack of correspondence between the position of the centromere as shown by the table and those predicted by aid of translocations. It is true that only one centromere has been anything like precisely located but this happens to be for the chromosome bearing the *d* locus which is thought to be practically contiguous with the centromere on the affinity hypothesis. The analysis of the behaviour of one translocation has indicated that the centromere for linkage group III is quite close to the *cw* locus (about two crossover units) which is approximately 35 units from *d*. Unless the translocation turns out to be extraordinarily complicated this value is difficult to reconcile with the continguity of *d* and the centromere. Locus *Sd* is a terminal locus of linkage group V but *fi* is located in the middle of the group some 20 units from *Sd*. The translocation analysis has indicated that the centromere of the chromosome bearing the group is located beyond *Sd* away from *fi*. This deduction is difficult to reconcile with the conclusion of the affinity data that the centromere lies between *Sd* and *fi*.

A translocation analysis with linkage group VI has indicated that the centromere lies beyond *uw* on the side opposite to *Ca*, the latter being some 44 units distant from *uw*. This is in conflict with the affinity conclusion that *Ca* is about 17.2 units from the centromere. A similar conflict of distance arises from *W* in group XVII. The position of the centromere has not been determined for this group by the translocation method, but the eventual position is likely to be outside of the linkage group as presently constituted. However, the *W* locus is sited 27 units from one terminal marker and 29 from the other. These distances are about threefold the estimate of 7.5 units between *W* and the centromere according to the affinity analysis.

Not all of the above discrepancies may prove to be significant but, taken together, they point to the conclusion that affinity may not be due in every case to attraction between centromeres. The simplest explanation is that one or more sites of attraction may occur along the chromosome. For this reason, and also to leave the matter as open as possible, affinity may be represented as arising from sites or regions of attraction between chromosomes, without the restriction that these need specifically involve the centromere region. The affinity effect is created because a mutant locus happens to be within a measurable distance.

Although Wallace may be the only worker to tackle affinity in a strictly analytical manner, reports of other reputative cases deserve to be examined. The reason is that it may be possible to discern a web of associations/disassociations involving a limited number of genes. Certain mutants, because of their proximity to a point of affinity, should display recurrent affinity-like associations. Given sufficient observations, these should be detectable. In fact, the expression of weak linkage between *b* and *d*, not only in Gates' (1926) data but also in those of Little (1927) and Green (1931), was a factor in giving credence to the affinity concept (Michie 1955d). Affinity, of course, does not demand that these genes should always display an association but that they will do so on an above-average number of occasions and that disassociations will occur as frequently as associations. This latter will differentiate affinity from weak chromosomal linkage.

Wallace's work revealed five genes displaying affinity. The locus of one of these, *se,* appears to be quite near to a point of affinity. The *se* gene is extremely closely linked to *d* and, in fact, the latter gene has behaved oddly in many segregations. Michie (1955d) pin-pointed the weak but significant associations of *d* with *b, s,* and *v.* The data of Gruneberg (1936a) and Fisher and Mather (1936b) indicated consistent disassociation of *d* and *wa-1.* The gene also shows disassociation with *c* (Durham, 1908; Carter and Falconer, 1952) and *sh-2* (Gates, 1934); an association with *Ra* (Carter and Phillips, 1954) and *Al* (Dickie, 1955); and suspicious behaviour in joint segregations with *p* (association in data of Little and Phillips (1913) and disassociation in Castle *et al.* (1936)). Gene *c* has exhibited association with *v* (Darbishire, 1904; Gates, 1926), *wa-1* (Burhoe, 1936; Carter and Falconer, 1952), *W* (Carter and Falconer, 1952) and disassociation with *Mi* (Bunker and Snell, 1948) – in a sense completing a discernible pattern.

Fisher and Mather (1936b) also reported an association of *d* with sex. However, the association was barely significant and would not be meaningful outside of the present context. Fisher and Mather's paper is noteworthy because it tabulates extensive data from five different males. It may turn out to be portentous that significant inter-male heterogeneity was apparent for random assortment of the genes *a - b, a - d,* and *b - wa-1,* although the totals for the three pairs do not show significance. Gene *b* has shown irregularities of assortment: an association with *W* (Wachter, 1927; Michie, 1955c) and *fi* (Carter and Gruneberg, 1950; Carter and Falconer, 1952; Wallace, 1961).

Parsons (1959; also Wallace, 1961) has commented that affinity doubtless occurs between *Sd* and the genes *ln* and *Sp,* of group XIII. In one segregation, there was disassociation between *Sd* and *Sp*, and assortment heterogeneity was present in all of the sub-groups of data on the genes. In some of Wallace's material, *fi* and *Sp* show association.

Gene *W* has manifested an association with *hr* (Gates and Pullig, 1945a, b; Lane 1967) and with *Va* (Cloudman and Bunker, 1945). *Ra* has displayed disassociation with *f* (Carter and Phillips, 1954; but with a small sample) and association with *Hk* (Green *et al.*, 1963). The gene *wa-1* has shown association with *dw* (Burhoe, 1936). Gene *hy-1* has displayed associations with *pa* (Clark, 1936) and with *sh-2* (Clark 1935, 1936). The *sh-2* gene has displayed disassociation with *f* (Heston *et al.*, 1952) and has behaved very curiously with sex (Wright, 1947b; Fisher *et al.*, 1947; Carter and Phillips, 1953). The gene has assorted very oddly altogether (together with the linked gene *wa-2*) as noted particularly by Carter and Phillips (1953) and by Michie (1955c). The weak linkage between *s* and *W* — first noted by Fisher (1946) but later shown to be false by Lane (1967) — may be an affinity effect.

Wallace (1958a) has given good grounds for thinking that *je* and *ru* can display affinity. Assortment data showing clear evidence of linkage are contrasted with others showing equally clear independence. Forsthoefel and Shenk (1970) have also noted a loose linkage between the two loci. Gene *je* has shown an association with *ch* (Phillips, 1956) and *ru* with *de* (Curry, 1959).

Ca was observed by Wallace (1961) to show weak and erratic affinity with *fi*, *Sd*, and *W*. No sign of affinity is evident between *Ca* and *b* for laboratory mice but Wallace (1958b) mentions that *Ca* and *b* display association when the two genes are involved in an interspecies cross but no numerical details seem readily available. Michie (1955c) records data showing disassociation of *Ca* with *a*, *vt*, and *Re*. Dickie and Woolley (1950) report a notable disassociation between *Ca* and *fz*. Gene *fz* has shown an association with *Dc* (Deol and Lane, 1966) and *Dc* has shown disassociation with *s* (Deol and Lane, 1966).

A remarkable case of probable affinity is reported by Dickie (1955). The *Al* gene showed significant associations with *b*, *d*, and *ln;* three genes belonging to the linkage groups II, VIII, and XIII. In essence, the gene is behaving similarly to *d*, with the implication that it is located close to an affined point. Dickie also observed that *Al* displayed weak linkages with genes *Re*, *sh-2*, and *wa-2* — but not with *vt* — of linkage group VII. These weak linkages resembled affinity in their respective magnitudes and were interpreted in this manner in company with the other associations. Lane (1970, appendix) has since obtained evidence that *Al* is fairly closely linked to *re*. Hence, some of the associations observed by Dickie represent genuine linkage, although the realized crossover values seem to be distorted by the affinity effects rife in the experiments.

Gene *oe* has shown association with *Pt* (Kelton and Rauch, 1968) and disassociation with *Tw* (Kelton and Rauch, 1968). Hutton and Roderick (1970) have reported disassociative assortment between the genes *bt* and *Es-3* and *Es-3* and *s*. The *s* has displayed sporadic non-random assortment on other occasions.

The above survey is probably incomplete if only because of the fitful nature of affinity. There is also the danger that the wider the search the greater the chance of picking up spurious associations. Moreover, negative data cannot be cited since these would be indecisive in refuting affinity — unlike linkage. The survey may, however, be useful as a guide for those genes most likely to lead to positive results for a thoroughgoing affinity analysis. It is patent that a suggestive network of associations/disassociations is fundamental for building up a preliminary case. Simple citation of cases of less or greater than expectation of recombination is not sufficient *per se*.

In brief, much will depend upon ·the number and position of points of affinity present in the chromosome pool of inbred strains and stocks of mice capable of displaying the phenomenon, and the occurrence of mutant loci in close proximity to these points. Genes *Al* and *d* may be two such loci. Also, it may not be a coincidence that so many genes of linkage groups V (*a, fi, pa, Ra,* and *Sd*) and VII (*Al, oe, Re, sh-2, vt,* and *wa-2*) recur in reported cases of deficient or excess of recombination. These chromosomes may have a disposition for affine assortment, possibly of as yet unknown complexity.

Direct observations of affinity as typified by non-random association of the chromosomes are not numerous at present. Even so, a few interesting aspects have emerged. These contributions have been outlined in the section on karyology. It is of interest that as many as six to eight associations of from two to six chromosomes may occur quite regularly. Furthermore, the associations may consist of the same chromosomes, although this aspect has not been directly demonstrated (Bennett, 1966; Douglas, 1966). If these observations can be more fully substantiated and extended, it is obvious that affinity may be more commonplace than is perhaps generally supposed. There appears to be evidence for several primary affined groups of centromeres or affinity points; probably accompanied by secondary affinities, depending on the likelihood of the same chromosome preferentially associating in the same group or exchanging between groups. At this time, such thoughts are largely speculative but, hopefully, not wildly so.

The results of Wallace have been re-examined by Douglas and Geerts (1966) and interpreted in terms of a theoretical model suggested by certain associations of three chromosomes (Douglas, 1966). The model deserves attention on two counts; (a) it is based on an actual karyological configuration and (b) it is an elegant demonstration of the mechanism by which the essentially acrocentric chromosomes of the mouse can associate to engender affinity. Any implicit assumption that affined chromosomes need be mesocentric is now unnecessary.

INDEPENDENT ASSORTMENT

Those published tests between gene pairs which have failed to reveal linkage are shown by Table 11.8. It may be presumed that the majority of loci are truly independent. Against this is the possibility that a number of genes could be loosely linked. Only future work will disclose the extent of this. The tabulation should be self-explanatory. The penultimate column headed 'closest linkage' is the closest linkage value in harmony with the available data. This value is approximated by multiplying the standard error by 196 and subtracting the product from the observed recombination value.

It is likely that the present compilation may be the last which is worthy of the effort. Ordinarily, this sort of tabulation would introduce a degree of order in what may be unco-ordinated research. That is to say, tests are often performed with those genes readily available, rather to some systematic plan. However, the use of special linkage testing stocks is changing this situation. On the other hand, there is the possibility that, at the present level of knowledge, some chromosomes may be capable of supporting two (exceptionally, perhaps more) linkage groups — quasi-independent. For this reason, it is still of interest to know how the terminal genes of each group are behaving with each other and, particularly, of course, with the numerous genes as yet unassigned.

Over 800 tests are shown in the table. At the conventional five per cent level, some 40 gene pairs would be expected to display significant deviation; say, 20 below and 20 above purely random assortment. Actually, 20 pairs show significant deviation below free assortment and 13 above. These observations are not strictly compatible with chance occurrence, specifically because of the lack of pairs showing excess recombination. Even the number of pairs showing poor recombination would not be so close to expectation were it not for the affinity phenomenon. It is probable that any suspicious deviation is followed up to ascertain if the deviation is real. Dickie (1955), for instance, commented that she obtained more than the ordinary amount of data on *Al* because of the gene's curious behaviour.

All of the 33 significant pairs of genes warrant at least a passing consideration. It is doubtful if many (or any) are indicative of weak linkage. Gene *Al* has finally been located in group VII, thus removing a range of possibilities. The most plausible explanation.for most of the associations at this time is affinity because the phenomenon is capable of explaining both poor and excess recombination. Most of the gene pairs known to be involved with affinity-like associations are cited in the section on affinity and there is no point in repeating the discussion. There is no intention to present affinity as an adequate or sole explanation for all gene pairs which happen to display interesting departures from expectation. As a cross-check, the following pairs are credited with significant deviation:

Al - b	*Al - ln*
Al - d	*bt - Es-3*

c - d	f - Ra
c - v	f - sh-2
Ca - fi	fi - Sp
Ca - Sd	fi - W
Ca - vt	Fu - Mi
ch - je	Hk - Ra
d - wa-1	hr - W
dse - Ra	hy-1 - sh-2
Dc - fz	Mdh-1 - Pgm-1
Dc - s	oe - Pt
de - ra	oe - Tw
du - f	Sd - W
dw - wa-1	sh-2 - δ
Es - Ldr-1	Va - W
Es-3 - s	

Weak linkage and affinity in its associative phase can be confused. Gene Al could have been credited with spurious false linkage if the multiplicity of associations had not been detected. Similarly, the depicted weak linkages between s and W (Fisher, 1946) and between je and ru (Fisher and Snell, 1948) turned out to be false (Lane, 1967; Wallace, 1958a). This sort of error is legitimate, of course, but it serves to remind that a weak linkage can be unreliable.

Bunker and Snell (1948) inadvertently used an unknown number of +r animals in the test of Mi and r. This made formal analysis of the joint segregation impossible but a simple 2 x 2 χ^2 revealed no evidence of linkage.

Statements recur in the literature that certain genes are inherited independently but lack publication of the supporting data. In several respects, this is highly unfortunate because this policy prevents data from different crosses from being suitably combined. Particularly, there is no excuse where the genes concerned have not been definitely assigned or stand revealed as the terminal members of a linkage group. For the mouse, possibly, this aspect is not of great moment in view of the many positive cases of linkage which are coming forward. For the record, the following may be stated.

Reed (1937) has published data which indicates that Fu is not linked to a, b, c, d, dw, ln, N, p, pa, r, s, sh-2, v, W, or wa-1.

However, the phase is not given for any of the crosses nor is the incomplete penetrance of *Fu* properly allowed for.

Snell (1941) states that *dw* assorted independently of *a, c, Ca, dse, hr, N, p, pa, s,* and *sh-1*.

Steinberg and Fraser (1943) state that *mc* gave no indication of linkage with *a, b, hr,* or *wa-2*.

Law, Morrow, and Greenspan (1952) state that *g* behaved as if independent of *a, b,* or *c* in both testcross and F_2 data.

Snell *et al.* (1954) quote W. Hollander that *Lp* segregates freely in conjunction with *a, b, c, p, s,* and *se*.

Snell (1955) states that *du* gave no indication of linkage with *a, b, c, Ca, f, Fu, fz, je, ln, Mi, p, Re, ru, s, sh-2, Sp, v, W, wa-1,* or *wa-2* (incomplete data given on *c, f,* and *p*.)

Bennett (1959) gives some incomplete data on the independent assortment of *bh* and *T*.

Paigen and Noell (1961) state that *g* and *rd* show no linkage with *a, b, d* or sex; nor that of *rd* with *pg* and *si*.

Kidwell, Gowen, and Stadler (1961) state that *pn* is apparently independent of *a, b, bt, c, f, fz, je, ln, Mi, p, Re, ru, se, si, To, v, Va,* and *wa-2*.

Sidman, Lane, and Dickie (1962) state that *sg* shows independence of *a, b, Mi, Os, Ra, T,* and *W*.

Popp (1962a) states that *Hba* segregated independently of *a, dse, p,* and sex.

Fraser and Schabtach (1962) state that *Cat* showed no linkage with *a, b, c,* or *N*.

Amos, Zumpft, and Armstrong (1963) make the point that *H-5* assorts independently of *a*.

Herzenberg, Warner, and Herzenberg (1965) state that *Ig-1* gave no evidence of linkage with *a, b, c, Ca, dse, Hc, H-2, p, s, Sl,* or sex (50 offspring examined for each gene). Subsequently, Herzenberg, McDevitt, and Herzenberg (1968) add the following to the list: *ax, Ea-1, ep, hr, Ir-1, Lp, Mi, Os, Ra, Re, ru, sa, Sd,* and *Va*.

Pelzer (1965) states that *Ee-1,* and *Ee-2* displayed no linkage with *b, c, H-2,* or *Hbb*; nor *Ee-1* with *a*.

Searle (1966) states that T^c, an allele of *T*, gave no evidence of linkage with the genes, *a, b, c, Ca, d, f, fz, je, ln, Mi, Ra, Re, ru, s, se, v, W, wa-1,* or *wa-2*.

Lilly (1966) states that the genes *a, c, H-6,* and *s* had no effect upon the incidence of induced Gross leukaemia and, in the circumstances, this could imply that none of the genes show significant linkage with either *rgv-1* and *rgv-2*.

McNutt (1967) states that 'porcine tail' is inherited independently of *tf*.

Snell, Hoecker, and Stimplfing (1967) note briefly that *H-14* is not closely linked to *a* or *c*.

Popp (1967a) states that *H-14* gave no signs of linkage with *a*, *b*, *c*, *Ca*, *d*, *Es-1*, *Es-2*, *f*, *Mi*, or *s*.

Biddle and Petras (1967) state that tests for linkage of *Pro-1* with *a*, *Es-1*, *Es-3*, and *Hbb* were negative.

Boyse *et al.* (1968) state that *H-2*, *Ly-A*, *Ly-B*, and '*0*' have not exhibited linkage with *c*.

Hauschka, Jacobs, and Holdridge (1968) state that *e* assorts independently of *a*, *b*, *bt*, *Mi*, *N*, *s*, and *T* (some data given).

Ruddle, Shows, and Roderick (1968) state that close linkage was not observed for the seven genes *c*, *Es-1*, *Es-3*, *Gpd-1*, *Id-1*, *Mdh-1*, and *rd* (two backcrosses of 13 and 19 offspring).

Shows and Ruddle (1968) state that *Ldr-1* is independent of *a*, *Es-1*, *Es-3*, *Gpd-1*, *Mdh-1*, and *Pgm-1* among 58 backcrossed offspring; also that linkage was not apparent between *Lrd-1* and *Hbb*.

Shows *et al.* (1969) state that close linkage can be excluded between *Pgm-1* and *a*, *Es-1*, *Es-2*, *Gpd-1*, *Id-1*, *Ig-1*, *Ldr-1*, or *Mdh-1*.

Platz and Wolfe (1969) state that no linkage is apparent between *Svp* and the genes *c* and *p*.

DeLorenzo and Ruddle (1969) state that linkage appeared to be absent between *Gpi-1* and genes *Es-1*, *Es-2*, *Es-3*, *Gpd-1*, *Gpi-1*, *Id-1*, *Mdh-1*, or *Pgm-1* (76 testcross progeny).

Hutton (1969b) states that no signs of linkage was found between the genes *b* and *d* with *Gpi-1*, *Hba*, *Id-1*, or *Ldr-1*; nor between *Es-1* and the genes *Es-3*, *Gpd-1*, *Hba*, *Ldr-1*, *Mdh-1*, or *Pgm-1*.

Ruddle, Shows, and Roderick (1969) state that tests failed to disclose linkage between *Es-2* and the loci *Gpd-1*, *Gpi-1*, *Id-1*, *Ig-1*, *Ldr-1*, *Mdh-1*, *Pgm-1* and *tf*.

Preliminary results from testing *ao* with *W* revealed no signs of linkage (Pelt, Cain, and Knorr 1969).

Russell *et al.* (1970) have shown that it is doubtful if *Hba* and *mk* are closely linked.

TABLE 11.8 Quasi-independent gene assortment in the house mouse

Loci	Recom- bination value	Phase balance	Closest linkage	References
a - ax	49±7	0	35	Lyon (1955)
a - b	51±1	99	49	Durham (1911), Little (1913, 1916, 1927), Little and Phillips (1913), Detlefsen and Roberts (1918), Dunn (1919), Green (1932), Gruneberg (1936a), Fisher and Mather (1936b), MacDowell (1950), Michie (1955c), Kreitner (1957)
a - bm	46±3	100	39	Lane and Dickie (1968)
a - bt	46±3	100	39	Murray and Snell (1945)
a - c	51±2	66	47	Dunn (1920b), MacDowell (1950), Carter and Falconer (1952), Michie (1955c), Kreitner (1957)
a - Ca	52±5	100	43	Michie (1955c), Wallace (1961)
a - ch	46±7	39	32	Phillips (1956)
a - cr	52±2	?	47	King (1956)
a - d	51±1	50	50	Durham (1911), Little (1913, 1916, 1927), Little and Phillips (1913), Green (1932), Gruneberg (1936a), Fisher and Mather (1936b), Michie (1955c)
a - Dc	56±5	100	46	Deol and Lane (1966)
a - dr	38±14	0	11	Lyon (1961a)
a - e	51±7	0	36	Bateman (1961)
a - ep	54±3	85	49	Lane and Green (1967)
a - f	57±4	100	49	Clark (1934b)
a - fz	58±8	100	42	Dickie and Woolley (1950)
a - H-1	54±6	-	42	Snell and Stevens (1961)
a - H-6	58±11	-	37	Amos et al. (1963)
a - Hbb	52±5	-	43	Cohen (1960)
a - hy-1	53±8	0	36	Clark (1936)
a - je	47±5	0	36	Gruneberg, Burnett, and Snell (1941)
a - ji	47±5	0	38	Snell (1945)
a - ln	50±4	100	41	Murray (1933)
a - lu	49±7	0	35	Green (1955)
a - lx	49±3	37	44	Carter (1951c)
a - Mi	55±5	45	46	Hertwig (1942), Kreitner (1957)
a - my	51±3	100	44	Carter (1956)

Loci	Recombination value	Phase balance	Closest linkage	References
a - N	40±3	100	35	Snell (1931)
a - oe	50±3	0	44	Kelton and Rauch (1968)
a - ol	40±7	100	26	Hertwig (1942)
a - p	51±1	84	49	Durham (1911), Little and Phillips (1913), Detlefsen and Roberts (1918), Dunn (1920b), Michie (1955c), Kreitner (1957)
a - pi	65±8	?	50	Dickie and Woolley (1946)
a - Pre	49±6	-	37	Shreffler (1964a)
a - Pt	54±3	0	48	Hollander and Strong (1951)
a - r	55±6	0	44	Keeler (1927, 1930)
a - Re	46±2	56	41	Crew and Auerbach (1940), Carter and Falconer (1952), Michie (1955c)
a - rl	55±7	0	42	Falconer (1951)
a - s	48±1	61	46	Dunn (1920b), Wachter (1921), Gates (1926), Fisher and Mather (1936b)
a - Sas-1	57±8	100	42	Wortis (1965)
a - Sd	53±7	44	40	Carter and Falconer (1952)
a - sh-2	52±3	26	46	Gates (1934), Carter and Falconer (1952)
a - shm	43±5	80	32	Green (1968)
a - Sp	48±3	78	43	Wallace (1961)
a - sr	55±10	0	35	Deol and Robins (1962)
a - T	49±3	0	43	Carter and Falconer (1952)
a - tf	54±5	100	45	Lyon (1956)
a - tg	45±5	100	35	Green, Snell, and Lane (1963)
a - Tr	51±6	0	39	Falconer (1951)
a - Trf	56±3	-	50	Shreffler (1960), Cohen (1960)
a - Tw	51±5	0	41	Lyon (1958)
a - v	56±5	100	47	Gates (1926)
a - Va	48±8	0	33	Carter and Falconer (1952)
a - vt	49±2	96	44	Michie (1955c)
a - W	50±1	49	48	Dunn (1920b), Wachter (1921), Michie (1955c), Wallace (1958b)
a - wa-1	51±1	95	48	Burhoe (1936), Gruneberg (1936a), Fisher and Mather (1936b), Crew and Auerbach (1940), Michie (1955c)
a - wa-2	53±3	72	47	Carter and Falconer (1952)
a - ♂	51±1	65	49	Fisher and Mather (1936b), Michie (1955c), Cohen (1960)

Loci	Recombination value	Phase balance	Closest linkage	References
ad - ln	41 ± 10	?	21	Falconer and Isaacson (1959)
ad - ♂	37 ± 10	?	17	,, ,, ,, ,,
Al - b	44 ± 2	100	40	Dickie (1955)
Al - Ca	54 ± 4	0	47	,, ,,
Al - d	44 ± 2	100	39	,, ,,
Al - f	56 ± 4	100	47	,, ,,
Al - Fu	53 ± 4	100	46	,, ,,
Al - fz	51 ± 4	100	44	,, ,,
Al - Hk	51 ± 4	0	43	Green, Snell, and Lane (1963)
Al - je	48 ± 3	100	42	Dickie (1955)
Al - ln	41 ± 2	100	37	,, ,,
Al - m	44 ± 4	100	37	,, ,,
Al - ru	42 ± 8	100	27	,, ,,
Al - s	50 ± 4	100	42	,, ,,
Al - sd	51 ± 5	0	41	,, ,,
Al - Sp	53 ± 4	0	46	,, ,,
Al - v	58 ± 4	100	50	,, ,,
Al - W	44 ± 4	0	36	,, ,,
ap - fs	49 ± 4	0	41	Tutikawa (1955)
asp - d	51 ± 1	100	49	Collins and Fuller (1968), Collins (1970)
ax - b	46 ± 8	0	30	Lyon (1955)
ax - c	44 ± 3	33	37	,, ,,
ax - f	46 ± 5	0	35	,, ,,
ax - fz	53 ± 5	0	43	,, ,,
ax - je	49 ± 5	0	39	,, ,,
ax - ln	55 ± 4	0	46	,, ,,
ax - lx	60 ± 8	100	44	,, ,,
ax - Mi	47 ± 6	100	35	,, ,,
ax - N	59 ± 5	100	49	,, ,,
ax - p	54 ± 5	0	44	,, ,,
ax - Re	50 ± 6	100	37	,, ,,
ax - ru	50 ± 5	0	39	.,, ,,
ax - s	47 ± 4	0	38	,, ,,
ax - Sd	54 ± 8	100	38	,, ,,
ax - se	51 ± 7	0	38	,, ,,
ax - T	42 ± 5	100	31	,, ,,
ax - v	47 ± 7	0	34	,, ,,
ax - Va	42 ± 6	100	29	,, ,,
ax - wa-2	54 ± 4	0	46	,, ,,
ax - ♂	45 ± 4	53	37	,, ,,
b - bt	53 ± 4	100	45	Murray and Snell (1945)
b - c	52 ± 2	100	48	Durham (1908), Feldman and

Loci	Recombination value	Phase balance	Closest linkage	References
				Pincus (1926), MacDowell (1950), Michie (1955c), Kreitner (1957)
b - Ca	50±2	70	47	Michie (1955c), Wallace (1961)
b - ch	63±8	46	48	Phillips (1956)
b - cl	54±7	79	40	Robins (1959)
b - cr	47±3	?	43	King (1956)
b - d	49±1	91	48	Durham (1908), Castle and Little (1909), Little (1913, 1916, 1927), Little and Phillips (1913), Gates (1926, Green (1932), Gruneberg (1936a), Fisher and Mather (1936b), Castle et al. (1936), Castle et al (1936), Michie (1955c)
b - de	45±8	0	43	Curry (1959)
b - dr	60±6	0	48	Lyon (1961a)
b - e	60±8	100	44	Bateman (1961)
b - ep	51±4	63	43	Lane and Green (1967)
b - f	55±4	0	48	Clark (1934b)
b - fi	43±3	53	39	Carter and Gruneberg (1950), Carter and Falconer (1952), Wallace (1961)
b - fz	48±3	35	42	Dickie and Woolley (1960), Carter and Falconer (1952), Wallace (1961)
b - Hk	46±3	100	40	Green, Snell, and Lane (1963)
b - hy - 1	52±2	0	24	Clark (1936)
b - Ig-1	52±7	-	39	Dray et al. (1936)
b - je	45±6	39	33	Gruneberg, Burnett, and Snell (1941)
be - le	49±6	0	39	Lane and Green (1967)
b - ln	52±2	100	49	Murray (1933), Castle (1941)
b - lx	48±3	67	43	Carter (1951c)
b - Mi	49±3	84	44	Hertwig (1942), Bunker and Snell (1948), Kreitner (1957)
b - my	45±3	100	42	Carter (1956)
b - N	54±3	100	49	Snell (1931)
b - oe	48±4	0	40	Kelton and Rauch (1968)
b - al	52±3	100	46	Hertwig (1942)
b - p	49±1	96	48	Durham (1911), Little (1913), Little and Phillips (1913), Detlefsen and Roberts (1918), Gates (1926), Castle et al.

Loci	Recombination value	Phase balance	Closest linkage	References
				(1936), Kreitner (1957)
b - pa	48±1	100	46	Castle (1941)
b - pi	52±6	?	39	Dickie and Woolley (1946)
b - Pre	53±6	-	43	Shreffler (1964a)
b - r	43±5	0	33	Keeler (1927, 1930)
b - Ra	49±3	0	43	Carter and Phillips (1954)
b - Re	51±2	53	46	Crew and Auerbach (1940), Carter and Falconer (1952), Michie (1955c)
b - rl	51±7	0	40	Falconer (1951)
b - s	50±1	18	49	Gates (1928), Wacher (1927), Fisher and Mather (1936b), Wallace (1961)
b - Sd	50±2	65	47	Wallace (1961)
b - se	53±3	61	47	,, ,,
b - sh-2	51±4	71	44	Clark (1935), Carter and Falconer (1952)
b - shm	46±5	100	36	Green (1968)
b - sr	43±6	0	31	Deol and Robins (1962)
b - stb	56±6	100	44	Lane and Dickie (1968)
b - T	50±5	100	40	Clark (1934a)
b - tf	53±4	100	46	Lyon (1956)
b - tg	50±5	100	39	Green, Snell, and Lane (1963)
b - Tr	44±7	56	31	Falconer (1951)
b - Tw	55±5	0	46	Lyon (1958)
b - v	50±1	100	47	Gates (1926), Carter and Falconer (1952)
b - Va	45±6	100	34	Carter and Falconer (1952), Michie (1955c)
b - vt	51±3	100	46	Michie (1955c)
b - W	50±1	56	47	Wachter (1927), Snell (1941), Michie (1955c)
b - wa-1	50±1	100	47	Burhoe (1936), Gruneberg (1936a), Fisher and Mather (1936b)
b - wa-2	49±3	100	43	Carter and Falconer (1952)
b - we	46±2	70	43	Hertwig (1942)
b - ♂	52±1	62	49	Fisher and Mather (1936b), Michie (1955c)
bg - Dc	53±5	100	43	Deol and Lane (1966)
bg - ep	50±8	0	34	Lane and Green (1967)
bg - bm	46±3	0	40	Lane and Dickie (1968)
bm - bt	44±5	0	34	Lane and Green (1968)
bm - Ca	47±4	100	39	Lane and Dickie (1968)
bm - fz	54±5	0	46	,, ,, ,, ,,

Loci	Recombination value	Phase balance	Closest linkage	References
bm - ln	50±4	0	42	Lane and Dickie (1968)
bm - Mi	50±2	100	46	,, ,, ,, ,,
bm - p	58±5	0	48	,, ,, ,, ,,
bm - pe	51±5	0	41	,, ,, ,, ,,
bm - Pt	45±4	100	37	,, ,, ,, ,,
bm - Ra	44±4	4	37	,, ,, ,, ,,
bm - Re	52±4	100	45	,, ,, ,, ,,
bm - Sd	49±4	100	42	,, ,, ,, ,,
bm - sh-2	51±5	0	42	,, ,, ,, ,,
bm - Sl	52±4	100	45	,, ,, ,, ,,
bm - T	56±4	100	48	,, ,, ,, ,,
bm - Tw	45±4	100	37	,, ,, ,, ,,
bm - Va	44±4	100	37	,, ,, ,, ,,
bm - W	46±3	100	41	,, ,, ,, ,,
bm - wa-1	58±5	0	48	,, ,, ,, ,,
bm - wa-2	54±5	0	44	,, ,, ,, ,,
bt - c	49±3	100	44	Carter and Falconer (1952)
bt - cr	49±3	?	43	King (1956)
bt - d	44±7	100	31	Murray and Snell (1945)
bt - Es-3	68±9	-	50	Hutton and Roderick (1970)
bt - f	48±7	0	33	Murray and Snell (1945)
bt - Fu	47±5	0	38	,, ,, ,, ,,
bt - Id-1	33±13	-	1	Hutton and Roderick (1970)
bt - lst	55±5	0	44	Forsthoefel and Shenk (1970)
bt - lx	56±6	0	43	Carter (1951c)
bt - Mi	50±7	100	39	Bunker and Snell (1948)
bt - oe	47±5	0	38	Kelton and Rauch (1968)
bt - p	53±6	0	42	Murray and Snell (1945)
bt - Ra	55±11	100	33	Carter and Phillips (1954)
bt - rl	44±6	0	31	Falconer (1951)
bt - sh-2	50±5	0	39	Murray and Snell (1945)
bt - Tr	47±7	100	24	Falconer (1951)
bt - v _	57±7	23	44	Carter and Falconer (1952)
bt - wa-2	58±7	0	46	Murray and Snell (1945)
c - Ca	48±3	80	47	Carter and Falconer (1952), Michie (1955c)
c - ch	47±7	0	33	Phillips (1956)
c - cr	50±3	?	51	King (1956)
c - d	56±2	95	52	Durham (1908), Carter and Falconer (1952)
c - dse	47±7	0	34	Michie (1955c)
c - de	46±6	0	34	Curry (1959)
c - dr	54±4	0	42	Lyon (1961a)
c - du	46±3	?	40	Snell (1955)

Loci	Recombination value	Phase balance	Closest linkage	References
c - e	51 ± 5	100	41	Bateman (1961)
c - f	49 ± 3	100	44	Clark (1934b)
c - fi	47 ± 3	100	42	Gruneberg (1943), Carter and Gruneberg (1950)
c - fz	50 ± 3	?	44	Carter and Falconer (1952), Dickie and Woolley (1960)
c - Hba	49 ± 3	-	44	Popp (1962b)
c - Hk	49 ± 4	100	41	Green, Snell, and Lane (1963)
c - hy-1	55 ± 6	0	42	Clark (1936)
c - je	59 ± 7	90	45	Gruneberg, Burnett, and Snell (1941), Carter and Falconer (1952)
c - jt	46 ± 6	100	35	Center (1966)
c - le	43 ± 6	0	32	Lane and Green (1967)
c - ln	50 ± 8	0	34	Murray (1933)
c - lr	45 ± 4	100	38	Fraser and Hever (1950)
c - lx	46 ± 3	55	41	Carter (1951c)
c - Mi	51 ± 3	56	45	Bunker and Snell (1948), Carter and Falconer (1953)
c - my	52 ± 6	0	40	Carter (1956)
c - N	44 ± 7	0	31	Snell (1931)
c - oe	49 ± 2	100	45	Kelton and Rauch (1968)
c - pi	47 ± 6	?	36	Dickie and Woolley (1946)
c - Pt	47 ± 5	100	39	Hollander and Strong (1951)
c - r	48 ± 3	100	43	Keeler (1927, 1930)
c - Ra	48 ± 3	100	42	Carter and Phillips (1954)
c - rd	46 ± 6	100	34	Sorsby et al. (1954)
c - Re	37 ± 2	98	42	Crew and Auerbach (1940), Carter and Falconer (1952), Michie (1955c)
c - rl	52 ± 7	0	39	Falconer (1951)
c - ru	46 ± 8	87	31	Carter and Falconer (1952)
c - s	52 ± 5	39	43	Durham (1908), Carter and Falconer (1952), Michie (1955c)
c - Sas-1	40 ± 8	100	25	Wortis (1965)
c - Sd	44 ± 9	100	27	Carter and Falconer (1952)
c - se	51 ± 1	49	49	Carter and Falconer (1953)
c - sh-2	53 ± 5	61	43	Clark (1934b), Gates (1934), Carter and Falconer (1952)
c - stb	50 ± 5	0	40	Lane and Dickie (1968)
c - T	46 ± 4	25	37	Clark (1934a), Carter and Falconer (1952)
c - tf	51 ± 4	100	43	Lyon (1956)

Loci	Recombination value	Phase balance	Closest linkage	References
c - Tr	46 ± 7	42	34	Falconer (1951)
c - Tw	45 ± 4	0	37	Lyon (1958)
c - un	49 ± 6	0	38	Wright (1947a), Michie (1955c)
c - v	46 ± 2	52	42	Darbishire (1904), Gates (1926)
c - Va	57 ± 1	100	37	Michie (1955c)
c - vt	50 ± 3	100	44	„ „
c - W	43 ± 1	43	40	Carter and Falconer (1952), Michie (1955c)
c - $wa-1$	45 ± 2	0	40	Burhoe (1936), Carter and Falconer (1952)
c - $wa-2$	54 ± 5	0	44	Carter and Falconer (1952)
c - we	51 ± 2	100	46	Hertwig (1942)
c - $\male$	50 ± 1	58	47	Carter and Falconer (1952), Michie (1955c)
Ca - ch	44 ± 6	83	33	Phillips (1956)
Ca - cr	51 ± 6	?	40	King (1956)
Ca - Dc	52 ± 6	0	41	Deol and Lane (1966)
Ca - Dr	46 ± 4	0	39	Lyon (1961a)
Ca - ep	47 ± 3	100	42	Lane and Green (1967)
Ca - $Es-3$	47 ± 8	-	31	Hutton and Roderick (1970)
Ca - fi	41 ± 5	100	31	Wallace (1961)
Ca - fz	54 ± 3	?	49	Dickie and Woolley (1950), Wallace (1961)
Ca - $Id-1$	40 ± 9	-	23	Hutton and Roderick (1970)
Ca - je	57 ± 5	100	48	Gruneberg, Burnett, and Snell (1941)
Ca - ji	57 ± 4	31	44	Snell (1945)
Ca - Lc	52 ± 4	88	45	Phillips (1960)
Ca - ln	49 ± 5	100	38	Wallace (1961)
Ca - lx	45 ± 7	100	31	Carter (1951c)
Ca - Mi	47 ± 5	85	38	Bunker and Snell (1948)
Ca - my	49 ± 5	0	38	Carter (1956)
Ca - oe	51 ± 14	100	23	Kelton and Rauch (1968)
Ca - pi	52 ± 1	100	50	Dickie and Woolley (1946)
Ca - Ra	48 ± 4	93	41	Carter and Phillips (1954)
Ca - Re	46 ± 3	0	34	Crew and Auerback (1939)
Ca - s	53 ± 3	0	48	Wallace (1961)
Ca - Sd	53 ± 1	11	51	Dunn et al. (1940), Carter and Falconer (1952), Wallace (1961)
Ca - se	51 ± 2	53	49	Wallace (1961)
Ca - tf	49 ± 5	100	39	Lyon (1956)
Ca - ti	56 ± 5	0	47	Searle (1961)
Ca - Tr	55 ± 8	0	40	Falconer (1951)

Loci	Recombination value	Phase balance	Closest linkage	References
Ca - Tw	48±5	0	39	Lyon (1958)
Ca - v	41±7	72	27	Carter and Falconer (1952)
Ca - vt	58±4	100	51	Michie (1955c)
Ca - W	47±3	100	41	Michie (1955c), Wallace (1961)
Ca - ♂	50±4	100	42	„　　　„　　　„　　　„
ch - dse	63±9	0	59	Phillips (1956)
ch - fz	42±14	0	14	„　　„
ch - je	30±9	0	12	„　　„
ch - ln	55±17	0	23	„　　„
ch - p	47±9	0	29	„　　„
ch - Re	40±10	100	20	„　　„
ch - ru	40±6	2	28	„　　„
ch - s	60±11	0	38	„　　„
ch - Sd	48±9	0	31	„　　„
ch - T	42±6	100	30	„　　„
ch - v	58±14	0	30	„　　„
ch - W	50±10	76	31	„　　„
ch - wa-2	45±17	0	19	„　　„
ch - ♂	53±4	33	45	„　　„
cl - d	61±8	100	46	Robins (1959)
cl - p	64±7	100	50	„　　„
cl - se	64±18	100	29	„　　„
cl - vt	54±7	50	39	„　　„
cl - wa-2	63±9	60	47	„　　„
cl - ♂	38±7	0	25	„　　„
cr - d	45±4	?	38	King (1956)
cr - fz	55±6	?	43	„　　„
cr - je	46±6	?	33	„　　„
cr - ln	48±5	?	38	„　　„
cr - lx	55±6	?	43	„　　„
cr - Mi	48±6	?	35	„　　„
cr - N	55±6	?	43	„　　„
cr - p	52±8	?	37	„　　„
cr - Re	53±6	?	41	„　　„
cr - ru	40±7	?	26	„　　„
cr - s	53±5	?	44	„　　„
cr - sh-2	41±10	?	22	„　　„
cr - Sd	59±6	?	46	„　　„
cr - T	56±5	?	47	„　　„
cr - v	47±7	?	33	„　　„
cr - Va	45±7	?	32	„　　„
cr - wa-2	59±3	?	41	„　　„
cr - ♂	47±2	?	43	„　　„
d - dr	59±6	0	47	Lyon (1961a)

Loci	Recombination value	Phase balance	Closest linkage	References
d - ep	45±4	63	37	Lane and Green (1967)
d - f	46±3	0	40	Clark (1934b)
d - fi	50±8	100	49	Gruneberg (1943)
d - fz	53±6	0	43	Dickie and Woolley (1950)
d - hy-1	59±8	0	44	Clark (1936)
d - Ig-1	52±7	-	39	Dray et al. (1968)
d - ln	56±4	0	48	Murray (1933)
d - lx	49±3	36	43	Carter (1951c)
d - Mi	69±11	100	48	Hertwig (1942)
d - Mup-a	46±6	-	33	Hudson et al. (1967)
d - N	50±5	100	41	Snell (1931)
d - ol	68±9	100	51	Hertwig (1942)
d - p	51±1	98	49	Little (1913), Little and Phillips (1913), Gates (1926), Castle et al. (1936)
d - pi	56±6	?	44	Dickie and Woolley (1946)
d - r	46±5	0	38	Keeler (1927, 1930)
d - Re	49±3	62	43	Crew and Auerbach (1940), Carter and Falconer (1952), Michie (1955c)
d - ru	61±9	100	44	Carter and Falconer (1952)
d - s	48±1	33	46	Gates (1926), Wachter (1927), Fisher and Mather (1936b)
d - sh-2	56±3	0	51	Gates (1934)
d - slt	47±5	19	36	Pierro and Chase (1963)
d - sr	45±6	0	33	Deol and Robins (1962)
d - T	49±5	100	40	Clark (1934a)
d - tf	37±11	100	16	Lyon (1956)
d - tg	45±6	100	34	Green, Snell, and Lane (1963)
d - ti	47±2	22	43	Searle (1961)
d - Tr	43±10	100	23	Falconer (1951)
d - Trf	47±5	-	37	Shreffler (1964a)
d - Tw	47±5	-	37	Lyon (1958)
d - v	53±1	5	50	Gates (1926), Carter and Falconer (1952)
d - vt	46±4	56	39	Michie (1955c)
d - W	50±2	40	46	Wachter (1927)
d - wa-1	53±1	48	50	Gruneberg (1936a), Fisher and Mather (1936b)
d - wa-2	53±5	100	44	Carter and Falconer (1952)
d - we	52±5	100	42	Hertwig (1942)
d - ♂	47±1	60	45	Fisher and Mather (1936b)
dse - Hk	54±3	100	49	Green, Snell, and Lane (1963)
dse - je	44±5	0	34	Gruneberg, Burnett, and Snell 1941

Loci	Recombination value	Phase balance	Closest linkage	References
dse - ji	58±4	0	50	Snell (1945)
dse - Mi	41±5	100	31	Bunker and Snell (1948)
dse - my	48±3	0	47	Carter (1956)
dse - Ra	42±3	80	36	Carter and Phillips (1954)
dse - rl	40±7	0	27	Falconer (1951)
dse - s	53±9	100	35	Michie (1955c)
dse - un	59±9	100	40	,, ,,
dse - ♂	47±4	0	39	,, ,,
Dc - fz	40±5	100	30	Deol and Lane (1966)
Dc - gr	51±7	100	36	,, ,, ,, ,,
Dc - ln	58±5	100	49	,, ,, ,, ,,
Dc - ma	51±6	100	39	,, ,, ,, ,,
Dc - Mi	50±6	0	38	,, ,, ,, ,,
Dc - p	53±7	100	39	,, ,, ,, ,,
Dc - Pt	51±8	0	36	,, ,, ,, ,,
Dc - Ra	55±10	0	36	,, ,, ,, ,,
Dc - Re	60±8	0	44	,, ,, ,, ,,
Dc - s	66±8	100	50	,, ,, ,, ,,
Dc - sa	53±5	100	43	,, ,, ,, ,,
Dc - se	51±5	100	41	,, ,, ,, ,,
Dc - Sl	59±5	41	50	,, ,, ,, ,,
Dc - T	52±6	0	41	,, ,, ,, ,,
Dc - W	48±6	0	37	,, ,, ,, ,,
de - f	50±6	0	38	Curry (1959)
de - fz	42±6	0	30	,, ,,
de - je	40±6	0	28	,, ,,
de - ln	58±7	0	30	,, ,,
de - Mi	50±5	0	49	,, ,,
de - N	55±5	0	45	,, ,,
de - p	55±6	0	43	,, ,,
de - Ra	50±5	0	41	,, ,,
de - Re	48±4	0	40	,, ,,
de - ru	37±6	0	25	,, ,,
de - s	44±7	0	30	,, ,,
de - Sd	50±5	0	40	,, ,,
de - se	55±7	0	43	,, ,,
de - T	50±5	0	40	,, ,,
de - v	41±6	0	29	,, ,,
de - W	50±5	0	49	,, ,,
de - wa-2	59±7	0	45	,, ,,
de - ♂	52±4	0	45	,, ,,
dr - Mi	48±4	100	41	Lyon (1961a)
dr - p	48±7	0	34	,, ,,
dr - Ra	54±4	100	46	,, ,,

Loci	Recombination value	Phase balance	Closest linkage	References
dr - Re	45±3	65	39	Lyon (1961a)
dr - s	49±6	0	37	,, ,,
dr - Sd	49±5	100	39	,, ,,
dr - T	52±8	46	40	,, ,,
dr - Tw	56±6	100	45	,, ,,
dr - vt	53±7	0	39	,, ,,
dr - W	54±4	100	48	Falconer and Sierts-Roth (1951)
dr - wa-2	47±7	0	33	Lyon (1961a)
du - f	41±4	?	34	Snell (1955)
du - p	43±6	?	32	Snell (1955)
dw - f	56±8	0	41	Clark (1934b)
dw - hy-1	61±8	0	47	Clark (1936)
dw - ln	63±8	0	47	Murray (1933)
dw - sh-2	47±8	0	45	Clark (1935)
dw - T	45±13	100	21	Clark (1934a)
dw - v	50±12	100	26	Snell (1941)
dw - W	44±14	0	18	,, ,,
dw - wa-1	59±2	100	55	Burhoe (1936)
e - fz	47±7	0	40	Hauschka et al. (1968)
e - ln	51±14	0	23	,, ,, ,, ,,
e - ♂	45±4	40	46	Bateman (1961)
Ee-1 - Ee-2	54±4	-	47	Pelzer (1965)
ep - fr	49±6	0	38	Lane and Green (1967)
ep - je	51±3	0	44	,, ,, ,, ,,
ep - le	47±17	0	30	,, ,, ,, ,,
ep - Mi	54±6	100	47	,, ,, ,, ,,
ep - Os	44±4	100	36	,, ,, ,, ,,
ep - pe	55±4	100	48	,, ,, ,, ,,
ep - Pt	43±4	100	35	,, ,, ,, ,,
ep - Re	56±4	100	48	,, ,, ,, ,,
ep - Sd	53±4	100	45	,, ,, ,, ,,
ep - Sl	52±5	100	43	,, ,, ,, ,,
ep - T	59±5	100	49	,, ,, ,, ,,
ep - Va	51±5	100	42	,, ,, ,, ,,
Es-1 - Es-3	53±4	-	45	Ruddle and Roderick (1965a, 1966)
Es-1 - Ldr-1	68±7	-	52	Hutton (1969b)
Es-1 - Pre	49±6	-	38	Shreffler (1964a)
Es-3 - Fu	46±10	-	28	Hutton and Roderick (1970)
Es-3 - ln	65±12	-	42	,, ,, ,, ,,
Es-3 - ru	45±9	-	28	,, ,, ,, ,,
Es-3 - s	66±8	-	51	,, ,, ,, ,,
Es-3 - Sl	64±9	-	47	,, ,, ,, ,,

Loci	Recombination value	Phase balance	Closest linkage	References
Es-3 - Sp	65±12	-	42	Hutton and Roderick (1970)
Es-3 - Va	61±10	-	42	„ „ „ „
f - fz	47±4	?	39	Dickie and Woolley (1950)
f - Hk	47±4	100	40	Green et al. (1963)
f - hr	47±6	0	34	Clark (1934b)
f - hy-1	56±8	0	40	Clark (1936)
f - je	55±8	0	40	Gruneberg et al. (1941)
f - ln	50±4	100	42	Clark (1936b)
f - lr	49±4	100	41	Fraser and Herer (1950)
f - lst	50±4	0	42	Forsthoefel and Shenk (1970)
f - lx	57±5	9	47	Carter (1951c)
f - Mi	46±6	100	34	Bunker and snell (1948)
f - my	51±8	0	36	Carter (1956)
f - N	49±6	100	37	Clark (1934b)
f - pe	42±4	0	35	Kelton and Rauch (1968)
f - p	49±5	100	39	Clark (1934b)
f - pa	55±8	0	40	„ „
f - r	43±8	0	29	„ „
f - Ra	78±12	100	55	Carter and Phillips (1954)
f - s	49±11	0	28	Clark (1934b)
f - sh-2	56±3	0	95	Clark (1934b), Heston et al.(1952)
f - T	47±3	100	40	Clark (1934a, b)
f - v	59±9	0	41	Clark (1934b)
f - W	39±11	100	18	„ „
f - wa-1	54±6	0	42	Burhoe (1936)
f - wa-2	46±3	100	41	Heston et al. (1952)
f - ♂	51±6	0	38	Clark (1934b)
fa - p	33±10	0	14	Yosida (1960a, b)
fa - s	60±10	0	41	„ „
fi - my	57±9	0	40	Carter (1956)
fi - Sp	42±3	100	42	Wallace (1961)
fi - W	48±1	86	46	Carter and Gruneberg (1950), Wallace (1958b, 1961)
fi - ♂	53±3	0	48	Carter and Falconer (1952)
fr - Hk	53±3	100	47	Green et al. (1963)
fr - oe	43±4	0	35	Kelton and Rauch (1968)
fs - hr	47±6	11	35	Green (1954)
fs - ♂	51±6	100	40	„ „
Fu - fz	52±3	?	45	Dickie and Woolley (1950)
Fu - je	48±25	100	0	Gruneberg et al. (1941)
Fu - ji	43±9	0	26	Snell (1945)
fz - lst	42±6	0	31	Forsthoefel and Shenk (1970)
Fu - lx	50±5	37	40	Carter (1951c)
Fu - Mi	21±11	0	8	Bunker and Snell (1948)

Loci	Recombination value	Phase balance	Closest linkage	References
Fu - pi	50±5	100	41	Dickie and Woolley (1946)
Fu - Re	47±4	100	39	Carter and Falconer (1952)
Fu - un	50±9	100	33	Wright (1947a)
fz - Hba	62±11	-	41	Hutton *et al.* (1964)
fz - Hk	54±5	100	45	Green *et al.* (1963)
fz - hr	61±4	?	52	Dickie and Woolley (1950)
fz - lx	47±7	0	48	Carter (1951c)
fz - Mi	52±5	100	41	Bunker and Snell (1948)
fz - my	43±8	0	28	Carter (1956)
fz - oe	51±3	0	45	Kelton and Rauch (1968)
fz - p	42±9	0	25	Carter and Falconer (1952)
fz - pi	54±6	0	42	Dickie and Woolley (1950)
fz - Ra	40±16	100	31	Carter and Phillips (1954)
fz - Sd	53±3	?	47	Dickie and Woolley (1950)
fz - se	48±4	100	40	Wallace (1961)
fz - Tw	50±5	100	41	Lyon (1958)
fz - W	49±3	?	42	Dickie and Woolley (1950)
fz - ♂	51±4	56	44	Carter and Falconer (1952)
H-2 - H-5	44±10	-	25	Amos *et al.* (1963)
H-2 - H-6	43±9	-	26	,, ,, ,, ,,
H-2 - H-14	45±4	?	38	Popp (1967a), Snell *et al.* (1967)
H-2 - Ly-A	50±5	-	41	Boyse *et al.* (1968)
H-2 - Ly-B	52±3	-	45	,, ,, ,, ,,
H-2 - 'θ'	55±4	-	48	,, ,, ,, ,,
H-14 - T	53±7	?	39	Snell *et al.* (1967)
Hba - Hbb	53±2	-	48	Popp (1962c, d, 1969a, b, c)
Hba - ln	67±10	-	47	Hutton *et al.* (1964)
Hba - Mi	68±10	-	49	,, ,, ,, ,,
Hba - mk	48±8	-	33	Russell *et al.* (1970)
Hba - Ph	53±8	-	37	Hutton *et al.* (1964)
Hba - Sl	37±8	-	22	,, ,, ,, ,,
Hba - Va	52±11	-	31	,, ,, ,, ,,
Hbb - Pre	50±6	-	38	Shreffler (1964a)
Hbb - Trf	53±3	-	41	Cohen (1960), Shreffler (1960)
Hbb - ♂	44±6	-	32	Cohen (1960)
Hc - Ig-1	53±4	-	46	Cinader *et al.* (1966)
Hc - Sas-1	56±7	100	42	Wortis (1965)
Hk - ln	51±5	100	42	Green *et al.* (1963)
Hk - Mi	45±4	0	37	,, ,, ,, ,,
Hk - N	52±4	0	38	,, ,, ,, ,,
Hk - p	46±2	100	42	,, ,, ,, ,,
Hk - Ra	41±4	0	32	,, ,, ,, ,,
Hk - Re	53±4	46	46	,, ,, ,, ,,
Hk - s	49±5	100	39	,, ,, ,, ,,

Loci	Recombination value	Phase balance	Closest linkage	References
Hk - sh-1	53±3	100	47	Green et al. (1963)
Hk - sh-2	48±4	100	41	,, ,, ,, ,,
Hk - Sl	59±10	0	40	,, ,, ,, ,,
Hk - Sp	50±5	0	42	,, ,, ,, ,,
Hk - v	57±5	100	50	,, ,, ,, ,,
Hk - Va	45±5	100	35	,, ,, ,, ,,
Hk - wa-2	47±4	100	40	,, ,, ,, ,,
Hm - shm	46±5	100	36	Green (1968)
hr - le	47±2	82	43	Lane (1967)
hr - ln	49±4	0	41	Murray (1933)
hr - lx	48±4	3	41	Carter (1951c)
hr - N	60±6	100	48	Snell (1931)
hr - pi	45±3	91	40	Dickie and Woolley (1948), Lane (1967)
hr - sh-2	57±7	0	44	Clark (1935)
hr - stb	45±8	100	44	Lane and Dickie (1968)
hr - T	44±6	100	32	Clark (1934a)
hr - W	45±2	100	42	Gates and Pullig (1945a, b), Lane (1967)
hr - ♂	42±5	0	32	Green (1954)
hy-1 - ln	54±7	0	40	Clark (1936)
hy-1 - N	48±8	100	31	,, ,,
hy-1 - r	43±8	0	30	,, ,,
hy-1 - s	53±7	0	39	,, ,,
hy-1 - sh-2	40±4	0	32	,, ,,
hy-1 - T	43±7	100	13	Clark (1934a, 1936)
hy-1 - v	43±8	0	26	Clark (1936)
hy-1 - W	52±12	100	28	,, ,,
Id-1 - lt	48±10	-	28	Hutton and Roderick (1970)
Id-1 - ru	52±10	-	32	,, ,, ,, ,,
Id-1 - s	54±10	-	34	,, ,, ,, ,,
Id-1 - Sl	44±10	-	26	,, ,, ,, ,,
Ig-1 - Ir-1	53±12	-	30	MacDevitt (1968)
Ig-1 - Sas-1	60±10	-	40	Wortis (1965)
je - lst	53±7	0	39	Forsthoefel and Shenk (1970)
je - lx	49±7	0	35	Carter (1951c)
je - my	49±8	0	34	Carter (1956)
je - oe	56±4	0	48	Kelton and Rauch (1968)
je - p	51±5	0	42	Gruneberg et al. (1941)
je - Ra	50±11	100	28	Carter and Phillips (1954)
je - Re	52±6	100	40	Gruneberg et al. (1941)
je - ru	49±2	85	46	Fisher and Snell (1948), Wallace (1958a)
je - s	44±4	69	36	Gruneberg et al. (1941)

Loci	Recombination value	Phase balance	Closest linkage	References
je - sh-2	48±8	100	32	,, ,, ,, ,,
je - Tw	56±9	100	39	Lyon (1958)
je - v	47±8	0	32	Gruneberg et al. (1941)
je - W	55±8	100	40	,, ,, ,, ,,
je - wa-1	44±8	0	30	,, ,, ,, ,,
je - wa-2	53±8	0	37	,, ,, ,, ,,
je - ♂	50±4	0	42	Carter and Falconer (1952)
ji - ln	56±7	0	43	Snell (1945)
ji - oe	52±5	0	42	Kelton and Rauch (1968)
ji - p	52±4	0	44	Snell (1945)
ji - r	46±8	0	29	,, ,,
ji - s	53±7	0	39	,, ,,
ji - W	43±4	79	29	,, ,,
ji - wa-1	53±7	0	40	,, ,,
jt - lu	51±10	10	32	Center (1966)
le - Os	48±4	100	42	Lane and Green (1967)
le - Ra	54±4	100	46	,, ,, ,, ,,
le - stb	51±5	100	42	Lane and Dickie (1968)
ln - je	42±8	0	27	Gruneberg et al. (1941)
ln - lst	57±6	0	44	Forsthoefel and Shenk (1970)
ln - lx	50±6	88	38	Carter (1951c)
ln - Mi	51±3	100	45	Bunker and Snell (1948)
ln - my	50±7	0	45	Carter (1956)
ln - oe	53±3	0	47	Kelton and Rauch (1968)
ln - Ra	57±4	100	50	Carter and Phillips (1954)
ln - Re	53±7	100	39	Crew and Auerbach (1940), Carter and Falconer (1952)
ln - s	47±7	0	34	Wallace (1961)
ln - Sd	51±2	59	48	Parsons (1959), Wallace (1961)
ln - se	50±7	0	36	Wallace (1961)
ln - sh-2	59±7	0	45	Clark (1935)
ln - T	51±5	100	41	Clark (1934a)
ln - Tw	50±5	100	41	Lyon (1958)
ln - v	45±6	0	33	Murray (1933)
ln - W	49±3	100	42	,, ,,
ln - wa-1	56±5	23	47	Burhoe (1936), Crew and Auerbach (1940), Carter and Falconer (1952)
ln - ♂	60±5	60	50	Carter and Falconer (1952),
Lp - oe	43±7	100	42	Kelton and Rauch (1968)
lr - sh-2	56±6	0	45	Fraser and Herer (1950)
lr - wa-2	46±6	0	35	,, ,, ,, ,,
lr - ♂	50±5	76	40	,, ,, ,, ,,
lst - Os	37±7	100	23	Forsthoefel and Shenk (1970)

Loci	Recombination value	Phase balance	Closest linkage	References
lst - T	49±5	0	39	Forsthoefel and Shenk (1970)
lst - Pt	40±7	100	27	,, ,, ,, ,,
lst - ru	46±5	0	36	,, ,, ,, ,,
lst - s	51±7	0	36	,, ,, ,, ,,
lst - wa-1	44±5	0	34	,, ,, ,, ,,
lx - p	54±3	37	49	Carter (1951c)
lx - pa	43±5	0	33	,, ,,
lx - Re	45±5	58	38	,, ,,
lx - ru	59±7	0	46	,, ,,
lx - s	51±2	42	48	,, ,,
lx - Sd	44±7	100	31	,, ,,
lx - si	45±6	100	33	,, ,,
lx - v	55±6	37	45	,, ,,
lx - Va	46±6	100	34	,, ,,
lx - wa-1	41±6	0	29	,, ,,
lx - wa-2	48±6	5	35	,, ,,
lx - ♂	49±1	35	46	,, ,,
Ly-A - Ly-B	53±3	-	48	Boyse et al. (1968)
Ly-A - 'θ'	46±8	-	30	,, ,, ,, ,,
Ly-B - 'θ'	47±4	-	38	,, ,, ,, ,,
Mdh-1 - Pgm-1	30±7	-	16	Hutton (1969b)
Mi - my	48±8	100	33	Carter (1956)
Mi - oe	47±3	100	41	Kelton and Rauch (1968)
Mi - p	56±4	74	47	Bunker and Snell (1948), Kreitner (1957)
Mi - Ra	54±8	0	38	Carter and Phillips (1954)
Mi - Re	51±4	0	43	Bunker and Snell (1948)
Mi - s	54±3	100	48	,, ,, ,, ,,
Mi - se	49±1	49	47	Carter and Falconer (1953)
Mi - sey	55±4	?	47	Roberts (1967)
Mi - sh-2	45±4	100	37	Bunker and Snell (1948)
Mi - Sp	44±4	0	37	,, ,, ,, ,,
Mi - tf	51±5	90	42	Lyon (1956)
Mi - ti	44±5	0	34	Searle (1961)
Mi - Tw	56±5	0	46	Lyon (1958)
Mi - v	48±3	100	42	Bunker and Snell (1948)
Mi - W	44±5	0	34	,, ,, ,, ,,
Mi - wa-2	47±4	100	39	,, ,, ,, ,,
Mi - ♂	48±8	18	33	Carter and Falconer (1952)
my - p	49±12	0	24	Carter (1956)
my - Ra	53±5	100	43	,, ,,
my - Re	46±7	100	33	,, ,,
my - ru	58±5	0	48	,, ,,
my - s	51±8	0	36	,, ,,

Loci	Recombination value	Phase balance	Closest linkage	References
my - Sd	43±7	0	31	Carter (1956)
my - T	50±5	100	40	,, ,,
my - v	41±13	0	16	,, ,,
my - Va	57±16	100	25	,, ,,
my - W	53±5	100	42	,, ,,
my - wa-2	43±9	0	26	,, ,,
my - ♂	54±4	25	47	,, ,,
N - ln	47±4	100	39	Murray (1933)
N - p	54±4	100	47	Snell (1931)
N - r	65±8	0	49	Keeler (1930), Snell (1931)
N - s	60±9	100	42	Snell (1931)
N - se	44±3	100	38	,, ,,
N - sh-2	53±5	100	44	Clark (1935)
N - T	49±3	43	42	Clark (1934a), Carter and Falconer (1952)
N - v	60±9	100	42	Snell (1931)
N - W	47±7	0	33	,, ,,
N - wa-1	65±11	100	43	Burhoe (1936)
N - We	46±13	100	21	Hertwig (1942)
N - ♂	52±4	50	44	Carter and Falconer (1952)
oe - Os	47±3	100	42	Kelton and Rauch (1968)
oe - Pt	43±3	100	38	,, ,, ,, ,,
oe - Ra	56±3	100	51	,, ,, ,, ,,
oe - ru	50±5	0	41	,, ,, ,, ,,
oe - s	47±3	?	41	,, ,, ,, ,,
oe - Sd	50±4	100	43	,, ,, ,, ,,
oe - se	54±5	0	45	,, ,, ,, ,,
oe - Sl	47±4	100	30	,, ,, ,, ,,
oe - T	48±3	100	42	,, ,, ,, ,,
oe - Tw	60±4	100	53	,, ,, ,, ,,
oe - v	58±5	0	48	,, ,: ,, ,,
oe - W	51±4	100	44	,, ,, ,, ,,
Os - Pt	49±3	79	44	Green et al. (1963)
Os - Ra	47±2	89	44	,, ,, ,, ,,
Os - rd	56±12	100	32	Sidman and Green (1965)
Os - shm	45±7	100	31	Green (1931)
Os - sl	55±6	100	42	Green et al. (1963)
Os - Tw	53±6	0	41	,, ,, ,, ,,
p - Pt	50±9	0	33	Hollander and Strong (1951)
p - r	44±5	100	33	Keeler (1927, 1930)
p - Ra	47±9	100	29	Carter and Phillips (1954)
p - Re	47±6	46	43	Crew and Auerbach (1940), Carter and Falconer (1952)
p - ru	47±5	43	37	Dunn (1945)

Loci	Recombination value	Phase balance	Closest linkage	References
p - s	49±1	10	46	Dunn (1920b), Gates (1926), Carter and Falconer (1952)
p - Sd	52±3	72	49	Wallace (1961)
p - sh-2	48±3	0	43	Gates (1934)
p - slt	47±4	0	40	Pierro and Chase (1963)
p - sr	54±6	0	42	Deol and Robins (1962)
p - Tr	49±6	100	38	Falconer (1951)
p - v	52±2	30	48	Darbishire (1904), Gates (1926)
p - vt	38±7	0	25	Michie (1955c)
p - W	52±1	66	47	Dunn (1920b), Wachter (1921)
p - wa-1	51±3	88	45	Burhoe (1936), Gruneberg (1936a)
p - wa-2	48±3	60	41	Carter and Falconer (1952), Michie (1955c)
p - ♂	50±3	42	44	Carter and Falconer (1952), Michie (1955c)
pa - sh-2	57±8	0	41	Clark (1935)
pa - T	47±5	100	37	Clark (1939a)
Ph - s	48±11	0	26	Gruneberg and Truslove (1960)
pi - Re	48±3	100	41	Dickie and Woolley (1946)
pi - Sp	54±3	100	48	,, ,, ,, ,,
pn - W	48±3	0	42	Kidwell et al. (1961)
Pre - Ss	46±6	-	33	Shreffler (1964a)
Pre - Trf	48±5	-	38	,, ,,
Pt - s	52±4	100	44	Hollander and Strong (1951)
Pt - se	50±10	?	31	,, ,, ,, ,,
qv - wd	51±5	0	41	Yoon (1959)
qv - ♂	44±3	100	39	Yoon (1959, 1961)
r - s	49±7	0	35	Keeler (1927, 1930)
r - se	48±8	0	32	,, ,, ,,
r - sh-2	55±9	0	38	Clark (1935)
r - T	46±5	100	36	Clark (1934a)
r - v	29±27	100	0	Keeler (1927, 1930)
r - W	42±8	100	27	,, ,, ,,
r - wa-1	45±7	0	31	Burhoe (1936)
Ra - Re	57±4	27	48	Carter and Phillips (1954)
Ra - ru	43±11	100	21	,, ,, ,, ,,
Ra - s	54±3	100	48	,, ,, ,, ,,
Ra - shm	30±17	100	0	Green (1967)
Ra - T	54±4	0	47	Carter and Phillips (1954)
Ra - tf	61±9	100	44	Lyon (1956)
Ra - ti	53±4	0	46	Searle (1961)
Ra - Tw	51±5	0	42	Lyon (1958)
Ra - tg	46±4	100	38	Green et al. (1963)

Loci	Recombination value	Phase balance	Closest linkage	References
Ra - v	30±16	100	0	Carter and Phillips (1954)
Ra - Va	60±17	0	17	,, ,, ,, ,,
Ra - W	51±3	100	45	,, ,, ,, ,,
Ra - wa-2	55±11	100	33	,, ,, ,, ,,
Ra - ♂	50±3	55	46	,, ,, ,, ,,
rd - Mi	52±10	100	32	Sidman and Green (1965)
rd - si	51±3	0	46	Di Paola and Noell (1962)
Re - s	52±4	32	45	Crew and Auerbach (1940), Carter and Falconer (1952)
Re - Sd	57±5	32	48	Carter and Falconer (1952)
Re - stb	53±3	100	46	Lane and Dickie (1968)
Re - tf	61±6	100	50	Lyon (1956)
Re - tg	53±4	100	44	Green et al. (1963)
Re - Tw	49±5	0	40	Lyon (1958)
Re - v	45±10	100	26	Crew and Auerbach (1940)
Re - Va	44±10	100	24	Michie (1955c)
Re - W	45±6	0	33	Carter and Falconer (1952)
Re - wa-1	51±2	46	47	Crew and Auerbach (1940), Carter (1951b)
Re - ♂	52±1	62	51	Fisher, Lyon, and Owen (1947), Carter and Phillips (1953), Michie (1955a)
rl - ♂	53±6	0	42	Falconer (1951)
ro - ♂	49±4	0	41	Falconer and Snell (1952)
ru - slt	62±8	0	47	Pierro and Chase (1963)
ru - T	46±3	100	41	Dunn (1945)
ru - Tw	42±5	100	31	Lyon (1958)
ru - ♂	46±5	0	37	Carter and Falconer (1952)
s - se	52±3	20	45	Wallace (1961)
s - sh-2	49±4	30	41	Carter and Falconer (1952)
s - slt	46±4	100	38	Pierro and Chase (1963)
s - sr	59±10	0	39	Deol and Robins (1962)
s - Tr	48±6	100	36	Falconer (1951)
s - Tw	48±5	0	38	Lyon (1955)
s - un	50±8	100	34	Michie (1955c)
s - v	48±1	99	46	Gates (1926), Carter and Falconer (1952)
s - Va	47±4	100	40	Cloudman and Bunker (1945), Carter and Falconer (1952)
s - vt	47±7	0	34	Michie (1955c)
s - W	47±1	62	44	Little (1915, 1917), Dunn (1920b, c), Keeler (1931a), Fisher (1946)

Loci	Recombination value	Phase balance	Closest linkage	References
s - wa-1	48±1	21	47	Burhoe (1936), Crew and Auerbach (1940), Fisher and Mather (1936b)
s - wa-2	50±3	46	44	Carter and Falconer (1952)
s - ♂	51±1	50	48	Fisher and Mather (1936b), Michie (1955c)
Sd - se	50±1	8	47	Wallace (1961)
Sd - shm	55±9	100	38	Green (1968)
Sd - Sp	49±2	44	46	Parsons (1959), Wallace (1961)
Sd - T	54±8	0	38	Dunn et al. (1940)
Sd - tg	47±2	100	44	Green et al. (1963)
Sd - ti	44±5	0	34	Searle (1961)
Sd - Tw	48±5	0	38	Lyon (1958)
Sd - un	45±12	100	23	Wright (1947a)
Sd - Va	39±5	12	21	Carter and Falconer (1952)
Sd - W	49±1	73	49	Wallace (1958b)
Sd - ♂	45±3	45	39	Carter and Falconer (1952)
se - ln	48±4	100	41	Murray (1933)
se - sh-2	56±3	0	51	Gates (1934)
se - wa-1	57±7	0	43	Burhoe (1936)
sg - W	36±10	?	17	Sidman et al. (1962)
Sh-2 - T	53±5	100	43	Clark (1934a)
sh-2 - v	75±15	0	45	Clark (1935)
sh-2 - W	47±6	0	35	Gates (1934)
sh-2 - wa-2	53±6	0	41	Burhoe (1936)
sh-2 - ♂	54±1	49	51	Wright (1947b), Fisher et al. (1947), Carter and Phillips (1953), Michie (1955b)
shm - W	49±8	100	32	Green (1968)
Sl - W	52±2	30	48	Sarvella and Russell (1956)
sr - vt	54±6	0	42	Deol and Robins (1962)
sr - W	40±6	0	27	,, ,, ,, ,,
sr - wa-2	52±6	0	40	,, ,, ,, ,,
stb - Va	49±5	100	38	Lane and Green (1968)
T - Mi	45±6	65	34	Carter and Falconer (1952)
T - ti	52±5	0	42	Searle (1961)
T - Tw	50±5	0	40	Lyon (1958)
T - v	48±5	100	39	Clark (1934a), Carter and Falconer (1952)
T - W	52±4	83	38	Clark (1934a), Carter and Falconer (1952)
T - wa-1	46±4	100	40	Burhoe (1936)
T - ♂	43±4	58	35	Carter and Falconer (1952)
tf - W	53±7	100	39	Lyon (1956)

Loci	Recombination value	Phase balance	Closest linkage	References
tg - Va	71±15	100	43	Green et al. (1963)
ti - W	43±4	100	36	Searle (1961)
Tr - wa-1	45±6	100	33	Falconer (1951)
Tr - ♂	52±5	100	42	,, ,,
Trf - ♂	42±6	-	30	Cohen (1960)
Tw - W	46±5	0	36	Lyon (1958)
Tw - wa-2	50±6	100	39	,, ,,
un - ♂	51±8	0	35	Michie (1955c)
v - wa-1	48±6	34	37	Burhoe (1936), Crew and Auerbach (1940)
v - ♂	58±5	50	48	Carter and Falconer (1952)
Va - W	45±2	0	41	Cloudman and Bunker (1945)
Va - vt	36±10	100	10	Michie (1955c)
Va - ♂	44±5	44	26	Carter and Falconer (1952)
vt - W	48±8	100	33	Michie (1955c)
vt - ♂	52±2	55	49	Heston (1951), Michie (1955c)
W - wa-1	56±7	100	43	Burhoe (1936)
W - wl	43±4	100	35	Lane and Dickie (1961)
W - ♂	48±4	56	41	Carter and Falconer (1952), Michie (1955c)
wa-1 - ♂	50±1	69	47	Fisher and Mather (1936b)
wa-2 - ♂	51±1	48	48	Wright (1947b), Fisher et al. (1947), Carter and Phillips (1953), Michie (1955c)
wd - ♂	52±6	0	39	Yoon (1959)

CHROMOSOME MORPHOLOGY

The haploid chromosome number consists of 20 chromosomes. Many of the very early determinations were subsequently found to be in error but the correct number was finally established by Cox (1926), Painter (1928a, b) and Cutwright (1932). The rapid advances in karyological techniques of the last decade have rendered obsolete much of the early work although it remains of historical interest and some aspects still merit consideration. A good summary is provided by the appropriate chapter of Gruneberg (1952). The following list brings the relevant literature up to date: Slizynski (1949, 1955b), Griffen (1955, 1958a, b, 1960), Hungerford (1955), Levan (1956), Beatty (1957), Ohno et al. (1957b), Stich and Hsu (1960),

Bunker (1961, 1965), Fredga (1961), Levan, Hsu and Stich (1962), Ford and Woollam (1963), Stevens and Bunker (1964), Lagneau and Mewissen (1964), Bennett (1965), Leonard and Deknudt (1965), Watson and Oliver (1965), Ojima, Takayama and Kato (1965), Goljdman, Smertenko and Vilkina (1966) and Zeleny (1967). Two useful reviews are those of Russell (1962) and Griffen (1966).

The mouse karyotype is noteworthy for the absence of excessively large chromosomes and the absence of easily recognized morphological features by which individual elements can be distinguished. Speaking relatively, of course, since in first-class preparations experienced observers have been able to distinguish about four elements by virtue of size and the presence of secondary constrictions. The difficulties are two, (a) all of the chromosomes are effectively telocentric and (b) the majority show a smooth transition of size from the largest to the smallest. Because of this, a systematic classification, based on relative size and the position of the centromere (as attempted for other species), will not be rewarding. Instead, a general description will be presented based on conclusions of various observers. In view of the important position held by the mouse for mapping of mammalian chromosomes, it is obvious that a detailed charaterization of individual chromosomes would be of value.

The remarkable continuous graduation in size (length) of the chromosomes has been brought out by the ideograms prepared by Levan et al. (1962) and Crippa (1964); see also Ohno et al. (1957b). Despite this, several attempts have been made to designate elements of similar size and shape. Slizynski (1955a) has classed the autosomes as four long, six medium, and nine short, while Crippa (1964) recognizes five groups of autosomes (consisting of 2, 3, 10, 3, and 1 element(s) respectively), Watson and Oliver (1965) four groups (4, 9, 3, and 4, respectively) and Goljdman et al. (1966) four groups (5, 5, 5, and 4, respectively). There is apparently ample room for differences of opinion and classifications of this nature may represent the confidence of the observer as much as realistic grouping.

Identification of individual chromosomes is far more of a problem. Few observers care to go beyond the positive recognition of more than four chromosomes, although it does seem possible to proceed further with exceptional preparations and attention to fine detail. Levan *et al.* (1962) have discussed the fundamental difficulties which must be overcome. They consider that the largest and smallest of the autosomes can be identified in most cells; also, with less certainty, the second largest. The presence of secondary constrictions enables certain of the óther chromosomes to be recognized. Almost all of the elements are telocentric or, if second arms are present, these are exceedingly small. However, at least one chromosome (designated as No. 15) has a distinctly recognizable minute second arm.

The observations of Ford and Woollam (1963), Crippa (1964), and Bennett (1965) seem to be in fundamental agreement with the above. Ford and Woollam note that the largest and smallest chromosome can usually be identified; together with two others of different sizes with marked secondary constrictions. A fifth chromosome also shows secondary constriction but less constantly and hence less reliably. Bennett, however, takes the view that three pairs of chromosomes can be reliably observed with 'satellites'. These are described as large, medium, and small. She remarks that the remaining elements can be arranged in pairs, taking into account such factors as length, degree of contraction, angle of the arms and, especially, morphological differentiation of the centromere region but not with complete infallibility. Goljdman *et al.* (1966) state that the five largest and smallest chromosomes are the easiest to recognize, followed by the larger elements as a group. Most of the chromosomes appear as telocentrics but as many as five are said to be subtelocentric; one, in particular, being most noticeable.

Nakamura and Tonomura (1963) have featured the presence of one chromosome with a distinctive knob or head separated from the main body of the chromosome by a thin thread of chromatin. The chromosome falls into 15th place according to size. This element is evidently the one described by previous observers but perhaps with the satellite more prominently displayed – either as a technical artefact or as a strain characteristic. Chu and Monesi (1960) state that four autosomals can be individually identified but give no details.

A new approach to chromosome topography has been made

by Slizynski (1949) and Griffen (1955, 1958, and 1960). This takes the form of meticulous description of the more deeply staining areas (chromomeres) of the pachytene chromosome. The outcome are diagrams superficially resembling the banded salivary chromosomes of *Drosophila*. Three attempts have so far been completed, with fair agreement. In Slizynski's presentation, most of the chromosomes are shown as subtelocentric and at least six of the chromosomes are shown with medium centromeres. On the other hand, Griffen (1960) indicated that all of the chromosomes are telocentric, with the possible exception of the *X* and *Y* in which a small second arm appears probable. Griffen cautions that not all of the autosomes may be truly telocentric although second arms were not seen.

Slizynski denotes the individual chromosomes by Arabic numbers; while Griffen uses Roman capital letters but later substituting Roman numerals whenever it becomes possible to relate a genetic linkage group with a particular chromosome. Table 11.18 shows the resulting correspondences, based on the conclusions of Slizynski (1957) and Griffen (1960). These should be regarded as provisional since Slizynski (1967) states that the element taken as linkage group I may actually be the *X* chromosome.

Sachs (1955) states that he has not been able to detect any 'cross-lines' on the pachytene chromosome. Geyer-Duszynska (1963) found that the chromomeric banding pattern was too uniform and not sufficiently stable enough for individual chromosomes to be recognized. It would be unfortunate if this turns out to be true. Observations of very fine detail can become personified in the sense that consistent results can be produced by one worker which cannot be duplicated by others. Only the future will show whether or not the analysis of the chromomeric features of the mouse chromosome can become a practical tool.

SEX CHROMOSOMES

For some time, it was thought that the *X* chromosome was the largest of the complement, particularly among Japanese workers. For example, Ojima *et al.* (1965) give both numerical and illustrative data. Unfortunately, the strain of mice employed is not specified. However, other investigations do not support this view. In these, the X chromosome emerged as a large telocentric but not the largest. For example, Levan *et al.* (1962) and Goljdman *et al.* (1966) rank the X as the third in size. Both Ford and Woollam (1963) and Crippa (1964) place the X in the second group of large chromosomes (i.e., not the largest but the next to largest). Bennett (1965), moreover, concludes that the X could be one of the many medium sized chromosomes. Part of the problem is that it is not always obvious which is the X chromosome but, in this respect, Galton and Holt (1965) have demonstrated by means of autoradiography that the X is almost certainly one of the large chromosomes.

The Y chromosome is one of the smallest in the karyotype. In fact, the Y is identical to or just slightly smaller than the two smallest autosomes. With first-class preparations it is possible to observe two small chromosomes in female cells but three in male cells. This aspect has led Stich and Hsu (1960) to propose that the Y could serve as a convenient biological marker for a variety of purposes. At first sight, this idea seems to have much to recommend it but Eichwald *et al.* (1966) and Ford (1966) feel that it possesses a drawback which rather severely limits its ·usefulness. This lies in the high degree of technical perfection which is required for the unambiguous detection of the three small chromosomes. It is fair to point out that Bennett (1965) and Zeleny (1967) consider that the small Y can be identified consistently.

The Y displays a certain amount of variability in size. Levan *et al.* (1962) noted that the Y of C57B1 males was unusually long; being about 20 per cent larger than the Y of C3H and Swiss males. While the C3H or Swiss Y was approximately the same length as the smallest autosome, the C57B1 Y was intermediate in size between the two smaller autosomes. Nakamura and Tonomura (1964) present data that the Y of a polydactylous strain in their possession had a longer Y than usual. Similarly, Goldjman *et al.* (1966) commented on the size variation of the Y element, implying that the Y of a strain designated as AKR was larger than usual.

An undecided, if not an actual controversial, point is whether or not the X and Y associate at meiosis by means of chiasmata.

Conceptionally, the most satisfying theory from a genetic viewpoint, is that the XY bivalent is tripartite, consisting of a homologous segment common to both chromosomes, a differential segment perculiar to the X and a differential segment common to the Y. Thus, three forms of sex-linked heredity are possible, partial sex-linkage and complete X and Y linkage. Here, the question is whether the conception has a karyological basis. Crew and Koller (1932) aver that it has, although their observations lack precision. Koller and Darlington's (1934) description of the rat XY bivalent has often been taken as a model for the mouse but this is hazardous to say the least.

However, the tripartite structure of the mouse XY bivalent has received its most exact formation at the hands of Slizynski (1955b). He proposes that the centromere is located in the homologous region, close to the differential segments but not so close that chiasmata cannot occur between the centromere and these regions. The X and Y are not of equal length and it follows that the difference is due to relative lengths of the differential segments. The Y segment is depicted as minute while the X segment is longer. A chiasma occurring between the centromere and the differential segments will give rise to a first meiotic chromosome with unequal segments. Out of 119 clearly defined bivalents, 14 (12 per cent) were observed to be of this type. If it is assumed that the chance of chiasma is linearly dependent on length, the length ratio of the homologous region on each side of the centromere is about 7.5 : 1. Slizynski's paper should be consulted for full details; particularly for his rebuttal of criticism of the tri-partite concept.

It may be noted that the above description is at variance with the view that the X and Y are telocentric or subtelocentric bodies, although much would depend on the relative size of the second arm. It is also proposed that the Y is not one of the smallest chromosomes of the complement. Slizynski's depiction implies that the homologous regions of the X and Y could be extensive and could allow for the possibility of partial sex-linkage.

The formation of chiasmata by the XY has been disputed by Makino (1941), Matthey (1953), Sachs (1955), Ohno et al. (1957a, b) and Geyer-Duszynska (1963). The main objection seems to be that the X and Y are invariably associated in an end-to-end manner which effectively precludes chiasma attachment. Makino and Matthey make the point that the association is between the ends thought not to contain the centromeres. Ohno *et al.* (1959a) refer to end-to-end association and emphasize that the association cannot be disrupted by ribonuclear digestion nor by hypotonic treatment. It seems, in fact, to be as strong as an ordinary chiasma connection. The peculiar end-to-end association could result from extensive differential segments, the homologous regions being confined to extremely short arms proximal to the centromeres. This is a plausible suggestion and one which may not be easily set aside. If the association is held to be non-chiasmatic, then some positive alternative must be proposed.

The Ohno *et al.* depiction would lead to several predictions which seem generally consistent with current mouse genetics. The considerable size difference between the X and Y is upheld. This implies that, on size alone, the X could be more genetically active than the Y. Chiasma may occur only, or mainly, in the minute region beyond the centromere. This would either preclude partial sex-linkage or render it exceedingly unlikely. It would follow that average chiasma frequency for the XY will be expected to be less than that of the average autosome.

It is probably merely a matter of time before the uncertainties surrounding chiasmata formation by the XY bivalent are finally dispelled. Solari (1970), following an extensive three-dimensional reconstruction of the sex chromosomes, is of the opinion that the X and Y do have a small homogogous region. Furthermore, certain preliminary changes were observed which foreshadow the formation of chiasmata. It does not follow, of course, that chiasmata must ensue but it is definitely probable.

Pearson and Bobrow (1970), by the aid of a process of selective staining with a fluorescent dye, have shown that chiasmata very probably occur between the presumptive homologous short arms of the X and Y in man. It is still possible to argue that the evidence is inconclusive but selective staining offers a fresh means of tackling the problems. The method would seem to have possibilities for mammalian cytogenetics in general and for the more important laboratory species in particular. The mouse is well placed to take up any advantages possessed by the technique.

The Y chromosome occupies a unique position in the mammalian karyotype. Permanently hemizygous, with a minimum of homology with the X just sufficient to ensure regular meiosis, the element would seem to be exposed to rigorous selection. In terms of size, its genetic potential would seem to be limited, perhaps primarily to determining maleness. This is a vital function and sufficient, in itself, to assure that the Y is not lost. The strong male determining propensity of the Y is shown by Table 11.9. XO individuals are sub-viable females but XXY individuals are males (possibly not of fully normal viability but this quantitative aspect is unknown). The fertility and segregation of X borne genes means that the XO condition can be easily maintained. Morris (1968) has made a detailed study of the reproductive performation of XO females and has shown, moreover, that the YO constitution is a pre-implantation lethal. The modes of origin of these abnormal sex forms are discussed by Ohno, Kaplan, and Kinosita (1959b), McLaren (1960), Russell (1961), Russell and Chu (1961), and Russell and Saylors (1962).

The possible XXX individual has not yet been karyologically identified although it seems probable that the animal will be a superficially normal female. There may be inviability, but this is not certain, and there seems to be a somewhat greater probability of infertility but again not necessarily. The XYY condition has also not been found; with this form, sterility would seem probable and could act as a preliminary screen.

Several of the possible sex chromosome mosaics have been reported. A sexually normal fertile XX/XO was found by M. C. Green (1967; MNL 37, 33). Another XX/XO, described by Cattanach (1967), was sterile but, as the animal was heterozygous for a translocation, the sterility was probably not due to the mosaicism. Lyon (1969) has featured in detail a hermaphrodite of XY/XO constitution. Externally, he was male but sterile, with a poorly developed scrotum. Dissection revealed an abdominal testis, with arrested spermatogenesis (left side), and an ovary with immature follicles (right side). The two gonads were accompanied by the appropriate accessory structures. It would seem that sufficient XY and XO cells were

TABLE 11.9. Appearance and vital characteristics of abnormal sex chromosome constitutions.

Karyotype	Appearance	Viability	Fertility	Reference
XO	Female	Impaired	Impaired	Russell, Russell, and Gower (1959), Welshons and Russell (1959), McLaren (1960), Kindred (1961), Cattanach (1962), Morris (1968).
YO	—	Pre-implantation	—	Russell, Russell, and Gower (1959), Morris (1968).
XXY	Male	Normal (?)	Sterile	McLaren (1960), Russell and Chu (1961), Cattanach (1961a), Kindred (1961), Slizynski (1964).
XYY	Male	Normal (?)	Weakly fertile (?)	Cattanach and Pollard (1969)

present on the two sides of the body to determine sexual development according to immediate karyotype but that these could not bring about full development or there was subsequent hormonal interference. Lyon speculates that, on this evidence, the XX/XY individual would be a hermaphrodite of similar form. The XO/XYY mosaic reported by Evans et al. (1969) was a sterile male resembling XYY.

Evans, Ford, and Searle's (1969) examination of the $38+X/38+XYY$ mosaic has raised the possibility that the full development of a male gonad to the point of active spermatogenesis is dependent on the presence of the Y chromosome. There is reason to think that a small proportion of Y bearing cells is sufficient to produce a testis but that normal spermatogenesis cannot be completed unless the number of these cells is above a certain threshold. A working hypothesis would be that Y bearing cells must be in the majority or at least scattered extensively among the spermatogenic tissue. Hence, the influence of the Y may extend beyond that of being simply male determining. The conclusions of Krzanowska (1969) that the Y carries a locus governing sperm head morphology may have relevance in this context.

CHIASMA FREQUENCY

The frequency distribution of chiasmata is of considerable interest to geneticists because of its functional relationship to genetic crossingover. Several thoroughgoing studies have been made on chiasma frequency and these are summarized in the accompanying tables. Some of the earlier results were found by sampling of the bivalents over many cells and rejection of indistinct bivalents. This procedure could lead to a bias and more recent studies have attempted to count chiasmata for every bivalent in a nucleus. However much care is taken, it seems that the counts must be regarded as minimum estimates, particularly for bivalents with multiple chiasmata.

Table 11.10 summarizes the observations of chiasma frequencies for the male. The counts show a typical decrease for successive phases of meiosis, apparently as a loss of distinctive chiasmatic configurations due to terminalization. The distribution of chiasmata might be different for chromosomes

TABLE 11.10. Frequency distribution of chiasmata per bivalent in the male mouse.

Stock	Meiotic phase	No. of bivalents	Chiasma frequency					Mean	Reference
			1	2	3	4	5		
Unknown	Metaphase	40	21	17	2	-	-	1.53 ± 0.10	Cox (1926)
Unknown	Metaphase	40	31	7	2	-	-	1.28 ± 0.09	Painter (1927)
Unknown	Diplotene	85	7	38	35	5	-	2.45 ± 0.08	Crew and Koller (1932)
Unknown	Metaphase	220	108	101	10	1	-	1.56 ± 0.04	Crew and Koller (1932)
A	Diakinesis	500	304	181	15	-	-	1.42 ± 0.03	Huskins and Hearne (1934)
CBA	Diakinesis	500	247	185	61	7	-	1.64 ± 0.03	Huskins and Hearne (1934)
CBA	Diplotene	179	38	93	40	6	2	2.11 ± 0.06	Slizynski (1955a)
CBA	Diakinesis	218	114	91	11	1	1	1.55 ± 0.04	Slizynski (1955a)
Cross-breds	Diplotene	839	236	446	145	11	1	1.92 ± 0.02	Slizynski (1955a)
Cross-breds	Diakinesis	896	457	403	36	-	-	1.53 ± 0.02	Slizynski (1960)
Cross-breds	Metaphase	200	79	100	21	-	-	1.71 ± 0.05	Slizynski (1960)
C3H × 101	Diakinesis	4750	3729	1018	-	-	-	1.21 ± 0.01	Searle et al. (1970)

of different length and this is shown by Table 11.11. All bivalents seem to have at least one chiasma; the number rising with the length of chromosome but not necessarily proportionally. In those species (non-mammalian), where favourable material has enabled this aspect to be studied in depth, the number of chiasmata is often less than proportional for the longer chromosomes. Slizynski (1955a) has shown that this relationship holds for the mouse, but not obviously, and then probably only for the longest elements. This result would be in harmony with the absence of marked size differences for the chromosomes, with a consequential reduction of this feature.

Fewer observations have been made on chiasma frequency in the female and these are shown by Table 11.12. Comparison of mean frequencies obtained by the same observer reveals that chiasma is higher for the female. This implies that the average frequency of crossingover would be expected to be greater in the ' female than in the male, although certain intercepts of chromosome may be exceptional. By and large, this expectation is borne out by those experiments which have differentiated between the sexes in amount of crossingover.

A certain amount of information is available for mean chiasmata frequency for animals of different ages (Table 11.13). A decline with age is apparent; perhaps of greater magnitude for the female than for the male. This could be a reflection of the greatly different modes of germ-cell formation. A comparable reduction for both sexes may not occur until the male is much older than the female. A decrease in crossingover with age has been observed in most of the genetic studies on this phenomenon (Fisher, 1949; Bodner, 1961b, Reid and Parsons, 1963) but not conclusively for those of Wallace (1957a). In Henderson and Edward's (1968) oocyte material, the decline in chiasma frequency was accompanied by an increase in the production of univalents.

The genetic phenomenon of interference in crossingover is generally thought to arise (a) from a tendency of one chiasma to inhibit the formation of another close to it or (b) from a tendency of a second chiasma to involve chromatids other than those involved in the first. The first tendency is termed chiasma interference while the second is chromatid interference. The

TABLE 11.11. Frequency distribution of chiasmata at diplotene for three groups of bivalents in the male mouse.

Stock	Size of Bivalents	No. of Bivalents	Chiasma frequency 1	2	3	4	5	Mean	Reference
Unknown	Short	9	-	14	6	-	-	2.22 ± 0.15	Crew and Koller (1932)
	Medium	6	-	3	3	-	-	2.50 ± 0.22	Crew and Koller (1932)
	Long	5	-	2	2	1	-	2.80 ± 0.12	Crew and Koller (1932)
CBA	Short	81	26	49	6	-	-	1.75 ± 0.07	Slizynski (1955a)
	Medium	54	9	26	15	4	-	2.26 ± 0.11	Slizynski (1955a)
	Long	44	3	15	16	2	-	2.63 ± 0.11	Slizynski (1955a)
Cross-breds	Short	369	207	154	8	-	-	1.46 ± 0.03	Slizynski (1955a)
	Medium	246	25	174	46	1	-	2.09 ± 0.04	Slizynski (1955a)
	Long	226	4	79	70	10	1	2.55 ± 0.04	Slizynski (1955a)

TABLE 11.12. Frequency distribution of chiasmata per bivalent in the female mouse.

Stock	Meiotic phase	No. of bivalents	Chiasma frequency						Mean	Reference
			1	2	3	4	5	6		
Unknown	Diplotene	42	8	10	11	8	3	2	2.86 ± 0.21	Crew and Koller (1932)
Unknown	Metaphase	100	29	54	8	9	-	-	1.97 ± 0.08	Crew and Koller (1932)
Mixed	Metaphase	779	174	403	178	24	-	-	2.11 ± 0.03	Slizynski (1960)

TABLE 11.13. Variation of chiasma frequency with age in the mouse.

Sex	Strain	Age (Months)	Meiotic phase	No. of bivalents	Mean	Reference
Male	Unknown	2.5	Metaphase	120	1.64 ± 0.06	Crew and Koller (1932)
		Adult	Metaphase	100	1.47 ± 0.05	Crew and Koller (1932)
	CBA	2	Diakinesis	25	1.63 ± 0.02	Huskins and Hearne (1936)
		8	Diakinesis	25	1.59 ± 0.02	Huskins and Hearne (1936)
	C57B1	1	Diakinesis	25	1.52 ± 0.02	Huskins and Hearne (1936)
		8	Diakinesis	25	1.49 ± 0.02	Huskins and Hearne (1936)
	CBA	2	Metaphase	?	$1.20 \pm ?$	Henderson and Edwards (1968)
		9	Metaphase	?	$1.30 \pm ?$	Henderson and Edwards (1968)
Female	CBA	2 - 2.5	Metaphase	840	1.28 ± 0.03	Henderson and Edwards (1968)
		4 - 6	Metaphase	840	1.22 ± 0.03	Henderson and Edwards (1968)
		9 - 15	Metaphase	480	1.06 ± 0.02	Henderson and Edwards (1968)
	C57B1	2	Metaphase	600	1.24 ± 0.02	Henderson and Edwards (1968)
		6	Metaphase	260	1.20 ± 0.02	Henderson and Edwards (1968)
		11 - 13	Metaphase	760	1.15 ± 0.01	Henderson and Edwards (1968)

two processes are not mutually exclusive. Karyologically, the two tendencies cannot be distinguished for mammaliam chromosomes at this time but Carter (1954) has shown that the problem is not insoluble. If chromatid interference is of frequent occurrence, theoretically it should be possible to detect its expression in genetic maps of over 60 units and, less precisely, in the behaviour of double crossingover. Carter's analysis uncovered no evidence for chromatid interference. If the tendency is present, it is weakly expressed and masked by chiasma interference.

LENGTH OF GENETIC CHROMOSOME

The earlier the phase of meiosis for which the chiasma frequencies can be determined, the more meaningful will be the derivation of estimated genetic map length. Thus, diplotene is preferable to metaphase and estimates will be based on this stage whenever possible. Variation of size between chromosomes could alter the estimated mean map length quite materially and this aspect has been taken into account to some extent by grouping the chromosomes as short, medium, and long as shown by Table 11.14. The lack of pronounced differences between the chromosomes for the mouse has one virue. It implies that the map lengths, though only means, will be more meaningful for the chromosomes of the mouse than for those species with considerable differences of chromosome size and morphology (acrocentric versus telocentric elements).

While each bivalent may be assumed to have at least one chiasma, it is probable that some will be undetected because of the process of terminalization. This is revealed by the reduction in frequency from diplotene to diakinesis to metaphase (Tables 11.10 and 11.12). Karyologically, the frequency distribution of chiasmata per bivalent for each phase of meiosis and their changes from one phase to the next are of interest but, cytogenetically, the prime interest is that of ascertaining the maximum frequency. It is usual to assume that one chiasma equals 50 crossover units of a genetic map. Therefore, the mean chiasma frequency multiplied by 50 gives a crude estimate of the mean genetic map length.

Henderson and Edwards (1968) have noted that chiasma and crossover frequencies may not correspond similarly in all situations because the position of the chiasmata must be considered. Terminally located chiasmata would give lower crossover values on the average than would interstitial chiasmata. In other words, chiasma localization would give rise to alternative crowding and spacing of loci on the genetic chromosome map. Direct observation of chiasmata has yet to be demonstrated for the mouse, although Henderson and Edwards have recorded that the frequency of terminal chiasmata is greater for old females than for young. This increase is correlated with a decrease in mean chiasma frequency with age. It is unknown if the increase is due to more rapid terminalization or to a change in location of chiasmata.

Table 11.14 presents a convenient summary of genetic map lengths as indicated by chiasma frequency. Both mean per bivalent and total length are shown. The first four entries relate to mean length, regardless of chromosome size, while the subsequent entries attempt to present a clearer picture by allowing for the size variation. The chiasma frequencies observed by Crew and Koller were significantly larger than those observed by Slizynski and the difference is carried over to the derived genetic maps. It is not difficult to imagine that mouse stocks could differ in chiasma frequency and hence in crossover behaviour. In any event, even without this complication, the estimates of the mean of map length should only be visualized as a representation. The overall estimate for all male diplotene data is 99 ± 1 (total length, 1986 ± 21). This estimate exceeds the length of most known maps based on crossover values and is, therefore, on the 'right side'. A map length of this magnitude is capable of supporting at least two short linkage groups. This is not to imply that this situation actually exists although it is evidently worth noting.

Comparable estimates of map length for the female are less easy to obtain. Crew and Koller give data on diplotene chiasma frequencies which are higher than those for the male. This produces a correspondingly longer genetic map (Table 11.14). Slizynski (1960) could not obtain reliable counts of diplotene chiasmata but he derives an approximation by assuming that the rate of change from metaphase to diplotene is the same for

TABLE 11.14. Mean and total estimated genetic map lengths, based on observed diplotene chiasma frequencies in the mouse.

Sex	Strain	Type of chromosome	Mean of map length	Total map length	Reference
Male	Unknown	All	122 ± 4	2450 ± 79	Crew and Koller (1932)
Female	Unknown	All	143 ± 11	2860 ± 214	Crew and Koller (1932)
Male	CBA	All	106 ± 3	2120 ± 57	Slizynski (1955a)
Male	Cross-breds	All	96 ± 1	1920 ± 24	Slizynski (1955a)
Male	Unknown	Short	111 ± 8		
Male	Unknown	Medium	125 ± 11	2449	Crew and Koller (1932)
Male	Unknown	Long	140 ± 6		
Male	CBA	Short	87 ± 4		
Male	CBA	Medium	113 ± 6	2119	Slizynski (1955a)
Male	CBA	Long	132 ± 6		
Male	Cross-bred	Short	73 ± 2		
Male	Cross-bred	Medium	105 ± 2	1922	Slizynski (1955a)
Male	Cross-bred	Long	128 ± 3		

females as for males. This gives a mean chiasma frequency of
2.37 ± 0.03 and a mean map of 119 ± 2 (total map length of
2370 ± 28). These figures differ slightly from those of Slizynski
because different data were used. The higher frequency of
chiasmata for the female must give longer map lengths. The
ratio of mean length for Crew and Koller's male and female data
is 1 : 1.17 and for Slizynski's adjusted data is 1 : 1.24. Little
can be made of this comparison because of the uncertainties
involved but the difference is not large.

Huskins and Hearne (1936) have determined the total
chiasma frequency at diakinesis per cell for some ten inbred
strains. One interesting result is that highly significant mean
differences were apparent between the strains. The authors
relate the differences in frequency to degree of susceptibility to
spontaneous tumours. In this respect, the correlation is
imperfect and whether or not tumour susceptibility is a relevant
factor, it may be argued that the interstrain factor certainly is.
The inference is that the rate of crossingover could vary
between strains, perhaps not for all intercepts and not
significantly unless appreciable numbers are analysed.

Huskins and Hearne's results are arranged in Table 11.15,
after conversion to mean chiasmata per bivalent. The mean
genetic map is also tabulated, based on diakinetic chiasma
frequencies. Slizynski's comparative data on diakinetic and
diplotene frequencies could be used to convert the map lengths
of the table to diplotene map lengths. The conversion factor
may be taken as 1.27. The range of map lengths is then from
114 to 69, with a mean of 95 (total map length of 1900). This
is a little below the overall mean of 99 found earlier for the
male. These derivations are probably consistent because Huskins
and Hearne's diakinetic counts are similar to Slizynski's
(although it need not follow that the rate of change from
diakinesis to diplotene is similar) and because the overall mean
of 99 takes account of the relatively high chiasma frequencies
found by Crew and Koller.

It is of interest that a very different approach by Carter
(1955) has yielded comparable estimates of the total genetic
map length. This method makes use of the chances of locating
linkage with new genes to a set of markers distributed among
the chromosomes. Two estimates of 1407 ± 45 and 1688 ± 48

were obtained, using different data. The combined estimate is 1620 ± 35. These estimates are lower than those obtained from chiasma counts but not significantly so. Carter was at pains to guard against bias, since the type of bias which could have produced an underestimation of the map length was the preferential reporting of linked versus independent segregation. Unfortunately, this sort of reportage seems almost indigenous to linkage investigation; only the early literature may be expected to be devoid of it and then only in a relative sense.

TABLE 11.15. Mean frequencies at diakinesis of inbred and cross-bred stocks in the male mouse.

Strain	Mean Chiasmata	Mean genetic map length	Reference
N (abds)	1.80 ± 0.02	90 ± 1	Huskins and Hearn (1936)
I (abdps)	1.70 ± 0.02	85·± 1	Huskins and Hearn (1936)
CBA	1.61 ± 0.02	81 ± 1	Huskins and Hearn (1936)
CBA	1.55 ± 0.04	76 ± 2	Slizynski (1955a)
C57B1 (a)	1.50 ± 0.01	75 ± 0.5	Huskins and Hearn (1936)
Y (A^y)	1.48 ± 0.01	74 ± 0.5	Huskins and Hearn (1936)
C3H	1.46 ± 0.02	88 ± 1	Huskins and Hearn (1936)
A (c)	1.44 ± 0.02	72 ± 1	Huskins and Hearn (1936)
C57Br (ab)	1.39 ± 0.02	70 ± 1	Huskins and Hearn (1936)
D (abd)	1.37 ± 0.02	69 ± 1	Huskins and Hearn (1936)
M (abln)	1.36 ± 0.02	68 ± 1	Huskins and Hearn (1936)
C57B1 (a) (Carrel's)	1.34 ± 0.02	54 ± 1	Huskins and Hearn (1936)
Cross-bred	1.53 ± 0.02	77 ± 1	Slizynski (1955a)
Grand mean	1.494 ± 0.002	75 ± 0.1	

TRANSLOCATIONS

Among the major alterations to the chromosomes, translocations probably have the greatest interest for geneticists. Not the least important reason is the relative ease by which translocations can be recovered. It seems highly probable that work on translocations will be extended as more are produced involving genes in known linkage groups. It has been realized that useful information can be obtained complementary to that forthcoming from linkage studies with intact chromosomes. Most translocations appear to be simple segmental interchanges, without appreciable gain or loss of material. These are probably the viable forms since the total genome is effectively preserved. In most cases, the translocation is terminal, involving only one break on each chromosome.

The manifestation of partial sterility is one of the classic means of detection of translocation, especially when accompanied by meiotic chromosome configuration characteristic of translocated elements. Most of the early work proceeded along these lines (Snell, 1933, 1935; Snell, Bodemann, and Hollander, 1934; Snell and Ames, 1939; Hertwig, 1940; Koller and Auerbach, 1941; Koller, 1944). The intricate karyological configurations are being analysed in detail as more material becomes available (Slizynski, 1952, 1957a, b, 1967); Ohno and Cattanach, 1962; Ford and Evans, 1964; Ohno and Lyon 1965). The more common or 'orthodox' configurations are discussed by Slyzinski (1957b) and Russell (1962). This theme is continued by Griffen (1967) but with emphasis on the theoretical separations of the chromosomes which can result in the rare aneuploid and other complex zygotes.

The detection of partial sterility is a useful screen for translocations and has been the method commonly employed, even if this means that those translocations most disruptive to normal bivalent assortment will be preferentially selected. Genetically, the discovery of linkage between genes hitherto known to assort independently is equally good evidence. In fact, it is decisive. At the same time, the discovery can establish that the linkage groups to which these genes belong do actually reside in different chromosomes and not merely in different regions of a common chromosome. In this respect, Snell (1946) showed that linkage groups V and VIII are in different chromosomes, while Carter et al. (1955, 1956), in conjunction with Slizynski (1957a), showed that groups I, II, III, V, VIII, IX, XI, and XIII are in different chromosomes.

A considerable literature exists on the induction of translocations but not all of this is relevant for present purposes. Table 11.16 presents a summary of those reports which are of unusual interest or have dealt primarily with the genetic aspects. The pioneer work is that of Snell (1946) who, in two studies, has laid much of the foundation for later

TABLE 11.16. Summary of features of the more genetically interesting translocations.

Designation	Loci involved In crosses	Break point located within 5% of loci	Translocation homozygote	References
T (1 ; ?) 12 Sn	c, p, sh-1;	c; ?	Viable	Snell (1946)
T (2 ; ?) 58 Sn	d;	?	?	Snell (1946)
T (5 ; 8) 146Sn	a, pa; b, m	a; ?	Viable	Snell (1946)
T (5 ; 8) 1 Ca	a, fi, Ra, Sd; b	?; b	?	Carter et al. (1955, 1956)
T (5 ; ?) 2 Ca	a, pa, Ra, we;	a, we; ?	Viable	Carter et al. (1955, 1956)
T (5 ; 13) 5 Ca	a; fz, ln,	a; ?	Lethal	Carter et al. (1955, 1956)
T (3 ; ?) 6 Ca	s;	s; ?	Viable	Carter et al. (1955, 1956)
T (5 ; 11) 7 Ca	a, pa, Ra; Lc, Mi,	pa; Lc, Mi	Viable	Carter et al. (1955, 1956)
T (1 ; 7) 8 Ca	c; Re, Sh-2, wa-2	c; ?	?	Carter et al. (1955, 1956)
T (5 ; 13) 83 Ca	a; fz, ln	a; ?	Viable	Carter et al. (1955, 1956)
T (2 ; 9) 138 Ca	d, sc, tk; T	?	Viable	Carter et al. (1955, 1956), Green and Lane (1967)
T (9 ; 13) 190 Ca	T; fz. ln	T; ?	?	Carter et al. (1955, 1956)
T (14 ; 17) 264 Ca	W; f	W; ?	Viable	Carter et al. (1955, 1956), R. J. S. Phillips (1966) MNL, 24, 34
T (11 ; ?) 281 Ca	Mi;	Mi; ?	Viable	Carter et al. (1955, 1956)
T (? ; Y) 8 Fa	—	?	—	Falconer et al. (1952)
T (8 ; X) 1 Rl	b, m, wi; Ta	?	?	Russell and Bangham (1961)
T (7 ; X) 1 ?	Re, wa-2; Mo	?	?	Schaible and Gowen (1961)
T (1 ; X) Ct	Hbb, pu, qv, ru-2, Sh-1; Bn, jp, Mo, spf, Ta	Hbb, Sh-1; jp, Mo, Ta	?	Cattanach (1961b), Wolfe (1967), Eicher (1967, 1970)

	Bn, Mo, Ta;	Bn, Mo, Ta; ?	?	
T (?; X) 16 H	cw, d, se, tk;	cw ; ?	?	Lyon et al. (1964), Ohno and Lyon (1965)
T (2 ;?) 163 H		?	Viable	Evans et al. (1967), Lyon et al. (1968)
T (11 ;?) 1 Ald	—	?	Viable	Leonard and Deknudt (1967a, b, 1969b)
T (? ;?); 1 Wh	—	?	?	White and Tjio (1967)
T (?; Y) 37 Ald	—		—	Leonard and Deknudt (1969a)
T (1 ; X) 2 Rl	c, p, pu, tp; Ta	c ; ?	?	Russell and Montgomery (1969, 1970)
T (1 ; X) 3 Rl	c, p, pu, tp; Ta	c ; ?	?	Russell and Montgomery (1969, 1970)
T (1 ; X) 4 Rl	c, p, pu, tp; Ta	c ; ?	?	Russell and Montgomery (1969, 1970)
T (1 ; X) 5 Rl	c, p, pu, tp; Ta	p ; ?	?	Russell and Montgomery (1969, 1970)
T (1 ; X) 6 Rl	c, p, pu, tp; Ta	p ; Ta	?	Russell and Montgomery (1969, 1970)
T (10 ;?) 18 H	v, gr;	v ; ?	?	A. G. Searle (1970), MNL, 42, 27

research. In general, translocations are analysed by determination of the linkage groups involved and by location of the position of the breakage point. This usually delimits the extent of the translocation. Just how precisely this can be done is clearly a function of the loci content of the chromosome region. That is, upon the known genetic map of the region. In some cases, a break has been located close to the loci of known mutants. Such loci serve as a convenient tag for the translocation in crosses. With this aspect in mind, mutant loci within five units of the break have been specially noted.

Table 11.17 attempts to discover the extent to which the presence of a translocation could interfere with the frequency of crossingover. In general, the effect is either small or non-existent unless one of the gene loci is quite close to the break point (distance in the table) and the intercept interval is small. Even then, the outcome is not straightforward, indicating, probably, that the nature of the translocation and its position in the chromosome are relevant factors. Out of the six significant cases, five are significant in only one sex. This differential is probably due to sample sizes as much as other factors because the studies were directed towards location of the break points. This meant that the sample sizes are usually inadequate to detect differences of crossover frequency. In four of the significant differences, the frequency is decreased while, in the other two, it is increased. The reality of the increase may be doubtful for fz - ln but it seems more substantiated for a - Ra. It is probable that the presence of a translocation would interfere with the distribution frequency of chiasma; inhibiting the frequency close to the break and perhaps causing a shift in frequency more distantly. In this manner, a region of low chiasma frequency could show an increase, which is reflected in a higher rate of crossingover in the translocated chromosome.

The group of translocations studied by Carter et al. (1955, 1956) appear to be terminals. Significant associations were found with known linkage groups for eleven translocations: in most cases, genes from two groups were involved. For various reasons (mainly that of availability of mutant genes), not all of the translocations could be investigated to the same depth. Even so, the breaks for several translocations were accurately pin-pointed. Briefly, breaks were found close to c in Group I, s

in III, a in V (twice), T in IX, and Mi in XI. For full details, the 1956 paper should be consulted.

All of the translocations notified by Russell and associates (Russell and Bangham, 1961; Russell, Bangham, and Saylors, 1962; Russell 1963; Russell and Montgomery, 1965, 1969, 1970) have involved the X chromosome. The first one involves the autosome carrying linkage group VIII. The break point is approximately 13 units from b and six units from Ta. Neither of these genes are thought to constitute a fully reliable tag. Five other translocations involve linkage group I. In three of these, the break is within seven units of c; two very close, 0 and 0.2 for T(I;X)2R1 and T(I;X)3R1, respectively. All show linkage with Ta, with the exception of T(I;X)4R1, but none closely. The break point of T(I;X)5R1 is the closest: about 6 - 15 units away. T(I;X)2R1 and T(I;X)3R1 are as much as 28 and 31 units away. If Ford and Evan's (1964; see later) deduction is correct that Ta lies near the middle of the X, then some of these translocations have involved substantial segments of the X while others may have involved only small pieces; depending how much of the physical chromosome is represented by 28-31 crossover units. According to Cachiero and Russell (1969; see Eicher 1970b), these translocations indicate that linkage group VIII is borne by one of the two largest chromosomes.

Two of the translocations have resulted in characteristic elements. T(3:?)6Ca was discovered to involve an unequal transference of material between two of the smallest chromosomes. The outcome is the appearance of an element much smaller than any of the normal karyotype (Ford $et\ al.$, 1956). Conversely, the T(9;13)190Ca heterozygote has an element much larger (and one much smaller) than any chromosome of the normal complement (Bennett, 1965). It was not possible to find which of the chromosomes were involved in the production of the abnormally large element. This would imply that these may be two of the numerous medium sized autosomes. Bennett also examined the chromosomes of T(2;9)138Ca but could not observe any unusual·configurations.

The T(X;?)16H translocation has contributed several interesting items of information (Lyon $et\ al.$, 1964). The break in the X is close to Ta (crossover value 0.85 ± 0.49). Crossingover was observed with Bn and Blo of the order of

TABLE 11.17 Comparison of crossover values for normal and translocated chromosome in the mouse.

Group	Translocation	Intercept	Distance	Sex	Crossover value		References
					Normal	Translocated	
I	T(1; ?p) 12Sn	c - p	0	♂	12.5±0.5	16.1±3.4	Snell (1946)
				♀	16.8±0.4	11.8±2.0*	,,
		c - sh-1	0	♂	2.6±0.5	1.1±1.8	,,
				♀	4.1±0.3	1.4±1.1*	,,
		p - sh-1	15.1±0.3	♂	12.6±2.4	17.2±3.9	,,
				♀	18.3±2.3	13.3±1.8	,,
II	T(2; 9) 138Ca	cw - se	1.8±1.2	♂	36.6±3.9	35.5±0.4	Lyon et al. (1968)
				♀	34.1±2.7	38.5±0.5	,, ,, ,,
		cw - tk	1.8±1.2	♂	42.7±0.2	38.7±6.2	,, ,, ,,
				♀	43.2±0.1	50.0±6.9	,, ,, ,,
		d - tk	24.4±2.2	♂	3.3±0.9	3.6±1.3	Green and Lane (1967)
		se - tk	24.4±2.2	♂	6.9±2.1	3.2±2.3	Lyon et al. (1968)
				♀	5.1±1.2	11.5±4.4	,, ,, ,,
V	T(5; 8) 146Sn	a - pa	1.2±5.7	♂	15.7±0.7	15.8±3.3	Snell (1946)
				♀	17.0±0.7	19.8±3.7	,,
	T(5; ?) 2Ca	a - pa	4±?	♂	15.7±0.7	3.4±1.5*	Carter et al. (1956)
				♀	17.0±0.7	13.1±7.0	,, ,, ,,
	T(5; 11) 7Ca	a - pa	5±?	♂	15.7±0.7	16.2±1.5	,, ,, ,,
				♀	17.0±0.7	20.0±5.1	,, ,, ,,
	T(5; ?) 2Ca	a - Ra	4±?	♂	23.6±0.1	32.2±3.6*	,, ,, ,,
				♀	19.7±1.1	39.1±10.3*	,, ,, ,,
	T(5; 8) 1Ca	fi - Sd	6±?	♂	19.4±1.7	23.3±6.4	,, ,,

			Distance	Sex			Reference
	T(5; ?) 2Ca	pa - Ra	5 ± ?	♀	20.5±1.7	13.8±4.5	"
				♂	45.2±7.6	30.2±3.8	"
		a - we	2 ± ?	♀	39.4±5.8	26.1±9.2	"
				♂	9.9±0.5	6.3±2.5	"
				♀	13.8±0.5	5.2±2.3*	"
VII	T(1; 7) 8Ca	sh-2 - wa-2	10.2±2.3	♀	23.7±1.0	23.3±13.8	"
VIII	T(5; 8) 146Sn	b - m	20.4±3.0	♂	6.6±1.5	6.7±1.9	Snell (1946)
				♀	7.7±0.6	8.0±6.3	"
XIII	T(5; 13) 65Ca	fz - ln	23.3±1.8	♂	35.3±0.8	38.5±4.0	Carter et al. (1956)
				♀	43.2±0.8	44.6±3.4	"
	T(5; 13) 83Ca	fz - ln	9.2±1.4	♂	35.3±0.8	46.0±3.8*	"
				♀	43.2±0.8	41.5±3.8	"
	T(9; 13) 190Ca	fz - ln	27.3±1.8	♂	35.3±0.8	37.9±3.3	"
				♀	43.2±0.8	39.5±3.5	"

Distance means the interval in crossover units between the break point and the nearest gene locus delineating the intercept

2.7 ± 1.9 and 2.5 ± 1.4 respectively. The order is assumed to be *Bn* - 2.7 - *break* - 0.85 - *Ta* - 1.65 - *Blo*. Some suppressing of crossingover is indicated since the usual map distance between *Bn* and *Ta* is about 15 units. Attempts to discover the autosome involved were unsuccessful, in spite of tests with genes from linkage groups I to V, VIII, X, XI, XIII, and XVIII. Ford and Evans' (1964) and Ohno and Lyon's (1965) karyological examinations suggest that the translocation is the result of a reciprocal exchange between the *X* and one of the smaller autosomes; the break points being approximately medial in both chromosomes. If the latter interpretation is correct, the *Ta* gene is located near the middle of the *X* chromosome.

    The translocation T(1;X)Ct first described by Cattanach (1961b) has proved to be the most complex yet reported. The translocation was first represented as an ordinary terminal exchange between an autosome and the *X*, although whether it was reciprocal or non-reciprocal was doubtful from the beginning. However, Ohno and Cattanach (1962) concluded that the situation was more complex. The quadrivalent configurations could best be explained by the insertion of a substantial segment of a medium sized autosome into the long arm of the *X*. The inserted segment was estimated to be as great as one third of the autosome. The length of the reconstituted *X* was such to imply that it had not lost any material.

    A more detailed examination by Slizynski (1967) confirmed that the transference was insertional. Further, peculiar large knob and loop formations of the complicated quadrivalent suggested (a) that the inserted piece of autosome had become reversed (relative to its centromere) and (b) a small piece of the *X* had been inserted into the autosome. Thus: there had been a two-break insertional translocation, accompanied by inversion and the production of potential duplication/deficiency heterozygotes.

    The translocation has initiated considerable interest. Part of the activity has been directed to delimiting the extent of the insertions. The autosome involved carries linkage group I, a conveniently well mapped group. Cattanach (1961b) showed that the *c* and *p* loci were involved. Later, Eicher (1967) showed that the *sh-1* locus was included while Wolfe (1967) produced evidence that the adjacent locus *Hbb* (about two units away) was not involved. This establishes one of the break points with fair accuracy. Eicher (1967) showed that the second break was beyond *sh-1* and in (1970a) that it was beyond *ru-2* but not as far as *qv*. Genes *ru-2* and *qv* are about nine units apart. A rough estimate of 14.4 crossover units is obtained between the second break and the locus *pu*, which is sited some 10 units beyond *qv*. This implies that the second break is nearer to *qv* than to *ru*, provided the crossover value of 14.4 is reasonable accurate and the break does not suppress crossingover. In terms of map units, the segment inserted in the *X* is about 26 units or about 46 per cent of the known linkage group.

The position of the inserted piece of autosome in the *X* has been determined by Cattanach (1961b, 1966). The data are extensive and indicate that the relative position is *Gy* - 20 - insertion - 2.2 - *Mo* - 4 - *Ta*. The *jp* locus is located between *Gy* and *Mo* and, therefore, must be close to the insertion. This is so, the crossover value is 0.65, but whether *jp* lies on the *Gy* or *Mo* side of the insertion is unknown. The very low value for the insertion - *jp* intercept strongly suggests that the insertion is suppressing crossingover in the region. Indeed, the above values for *Gy* - insertion and insertion - *Mo* are probably low relative to their spatial distance for the same reason. Should the *jp* locus lie on the *Gy* side of the insertion, its position will have been pin-pointed with fair precision. If Fqrd and Evan's (1964) deduction is correct that *Ta* is located medially in the *X*, then the closeness of the insertion to *Ta* would indicate that the insertion may also be medial. More research is required on these questions.

All of the mouse chromosomes are effectively telocentric, being either devoid of a second arm or having a secondary arm which is barely detectable. This means that single break translocations would not alter the appearance of the karyotype unless the segment exchange was grossly unequal. The majority of exchanges are between the long arms but, given the existence of very small second arms, an exchange between this and the long would produce a submetacentric or metacentric translocation. Fusion of centromeres could also produce a metacentric, accompanied by a step-down in the number of elements from 40 to 39 and 38 for heterozygotes and homozygotes, respectively. The exchange between unequal arms could also result in a reduction of the number of elements if the small complementary translocation could be lost without appreciable impairment of either viability or fertility. Several instances of metacentric bodies have been reported. Most have only been described karyologically but two have been subjected to both karyological and genetical examination.

Both Leonard and Deknudt (1967a, b, 1969b) and White and Tjio (1967) have described and illustrated metacentric translocations. Leonard and Deknudt observed chains and rings of four chromosomes in homozygotes. The homozygote was viable but small and of poor fertility. All animals were from an inbred strain and it is uncertain if the inviability aspects were a strain characteristic or due to the translocations. The metacentric was considered to be formed from the 4th and 13th autosomes in order of decreasing size. The translocation had been designated T(11;?)1Ald and the discovery that one of the chromosomes involved carries the XI linkage group is due to M. F. Lyon and S. Hawkes (1969; MNL, 41, 28). White and Tjio reported ring quadrivalents in their homozygotes but with no obvious impairment of fertility. In heterozygotes, trivalents of various configurations were noted (chains and 'frying-pan'). Non-disjunction occurred at a low frequency and there was a reduction in fertility. The submetacentric seemed to be formed from fusion of one of the largest autosomes with one of the smallest.

The T(2;?)138H submetacentric of Evans, Lyon, and Daglish (1967) is depicted as being formed from fusion of the 10th and 19th autosomes (medium and smallest). The females are of normal fertility but the males show some impairment. Ring bivalents were evident for the homozygote and trivalents for the heterozygotes. The translocation was found to share a common chromosome with T(2;9)138Ca with which it forms a large multivalent. The common chromosome was that bearing linkage group II. Lyon, Butler, and Kemp (1968) have shown that cw is closely linked (1.8 ± 1.2) to the break or fusion (hence to the centromere region) while se and tk are serially more distant (39 ± 5 and 46 ± 5, respectively). Crossingover for the cw - se and se - tk intercepts are not significantly different from that observed in the normal chromosome (37 ± 5 versus 35 ± 2 and 7 ± 3 versus 5 ± 1, respectively). It may be that centromeric fusions do not have the same disruptive effect in this respect as do joins eleswhere. The results indicate that cw could be sited close to the centromere in the normal chromosome unless an inversion has occurred in the production of T(2;?)138H. The authors discuss this possibility but conclude that the evidence is against it.

The exchange of material between chromosomes presents a direct method of relating linkage groups to individual chromosomes. If reliance is to be placed on this procedure, it is essential that each of the chromosomes can be individually defined — not just generally as members of a group but specifically and unambiguously. A useful start has been made in this direction by pachytene mapping (as noted earlier), culminating in several attempts to relate the linkage groups to individual chromosomes (Slizynski 1952, 1957a, 1967, Griffen 1960). The main conclusions are shown by Table 11.18. Unfortunately, it seems that these relationships require independent confirmation since Slizynski (1967) has commented that an autosome has been erroneously designated as the X in the maps of Slizynski (1949) and of Griffen (1955). However, it is to be hoped that these investigations will continue and particularly that greater refinement of pachytene mapping will be possible.

Translocations may presumably occur from any combination of chromosomes. Most of the translocations have embraced the autosomes since these are in the majority. Several have involved the X chromosome and even the small Y has not escaped. Two cases involving the Y have been described (Falconer et al., 1952; Leonard and Deknudt, 1969a). Complex translocations, formed from existing translocations, have not yet been reported.

Cattanach (1965) has brought forward evidence that the peculiar dominant abnormality 'snaker' could be associated with a translocation. In spite of the suggestiveness of the breeding results, direct karyological confirmation was not wholly convincing. Those quadrivalents which were observed were occurring at a low frequency, scarecely greater than the spontaneous rate. If a translocation is involved, it seems (a) that only small pieces of chromosome have been exchanged and (b) the exchange is insertional. Tests for association with groups III, VII, XI, and XVIII have been negative.

TABLE 11.18. Provisional correspondences of Slizynski's and Griffen's pachytene chromosomes and linkage groups.

Linkage group	Slizynski's Classification	Griffen's Classification
I	11	F
II	9	C
III	14	D
V	17	E
VII	13	A
VIII	18	H
IX	16	I
XI	15	B
XIII	19	K
XX	20	XY

CENTROMERE POSITION

In principle, there are several methods by which the position of the centromere may be hypothesized. Of these, the most reliable is probably by examination of the behaviour of linked groups of genes in translocation heterozygotes. The fact that the chromosomes are telocentric, or effectively so, simplifies the anlysis for the mouse in that the centromere will be located at the end or near the end of the linkage group.

Ford, Carter, and Hamerton (1956) observed that the T(5;8)146Sn translocation heterozygote tends to form quadrivalent configurations in about eight to 30 per cent of cells. These quadrivalents are invariably rings with four (rarely five) chiasmata. Failure to form quadrivalents seems to be due to lack of chiasma in the non-centromeric arms. Hence, the maximum amount of recombination in these arms would be expected to be in the order of 15 per cent. However, Snell (1946) has reported recombination of 27 ± 2.8 per cent between b and the break, notably in excess of the above. This implies that b is on the centromeric side of the break. Snell also showed that b and pa must lie in opposite centromeric arms. Thus, there is reason to think that the centromeres of the chromosomes bearing linkage groups VIII and V will be at the *asp* and *Sd* ends of the respective linkage groups. These locations have since been independently confirmed.

Where two telocentric elements have fused in the centromere region to form a novel metacentric, location of the centromere is straightforward. Those genes of the linkage groups involved which display the closest linkage to the break point will indicate the position of the centromere. In this manner, Lyon *et al.* (1968), by the aid of the metacentric T(2;?)163H, have shown that the centromere is located at the *cw* end of linkage group II. In fact, unless there is suppression of crossingover, the centromere is merely about two crossover units from *cw*. The only proviso which must be made is that the fusion has not also involved in a structural change (e.g., an inversion), for the chromosomes. This is briefly considered by the authors but, on balance, the evidence favours simple fusion.

The discovery that one of the chromosomes involved in the metacentric T(11;?)1Ald carries linkage group XI has led to studies with loci of the group. The following crossover values were found between the break point and the genes *Lc* and *Mi*: for *Lc*, 24 and 32 per cent for male and female heterozygotes, respectively, and for *Mi*, 31 and 47 per cent for male and female heterozygotes, respectively. This indicates the centromere is at the *Sig* end of linkage group (M. F. Lyon and S. Hawkes, 1970; *MNL*, 42, 27).

A more general method involves examination of the relative frequencies of assorting genes in translocation heterozygotes. The gametic output of these heterozygotes could be large and diverse but, for most practical purposes, is probably limited to a few main types. The majority of zygotes would probably die because of unbalance chromosome constitutions. The survivors represent the main gametic types and would be expected to occur in different proportions. These would doubtless reflect the relative frequency of gametes, themselves being dependent upon the viability of the gametes and their production. Fig. 11.18 is illustrative of the argument. Two imaginary T(5;8) heterozygotes are paired with the genotypes as shown. The six main types of gametes could determine 36 zygotes but only those with complete genomes are assumed to survive.

The four zygotes in the top left corner of the chequerboard would be expected to occur in the greatest numbers but not usually to the exclusion of others. To appreciate this, it is necessary to consider the formation of the gametes. Gametes 1 and 2 are fully balanced in that the total genome is intact. The zygotes resulting from union of these gametes will also be balanced. Gametes 3 and 4 may be expected to be produced at a similar frequency to the former but their constitutions are unbalanced and this could impair their viability. Two unions of these gametes, however, would restore the genome and these are usually viable. Gametes 5 and 6 are produced at a low frequency. Whereas, the first four gametes are formed from normal disjunction, these last two are formed from non-disjunction. Again, these are unbalanced as gametes but can produce balanced and usually viable zygotes. Due to the differential viabilities and modes of formation, the three gametic types produce zygotes in progressively lower frequencies. The last two often appreciably lower.

It is important to note that ordinarily the gametic types 3 and 4 and 5 and 6 form viable zygotes only by unions with their own kind. It is this aspect which can be utilized for the mapping of centromeres. If the translocation elements are suitable tagged with genes, these zygotes can be identified. Fig. 11.18 is illustrative of such a scheme, making use of a T(5;8) translocation based on the work of Snell and Searle. The *a* gene is included just to demonstrate the 3 : 1 ratio of the four zygotes in the top left corner. The other mutant phenotypes are uniquely determined by the gametic types 3 and 4 or 5 and 6. By the composition of the cross, the relative frequencies of the zygotes should indicate the position of the genes relative to the break point. In particular, since *pa* and *b* are known to be in different linkage groups, the occurrence of the *pab* combination means that the two genes lie on the centromere side of the break in their respective chromosomes.

Snell (1946) carried out a cross with T(5;8)146Sn as shown in the figure but without the *bp* gene. The progeny included *papabb* individuals at a frequency of $2/324 = 0.62$ per cent. He concludes that the low frequency implied that these individuals were the product of type 6 gametes. That is, they were produced by non-disjunction of homologous centromeres. In this he was correct, since subsequent work with other translocations has shown that type 3 or 4 gametes are usually

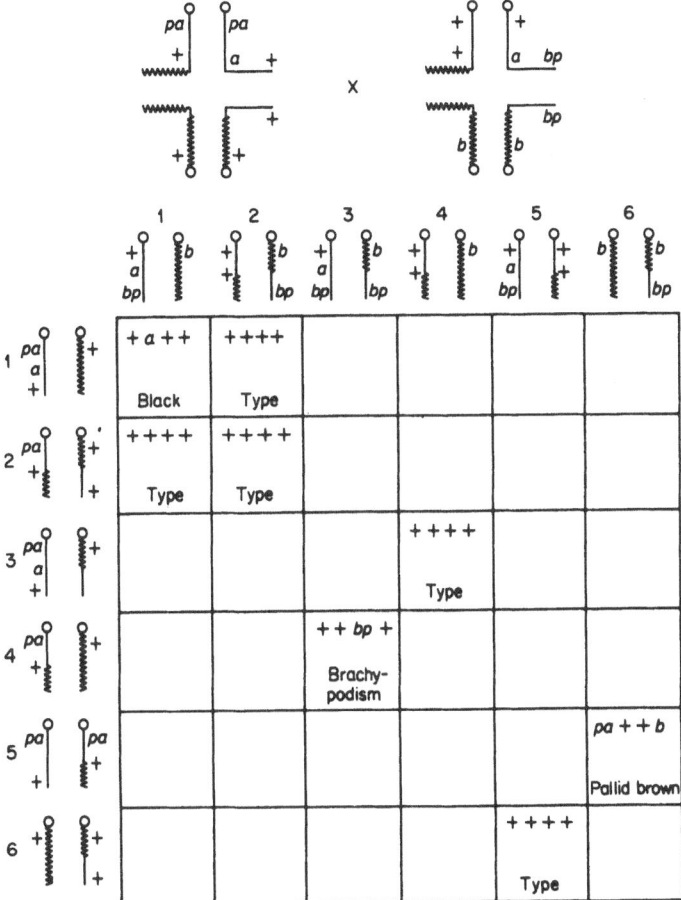

Fig. 11.18. The primary theoretical products of a T(5;8)146Sn type translocation in the house mouse.

produced at a higher frequency. This means that the *pa* and *b* genes lie on the centromere side of the break. Snell found that *a* was closely linked to the break (about one per cent crossingover) while *a* and *pa* showed 18 per cent crossingover. This indicates the order is centromere - *pa* - *a*. It follows that the centromere is located at the *Sd* end of linkage group V. Snell also found about 27 per cent crossingover between *b* and *a* (tagging for the break), 21 per cent between *m* and *a* and seven per cent between *b* and *m*. This indicates the order centromere -

b - m - break. Thus, the centromere is at the *asp* end of linkage group VIII.

To determine the position of the centromere explicitly for a single cross, it is necessary for genes of the linkage group to straddle the break. This is shown by the figure for the group V genes *pa* and *bp* whence the relative frequencies of the two mutant phenotypes should make it apparent which lies on the centromere side of the break. More usually, at least two crosses may be required to locate the centromere, each one being of the pattern Tx *y*/ + *y* x Tx +/++, where Tx is the translocation and *y* is the gene being tested. The only prerequisite is that suitable translocations be on hand. It would be useful to know beforehand where a break point occurs within a linkage group since this would decide the choice of genes for the test. Although Snell performed his cross as long ago as 1946, A. G. Searle (1968; *MNL*, **39**, 25) was probably the first person to exploit the cross as a routine method for mapping of centromeres.

Searle has located the relative position of the centromere for several linkage groups. In two tests with the genes *bp* and *pa* of linkage group V in conjunction with T(5;?)26H, he observed *bpbp* to assort at a frequency of 12 per cent while *papa* occurred at a frequency of 2.5 per cent (1968; *MNL*, **39**, 25). Thus, the rarer *papa* offspring are apparently engendered by non-disjunction and *pa*, therefore, is on the centromere side of the break. Subsequently, using the translocation T(2;5)11H, Searle and C. V. Beechey (1970; *MNL*, **43**, 29) observed *bpbp* young to assort at a frequency of 15.4 per cent. Thus, two different translocations have given essentially consistent results. That is, the centromere is evidently at the *Sd* end of linkage group V. These observations confirm those of Snell (1946); and the deductions of Ford *et al.* (1956).

In T(13;?)170H, one break is said to be between the loci *fz* and *ln* (Searle 1961; *MNL*, **40**, 25). Test matings with these genes gave 29 per cent of *lnln* but only 1.3 per cent of *fzfz*. The centromere, therefore, is indicated to be at the *fz* end of linkage group XIII. The two genes *pe* and *sa* were tested in conjunction with T(14;17)264H (Searle, 1969; *MNL*, **41**, 28). Among the resultant progeny were 12 per cent of *pepe* compared with 1.2 per cent of *bgbg*. The centromere is indicated to lie at the *bg* end of linkage group XIV.

Searle (1970; *MNL*, 42, 27) states that one break point for T(10;?)18H is close to the *v* locus and crosses between T18*gr*/+*gr* x T18+/++ have given 7.4 per cent of *grgr* progeny. This value is significantly higher than the frequency observed for the union of non-disjunctive gametes (1.2 to 2.5 per cent; see above). It is also much less, however, than the 12 to 29 per cent realized for the union of disjunctive unbalanced gametes, but *grgr* mice are known to be only about 50 per cent viable up to the time of classification. Doubling the observed frequency of *grgr* gives 15 per cent and this value is within the expected range. Searle concludes that the centromere for linkage group X is on the side of *v* away from *gr*. The close linkage of *v* to the break would probably exclude the reverse order since the *gr* - *v* intercept is about 13 units (at least) and could not be accommodated unless crossover suppression is present.

Green (1970; *MNL*, 43,32) has made use of T(3;?)6Ca to locate the centromere for linkage group III. Crosses of T6hr/++ x T6+/++ have given 4.9 per cent of *hrhr*. This frequency is higher than that so far reported for non-disjunction zygotes. However, there are two reasons for believing that the appearance of *hrhr* is due to non-disjunction; (a) there is no evidence of marked inviability of *hr* and (b) unpublished observations of C. S. Ford are suggestive that T6 has an unusually high rate of non-disjunctive separation of centromeres. This latter fact would raise the expected recovery of *hrhr*. The centromere, therefore, is seemingly at the *pn* end of linkage group III.

The validity of the two methods would be individually enhanced if each gave the same answer for centromere position in a linkage group. M. F. Lyon and S. Hawkes (1970; *MNL*, 42, 27) were able to examine this aspect for the genes *cw* and *se* of linkage group II. utilizing T(2;9)138Ca. The two tests gave a frequency of 22 per cent of *sese* but a frequency of only 1.3 per cent of *cwcw*. These frequencies indicate that the centromere is sited at the *cw* end of the linkage group in agreement with earlier results based on crossover values with the metacentric T(2;?)163H.

Lyon *et al.* (1968) deduced that the centromere is located at the *T* end of linkage group IX. This has now been confirmed to some extent by the present method, using T(2;9)138Ca (Lyon and Hawkes (*loc. cit.*). The Tt^{23} combination of genes were used for the test and were recovered among 134 offspring at a frequency of 0.75 per cent. A frequency of this magnitude would be more in keeping with that expected for non-disjunctive rather than disjunctive separation of the centromeres. Thus, the deduction is upheld.

The segregational method — as it may be called — of locating the approximate position of the centromere has much to recommend it. It is easy to perform and orientate the whole linkage group with respect to the centromere in one swoop. If the difference in frequency of the two types of gamete formation (3 and 4 versus 5 and 6) is consistently maintained (particularly, if non-overlapping limits exist), it may be practical to rely upon a test with a single gene. Conclusions based on a single test would certainly facilitate centromere location but it is possibly too soon to judge if this can be decisive without supplementary information. Yet, subject to unforeseen difficulties and the availability of suitably composed translocations, it should not be too difficult to complete the task of locating the centromere for all of the known linkage groups.

The segregational method may be instrumental in deciding the question whether or not all of the autosomes have second arms, and if gene loci are to be found thereon. On the other hand, precise positioning of the centromere will have to wait upon the chance discovery of close linkage to a break.

The location of the centromere for groups I and XX has been examined by Eicher (1970b). The evidence is very indirect but is suggestive that the centromere is at the *fr* end for group I and beyond *Gy* for group XX.

DEFICIENCY

The occurrence of deficiencies or deletions involving specific genes or linked groups of genes would be another method of relating genes to chromosomes, provided that the deficiency can be detected. A spontaneous deficiency was found by Gates (1927b) and confirmed karyologically by Painter (1927). This discovery is acknowledged as the first known case in laboratory animals. Painter concluded that the deficiency is due to loss of part of chromosome 17 on his classification. This is one of the smaller chromosomes and could be taken as a provisional (if vague) allocation of linkage group X.

Deficiencies are of common occurrence from X-irradiation and can be efficiently detected with the appropriate genetic techniques (Russell and Russell, 1960; Kidwell et al., 1961; Russell, Steele, and Thompson, 1966; Russell, 1967; Gluecksohn-Waelsch and Cori, 1970). In this manner, deficiencies involving loci in linkage groups I (c), II (d, se), and VI (bt, hl) have been found, and the method could be extended. No reports have been made on karyological aspects of these deficiencies. The genetic procedures will function as a screen for the more exacting karyological scrutiny.

The curious behaviour of the divers t alleles has prompted speculation that some form of chromosomal rearrangement is involved (Gruneberg 1952; Lyon and Phillips 1959; Dunn et al., 1962). In particular, new t alleles seem to arise from existing t alleles by a process of crossingover which could in actuality be further rearrangements. The t alleles are variously associated with lethality, sterility, partial fertility, modified segregation ratios, and suppression of crossingover. Chromosomal inversion, deficiency, and duplication have all been offered as likely explanations.

It was obvious that a detailed analysis of the chromosomes is desirable. Two early investigations were inconclusive (Bryson, 1944; Jaffe, 1952), largely because these were concerned with other matters or because they could not be brought to fruition. Geyer-Duszynska (1964) has since contributed an account which comes out in favour of deficiencies as the major factor involved. The bivalent configurations of instial loops suggest that the various t alleles are associated with deficiencies of various sizes. Further, the t alleles appear to have non-homologous regions which exaggerate the asynapsis. Some of the configurations indicated a failure of pairing beyond the presumptive t region (as if pairing was not resumed) or if the affected region extended over an appreciable area.

No deficiency seemed to be associated with either T or t^{w1} or, if so, was so small to be undetectable. The allele t^o appeared to be associated with a large deficiency than either t^{w6} or t^{w18};

$t^{1\,2}$ is depicted as being associated either with a small interstitial deficiency causing non-pairing of the remainder of the arm or with a large terminal deficiency. It may be noted that some of the loop configurations could be due to duplications. Indeed, it may be difficult to distinguish between the two possibilities. The presence of one could imply the presence of the other. Geyer-Dusynska appears to recognize this probability but she suggests that the assumption of variable deficiencies (leading to non-homologous regions) are sufficient to explain the observations.

The T locus (or t region, not to make too fine a point) belongs to the IX linkage group. The behaviour of the chromosome bearing the t region would, therefore, be a means of relating the IX linkage group to a particular chromosome. However, Geyer-Dusynska was not able to accomplish this. Should the problem be re-investigated, this is an aspect which deserves attention.

TRISOMY AND ANEUPLOIDY

Spontaneous trisomy is probably an exceedingly rare event but more likely as a remote consequence of X-irradiation or chemical disruption of meiosis. All of the trisomics so far observed fall into the latter category.

Eight cases of primary trisomy have been described by Cattanach (1964) and Griffen and Bunker (1964, 1967). Most were sterile and the remainder semi-fertile. Trivalent bodies could be observed, together with chains of three chromosomes. The majority of animals were trisomic for one of the smaller chromosomes. This is evidently not a *sine qua non* for viability since one trisomic had an extra medium sized and another an extra large chromosome (provisionally placed as number four in size). There was no indication that the additional elements for the above cases were other than a single chromosome.

In addition to the above trisomics, many other cases of a more complicated nature have been discovered (Lyon and Meredith, 1966; Griffen, 1967). These are the products of non-disjunction in translocation heterozygotes. The enhanced rate of non-disjunction is seemingly a consequence of the distorted meiosis of the quadrivalent pairing which give rise to a variety of aneuploid individuals. Perhaps only a few are viable but the data are insufficient for this to be assessed. By definition, the extra element in these trisomics is complex, being composed of two or more chromosomes. The above authors have described several cases in detail. Sterility and reduced fertility is common. Lyon and Meredith thought that one of their trisomics was associated with a phenotypic change (exceptionally broad skull between the eyes) but the evidence is ·not incontrovertible. A monosomic individual with 39 elements was observed; presumably the obverse of one of the trisomic forms.

The meiotic pairing of the odd element can have curious and significant features. Griffen (1967) has described the behaviour of one element which paired with different bivalents in different cells and, on one cell at least, paired with the two bivalents simultaneously. This is direct confirmation of the translocated or bichromosomal composition of the extra element. Griffen's paper should be consulted for a proposed terminology to represent the main aneuploid forms.

ASSOCIATIONS OF CHROMOSOMES

The observation of non-random spatial association of chromosomes is of importance because it provides a probable physical basis to the affinity phenomenon. Only a few reports are available at present but these have contributed several intriguing items. At this stage, the detection of associations have been mainly on a 'greater than chance' basis but if the associations are taken seriously it may be possible to tackle the problem more systematically.

The observations of Griffen (1960) are relevant in this context. He noted that several chromocentres may be present in cells at spermatogonial pachytene. These consisted of fused centromeres of up to seven chromosomes: the more common numbers being three, four, or five. There was no indication that particular chromosomes tended to be associated in the centres. Rather, he is of the opinion that the presence of any chromosome in a centre seemed to be a random process. Similar fusion of centromeres of different chromosomes during early first meiotic prophase has been commented upon by Ohno, Christian, and Stenius (1963) for oocyte material. The association consisted of from two to six chromosomes and few of the chromosomes escaped being involved. These authors consider the association to be a prelude to dictyotene, mainly because the associations occur more prominently in oocytes than in spermatocytes.

Bennett (1966) has contributed illustrations of typical non-random associations of mitotic chromosomes. These consisted of peculiar formations of chromosomes in a ring-like pattern, with their centromeric ends orientated towards a common centre. Upwards of six chromosomes may participate in these patterns. In two cells, it seemed as if two associations could be discerned − a ring of four and a ring of six. The larger the number of chromosomes, the more obvious the non-random arrangement. However, once the eye had been alerted, pairs of just two chromosomes '. . . showed an end-to-end relationship which seemed similar in general orientation to that of the larger associations'.

An attempt was made to ascertain if certain chromosomes were behaving in this manner, rather than random grouping. Three pairs of satellited chromosomes could be seen in cells, conveniently classified as short, medium, and long. Tabulation of the observed associations of these chromosomes in comparison with the depicted frequency for random grouping, led to the conclusion that these chromosomes featured in the ring-like patterns at a greater than chance frequency. This did not mean that the associations consisted mainly of satellited chromosomes or that at least one satellited chromosome must be present, since about half of the observed associations were made up of non-satellited elements. It was briefly speculated that some of the associations could be due to chromosomes which had been involved in the organization of a nucleolus during the previous interphase. The likelihood that certain chromosomes preferentially associate, whether due to centromere orientation or some other point of attraction, would constitute important evidence for affinity.

Independently, Douglas (1966) has given details of non-random associations for chromosomes in spermatogenetic material. In this material, non-homologous chromosomes were associated in groups of from two to five bivalents. These appear to be about six to eight associations per cell. The associations could be seen for most of the various cell stages. So far, it has not been possible to follow through these observations in order to ascertain if the number of associations or number of chromosomes per association remain constant. The associations are often joined together at pachytene and diplotene by a chromocentrum and certain pre-treatments are helpful in clarifying the various configurations. Douglas makes the point that some of the associations of chromosomes described by Koller (1944) were not perfectly correlated with infertility, as might be expected if all the associations were due to translocation quadrivalents. Some of the configurations may have been associations of the type discussed here. Douglas and Geerts (1966) have discussed the theoretical structure of some of the more simple configurations of three chromosomes, with special reference to their significance for affinity.

Associations of chromosomes have been observed by Omura (1968) in bone marrow cells. The association is termed 'temporary centric connection' and involved two chromosomes. The number of associations varied from none to several per cell. On occasion, the two chromosomes appeared to be connected by a short slender thread.

Bibliography

ABEELEN, J. H. F., Van and KROON, P. H. W., Van der (1967). Nijmegen waltzer, a new neurological mutant in the mouse. *Genet. Res.,* **10**, 117-118.

ALLEN, S. L. (1955a). *H-2ʲ*, a tenth allele in the histocompatibility locus in the mouse as determined by tumour transplantation. *Cancer Res.,* **15**, 315-319.

ALLEN, S. L. (1955b). Linkage relations of the genes histocompatibility-2 and fused tail, brachyury and kinky tail in the mouse, as determined by transplantation. *Genetics,* **40**, 627-650.

AMAND, W. St. (1970). Recombination frequency estimates for the satin-beige interval in the mouse. *J. Miss. Acad. Sci.,* **15**, 60.

AMOS, A. B., GOVER, P. A. and MIKULSKA, Z. B. (1955). An analysis of an antigenic system in the mouse (the *H-2* system). *Proc. Roy. Soc. B.,* **144**, 369-380.

AMOS, D. B., ZUMPFT, M. and ARMSTRONG, P. (1963). *H-5* and *H-6,* two mouse isoantigens on red cells and tissues detected serologically. *Transplantation,* **1**, 270.

AUERBACH, C., FALCONER, D. S., and ISAACSON, J. H. (1962). Test for sex-linked lethals in irradiated mice. *Genet. Res.,* **3**, 444-447.

BAILEY, D. W. (1962). Histocompatibility associated with the *X* chromosome in mice. *Genetics,* **47**, 939.

BAILEY, D. W. (1963a). Histocompatibility associated with the *X* chromosome in mice. *Transplantation,* **1**, 70-74.

BAILEY, D. W. (1963b). Mosaic histocompatibility of skin grafts from female mice. *Science,* **141**, 631-633.

BAILEY, D. W. (1964). Genetically modified survival time of grafts from mice bearing *X* linked histocompatibility. *Transplantation,* **2**, 203-206.

BAILEY, D. W. (1965). A search for genetic background influences on survival time of skin grafts from mice bearing *Y* linked histocompatibility. *Transplantation,* **3**, 531-534.

BATEMAN, N. (1961). Sombre, a viable dominant mutant in the house mouse. *J. Hered.,* **52**, 186-189.

BEATTY, R. A. (1957). Chromosome constancy in the corneal epithelium of the mouse. *Chromosoma,* **8**, 585-596.

BECK, S. L. (1963). The anophthalmic mutant of the mouse, II. An association of anophthalmia and polydactyly. *J. Hered.,* **54**, 79-83.

BENNETT, D. (1959). Brain hernia, a near recessive mutation in the house mouse. *J. Hered.,* **50**, 265-268.

BENNETT, D. (1965). The karyotype of the mouse, with identification of a translocation. *Proc. Nat. Acad. Sci. U.S.A.,* **53**, 730-737.

BENNETT, D. (1966). Non-random association of chromosomes during mitotic metaphase in tissue cells of the mouse. *Cytologia,* **31**, 411-415.

BENNETT, J. H. and GRESHAM, G. A. (1956). A gene for eyelids open at birth in the house mouse. *Nature,* **178**, 272-273.

BERNSTEIN, S. E., SILVERS, A. A. and SILVERS, W. K. (1958). An attempt to demonstrate a Y linked histocompatibility gene in the house mouse. *J. Nat. Cancer Inst.*, 20, 577.

BIDDLE, F. G. and PETRAS, M. L. (1967). The inheritance of a non-hemoglobin protein in *Mus musculus. Genetics*, 57, 943-949.

BILLINGHAM, R. E. and SILVERS, W. K. (1960). Studies on tolerance of the Y chromosome antigen in mice. *J. Immunol.*, 85, 14-26.

BILLINGHAM, R. E. and SILVERS, W. K. (1963). Sensitivity to homografts of normal tissues and cells. *Ann. Rev. Microbiol.*, 17, 531-564.

BITTNER, J. J. (1934). Linkage in transplantable tumours. *J. Genet.*, 29, 17-27.

BLOOM, J. L. (1961). Pulmonary tumours and group VII in the mouse. *Nature*, 191, 1124-1125.

BLOOM, J. L. and FALCONER, D. S. (1966). Grizzled, a mutant in linkage group X of the mouse. *Genet. Res.*, 7, 159-167.

BODNER, W. F. (1961a). Viability effects and recombination differences in a linkage test with pallid and fidget in the house mouse. *Heredity*, 16, 485-495.

BODNER, W. F. (1961b). Effects of maternal age on the incidence of congenital abnormalities in mouse and man. *Nature*, 190, 1134-1135.

BORGER, R. (1950), Order of genes on the fifth linkage group of the house mouse. *Nature*, 166, 697.

BOYSE, E. A., MIYAZAWA, M., AOKI, T. and OLD, L. J. (1968). Ly-A and Ly-B: two systems of lymphocyte isoantigens in the mouse. *Proc. Roy. Soc. Lond.*, B, 170, 175-193.

BOYSE, E. A., OLD, L. J. and LUELL, S. (1964). Genetic determination of the *TL* (thymus-leukemia) antigen in the mouse. *Nature*, 201, 779.

BRYSON, V. (1944). Spermatogenesis and fertility in *Mus musculus* as affected by factors at the *T* locus. *J, Morph.*, 74, 131-179.

BUNKER, H. and SNELL, C. D. (1948). Linkage of white and waved-1. *J. Hered.*, 39, 28.

BUNKER, L. E. (1959). Hepatic fusion, a new gene in linkage group I of the mouse. *J. Hered.*, 50, 40-44.

BUNKER, M. C. (1961). A technique for straining chromosomes of the mouse with Sudan Black B. *Canad. J. Genet. Cytol.*, 3, 355-360.

BUNKER, M. C. (1965). Chromosome preparations from solid tumours of the mouse: a direct method. *Canad. J. Genet. Cytol.*, 7, 78-83.

BURHOE, S. O. (1936). Linkage studies of waved coat in the house mouse. *J. Hered.*, 27, 119-120.

CACHEIRO, N. L. A. and RUSSELL, L. B. (1969). Cytological studies of *X*-autosomal translocations in the mouse. I. Mitotic chromosomes. *Genetics*, 61, s8.

CARTER, T. C. (1947). A new linkage in the house mouse: undulated and agouti. *Heredity*, 1, 367-372.

CARTER, T. C. (1949). Position of luxate in the third linkage group of the house mouse. *Nature*, 164, 1138.

CARTER, T. C. (1951a). The position of fidget in linkage group V of the house mouse. *J. Genet.*, 50, 264-267.

CARTER, T. C. (1951b). Wavy coated mice: phenotypic interactions and linkage tests between rex and (a) waved-1, (b) waved-2. *J. Genet.*, 50, 277-299.

CARTER, T. C. (1951c). The genetics of luxate mice. II. Linkage and independence. *J. Genet.*, 50, 300-306.

CARTER, T. C. (1954). A search for chromatid interference in the male house mouse. *Z. Ind. Abst. Vererbgslehre*, 86, 210-223.

CARTER, T. C. (1955). The estimation of total genetic map lengths from linkage test data. *J. Genet.*, 53, 21-28.

CARTER, T. C. (1956). Genetics of the Little and Bagg X-rayed mouse stock. *J. Genet.*, 54, 311-326.

CARTER, T. C. (1957). The use of linked marker genes for detecting recessive autosomal lethals in the mouse. *J. Genet.*, 55, 585-597.

CARTER, T. C. and FALCONER, D. S. (1951). Stocks for detecting linkage in the mouse and the theory of their design. *J. Genet.*, 50, 307-323.

CARTER, T. C. and FALCONER, D. S. (1952). A review of independent segregation in the house mouse. *J. Genet.*, 50, 399-413.

CARTER, T. C. and FALCONER, D. S. (1953). Independence of linkage groups I, II, and XI in the house mouse. *J. Genet.*, 51, 373-374.

CARTER, T. C. and GRUNEBERG, H. (1950). Linkage between fidget and agouti in the house mouse. *Heredity*, 4, 373-376.

CARTER, T. C., LYON, M. F. and PHILLIPS, R. J. S. (1954). Partial sex linkage in the mouse. *Nature*, 174, 309-310.

CARTER, T. 'C., LYON, M. F. and PHILLIPS, R. J. S. (1955). Gene tagged chromosome translocations in eleven stocks of mice. *J. Genet.*, 53, 154-166.

CARTER, T. C., LYON, M. F. and PHILLIPS, R. J. S. (1956). Further genetic studies of eleven translocations in the mouse. *J. Genet.*, 54, 462-473.

CARTER, T. C. and PHILLIPS, R. J. S. (1953). The sex distribution of waved-2, shaker-2 and rex in the house mouse. *Z. Ind. Abst. Vererbgslehre*, 85, 564-578.

CARTER, T. C. and PHILLIPS, R. J. S. (1954). Ragged, a semidominant coat texture mutant in the house mouse. *J. Hered.*, 45, 150-154.

CASTLE, W. E. (1919). Studies of heredity in rabbits, rats and mice. *Car. Inst. Wash. Pub.*, 288, 56pp.

CASTLE, W. E. (1925). A sex difference in linkage in rats and mice. *Genetics*, 10, 580-582.

CASTLE, W. E. (1941). Influence of certain colour mutations on body size in mice, rats and rabbits. *Genetics*, 26, 177-191.

CASTLE, W. E., GATES, W. H. and REED, S. C. (1936). Studies of a size cross in mice. *Genetics*, 21, 66-78.

CASTLE, W. E., GATES, W. H., REED, S. C. and LAW, L. W. (1936). Studies of a size cross in mice II. *Genetics*, 21, 310-323.

CASTLE, W. E., GATES, W. H., REED, S. C. and SNELL, G. D. (1936). Identical twins in a mouse cross. *Science*, 84, 581.

CASTLE, W. E. and LITTLE, C. C. (1909). The peculiar inheritance of pink eyes among coloured mice. *Science*, 30, 313-314.

CASTLE, W. E. and WACHTER, W. L. (1924). Variations of linkage in rats and mice. *Genetics*, 9, 1-12.

CATTANACH, B. M. (1961a). *XXY* mice. *Genet. Res.*, 2, 156-158.

CATTANACH, B. M. (1961b). A chemically induced variegated type position effect in the mouse. *Z. Veverbungslehre*, 92, 165-182.

CATTANACH, B. M. (1962). *XO* mice. *Genet. Res.*, 3, 487-490.

CATTANACH, B. M. (1964). Autosomal trisomy in the mouse. *Cytogenetics*, 3, 159-166.

CATTANACH, B. M. (1965). Snaker: a dominant abnormality caused by chromosomal imbalance. *Z. Vererbungslehre*, 96, 275-284.

CATTANACH, B. M. (1966). The location of Cattanach's translocation in the *X* chromosome linkage of the mouse. *Genet. Res.*, 8, 253-256.

CATTANACH, B. M. (1967). A test of distributive pairing between two specific non-homologous chromosomes in the mouse. *Cytogenetics*, 6, 67-77.

CATTANACH, B. M. and ISAACSON, J. H. (1967). Controlling elements in the mouse *X* chromosome. *Genetics*, 57 331.

CATTANACH, B. M., PEREZ, J. N. and POLLARD, C. E. (1970). Controlling elements in the mouse *X* chromosome. II. Location in the linkage map. *Genet. Res.*, 15, 183-195.

CATTANACH, B. M. and POLLARD, C. E. (1969). A *XYY* sex chromosome constitution in the mouse. *Cytogenetics*, 8, 80-86.

CATTANACH, B. M., POLLARD, C. E. and PEREZ, J. N. (1969). Controlling elements in the mouse *X* chromosome. I. Interaction with the *X* linked genes. *Genet. Res.*, 14, 223-235.

CELADA, F. and WELSHONS, W. J. (1963). An immunogenetic analysis of the male antigen in mice utilising animals with an exceptional chromosome constitution. *Genetics*, 48, 139-151.

CENTER, E. M. (1966). Genetical and embryological studies of the *jt* form of syndactylism in the mouse. *Genet. Res.*, 8, 33-40.

CHANG, S. S. and HILDEMANN, W. H. (1964). Inheritance of susceptibility to polyoma virus in mice. *J. Nat. Cancer Inst.*, 33, 303-313.

CHASE, H. B. (1946). A sex-linked recessive in the mouse. *Genetics*, 31, 214.

CHU, E. H. Y. and MONESI, V. (1960). Analysis of X-ray induced chromosome aberrations in mouse somatic cells *in vitro*. *Genetics* 45, 981.

CINADER, B., DUBISKI, S. and WARDLAW, A. C. (1966). Genetics of MuBl and a complement effect in inbred strains of mice. *Genet. Res.*, 7, 32-43.

CLARK, F. H. (1934a). Linkage studies of brachyury in the house mouse. *Proc. Nat. Acad. Sci., U.S.A.*, 20, 276-279.

CLARK, F. H. (1934b). The inheritance and linkage relations of a new recessive spotting in the house mouse. *Genetics*, 19, 365-393.

CLARK, F. H. (1935). Linkage relations of Zavadskaia shaker in the house mouse. *Proc. Nat. Acad. Sci. U.S.A.*, 21, 247-251.

CLARK, F. H. (1936). Linkage relations of hydrocephalus-1 in the house mouse. *Proc. Nat. Acad. Sci. U.S.A.*, 22, 474-478.

CLOUDMAN, A. M. and BUNKER, L. E. (1945). The varitint-waddler mouse. *J. Hered.*, 36, 258-263.

COHEN, B. L. (1960). Genetics of plasma transferrius in the mouse. *Genet. Res.*, 1, 431-438.

COLLINS, R. L. (1970). A new genetic locus papped from behavioural variation in mice: audiogenic siezure prone (*asp*) *Behav. Genet.*, 1, 99-109.

COLLINS, R. L. and FULLER, J. L. (1968). Audiogenic seizure prone: a gene affecting behaviour in linkage group VIII of the mouse. *Science*, 162, 1137-1139.

COLLINS, R. L. and HUTTON, J. J. (1970). The position of autosomal glucose-6-phosphate dehydrogenase on linkage group VIII of the mouse. *J. Hered.*, 61, 53-54.

COOPER, C. B. (1939). Linkage between naked and caracul in the house mouse. *J. Hered.*, 30, 212.

COX, E. K. (1926). The chromosomes of the house mouse. *J. Morph.*, 43, 45-54.

CREW, F. A. E. and AUERBACH, C. (1939). Rex: a dominant autosomal monogenic coat texture character in the mouse. *J. Genet.*, 38, 341-344.

CREW, F. A. E. and AUERBACH, C. (1940). Linkage date on the rex character in the house mouse. *J. Genet.*, 39, 225-227.

CREW, F. A. E. and KOLLER, P. C. (1932). The sex incidence of chiasma frequency and genetic crossingover in the mouse. *J. Genet.*, 26, 359-383.

CRIPPA, M. (1964). The mouse karyotype in somatic cells cultured *in vitro*. *Chromosoma*, 15, 301-311.

CURRY, G. A. (1959). Genetic and developmental studies on droopy eared mice. *J. Embryol. Exp. Morph.*, 7, 39-65.

CUTWRIGHT, P. R. (1932). The spermatogenesis of the mouse. *J. Morph.*, 54, 197-220.

DARBISHIRE, A. D. (1904). On the result of crossing Japanese waltzing with albino mice. *Biometrika*, 3, 1-51.

DAVID, C. S., SHREFFLER, D. C. and STIMPFLING, J. H. (1969) Analysis of four crossovers within the complex *H-2* region of the house mouse. *Genetics*, 61, s12-s13.

DE LORENZO, R. J. and RUDDLE, F. H. (1969). Genetic control of two electrophoretic variants of glucosephosphate isomerase in the mouse. *Biochem. Genet.*, 3, 151-162.

DEOL, M. S. and GREEN, M. C. (1966). Snell's waltzer, a new mutation affecting behaviour and the inner ear in the mouse. *Genet. Res.*, 8, 339-345.

DEOL, M. S. and LANE, P. W. (1966). A new gene affecting the morphogenesis of the vestibular part of the inner ear in the mouse. *J. Embryol. Exp. Morph.*, 16, 543-558.

DEOL, M. S. and ROBINS, M. W. (1962). The spinner mouse. *J. Hered.*, 53, 133-136.

DETLEFSEN, J. A. (1925). Linkage of dark-eye and color in mice. *Genetics*, 10, 17-32.

DETLEFSEN, J. A. and CLEMENTE, L. S. (1924). Linkage of a dilute color factor and dark-eye in mice. *Genetics*, 9, 247-260.

DETLEFSEN, J. A. and ROBERTS, E. (1918). On a backcross in mice involving three allelomorphic pairs of characters. *Genetics*, 3, 573-598.

DETLEFSEN, J. A. and ROBERTS, E. (1923). Linkage Studies in mice. *Anat. Rec.*, 24, 417.

DICKERMAN, R. C., FEINSTEIN, R. N. and GRAHM, D. (1968). Position of the acatalasemia gene in linkage group V of the mouse. *J. Hered.*, 59, 177-178.

DICKIE, M. M. (1954a). The tortoiseshell house mouse. *J. Hered.*, 45, 158, 190.

DICKIE, M. M. (1954b). The expanding knowledge of the genome in the mouse. *J. Nat. Cancer Inst.*, 15, 679-684.

DICKIE, M. M. (1955). Alopecia, a dominant mutation in the house mouse. *J. Hered.*, 46, 30-34.

DICKIE, M. M. (1964). New Splotch alleles in the mouse. *J. Hered.*, 55, 97-101.

DICKIE, M. M., KELTON, D. E., FIELDER, J. H., INGALLS, A. M. and SNELL, G. D. (1949). New mutations and linkage studies in the house mouse. *Anat. Rec.*, 105, 540.

DICKIE, M. M. and WOOLLEY, G. W. (1946). Linkage studies with the pirouette gene in the mouse. *J. Hered.*, 37, 335-337.

DICKIE, M. M. and WOOLLEY, G. W. (1948). Location of the pirouette gene in *Mus musculus*. *J. Hered.*, 39, 288.

DICKIE, M. M. and WOOLLEY, G. W. (1950). Fuzzy mice. *J. Hered.*, 41, 193-196.

DIPAOLO, J. A. and NOELL, W. K. (1962). Some genetic aspects of visual cell degeneration in mice. *Exp. Eye Res.*, 1, 215-220.

DOUGLAS, L. T. (1966). Meosis. I. Association of non-homologous bivalents during spermatogenesis in white mice. *Genetica*, 37, 466-480.

DOUGLAS, L. T. and GEERTS, S. J. (1966). Meiosis. II. A modified affinity model in mice. *Genetica*, 37, 511-542.

DRAY, S., LIEBERMAN, R. and HOFFMAN, H. A. (1963). Two murine gamma-globulin allotypic specifities identified by ascitic fluid isoprecipitins and determined by allelic genes. *Proc. Soc. Exp. Biol. Med.*, 113, 509.

DRAY, S. POTTER, M. and LIEBERMAN, R. (1965). Immunochemical and genetic studies of two distinct gamma-G-immunoglobulins in BALB/C mice. *J. Immunolog.*, 95, 823-833.

DUNN, L. C. (1919). Anomalous ratios in a family of yellow mice suggesting linkage between the genes for yellow and for black. *Amer. Nat.*, 53, 558-560.

DUNN, L. C. (1920a). Linkage in mice and rats. *Genetics*, 5, 325-343.

DUNN, L. C. (1920b). Independent genes in mice. *Genetics*, 5, 344-361.

DUNN, L. C. (1920c). Types of white spotting in mice. *Amer. Nat.*, 54, 465-495.

DUNN, L. C. (1945). A new eye color mutant in the mouse with asymmetrical expression. *Proc. Nat. Acad. Sci. U.S.A.*, 31, 343-346.

DUNN, L. C. (1956). Analysis of a complex gene in the house mouse. *Cold Spring Harb. Symp. Quant. Biol.*, 21, 187-195.

DUNN, L. C. and BENNETT, D. (1968). A new case of transmission ratio distortion in the house mouse. *Proc. Nat. Acad. Sci. U.S.A.*, 61, 570-573.

DUNN, L. C., BENNETT, D. and BEASLEY, A. B. (1962). Mutation and recombination in the vicinity of a complex gene. *Genetics*, 47, 285-303.

DUNN, L. C. and CASPARI, E. (1942). Close linkage between mutations with similar effects. *Proc. Nat. Acad. Sci. U.S.A.*, 28, 205-210.

DUNN, L. C. and CASPARI, E. (1945). A case of neighbouring loci with similar effects. *Genetics*, 30, 543-568.

DUNN, L. C. and GLUECKSOHN–SCHOENHEIMER, S. (1939). The inheritance of taillessness (anury) in the house mouse. II. Taillessness in a second balanced line. *Genetica*, 24, 587-609.

DUNN, L. C. and GLUECKSOHN–SCHOENHEIMER, S. (1943). Tests for recombination among three lethal mutations in the house mouse. *Genetics*, 28, 29-40.

DUNN, L. C. and GLUECKSOHN-SCHOENHEIMER, S. (1950). Repeat mutations in one area of a mouse chromosome. *Proc. Nat. Acad. Sci. U.S.A.*, 36, 233-237.

DUNN L. C., GLUECKSOHN-SCHOENHEIMER, S. and BRYSON, V. (1940). A new mutation in the mouse affecting spinal column and urogenital system. *J. Hered.*, 31, 343-348.

DUNN, L. C. and GLUECKSOHN-WAELSCH, S. (1953a). Genetic analysis of seven newly discovered mutant alleles at locus *T* in the house mouse. *Genetics*, 38, 261-271.

DUNN, L. C. and GLUECKSOHN-WAELSCH, S. (1953b). The failure of a *t* allele to suppress crossingover in the mouse. *Genetics*, 38, 512-517.

DUNN, L. C. and GLUECKSOHN-WAELSCH, S. (1954). Genetical study of the mutation 'fused' in the house mouse, with evidence concerning its allelism with a similar mutation 'kink'. *J. Genet.*, 52, 383-391.

DURHAM, F. M. (1908). A preliminary account of the inheritance of coat colour in mice. *Rep. Evol, Commit. Roy. Soc.*, 4, 41-53.

DURHAM, F. M. (1911). Further experiments on the inheritance of coat colour in mice. *J. Genet.*, 1, 159-178.

EGOROV, I. K. (1965). A new isoantigen of mouse erythrocytes. *Genetika, Mosk.*, 1965(6), 80-85.

EICHER, E. M. (1967). The genetic extent of the insertion involved in the flecked translocation in the mouse. *Genetics*, 55, 203-212.

EICHER, E. M. (1970a). The position of *ru-2* and *qv* with respect to the flecked translocation in the mouse. *Genetics*, 64, 495-510.

EICHER, E. M. (1970b). *X*-autosomal translocations in the mouse: total inactivation versus partial inactivation of the *X* chromosome. *Advanc. Genet.*, 15, 175-259.

EICHWALD, E. J., DAVIDSON, N. and MOORE, M. (1966). The murine *Y* chromosome as a marker. *Transplantation*, 4, 332-333.

EICHWALD, E. J. and SILMSER, C. R. (1955). Untitled communication. *Transpl. Bull.*, 2, 148-149.

EICHWALD, E. J., SILMSER, C. R. and WEISSMAN, I. (1958). Sex-linked rejection of normal and neoplastic tissue. I. Distribution and specificity. *J. Nat. Cancer Inst.*, 20, 563-575.

EICHWALD, E. J., SILMSER, C. R. and WHEELER, N. (1957). The genetics of skin grafting. *Ann. N.Y. Acad. Sci.*, 64, 737-740.

ERICKSON, R. P., GLUECKSOHN-WAELSCH, S. and CORI, C. F. (1968). Glucose-6-phosphatase deficiency caused by radiation induced alleles at the albino locus in the mouse. *Proc. Nat. Acad. Sci. U.S.A.*, 59, 437-444.

EPSTEIN, C. J. (1969). Mammalian oocytes: *X* chromosome activity. *Science*, 163, 1078-1079.

EVANS, E. P., FORD, C. E. and SEARLE, A. G. (1969). A 39, *X*/41,*XYY* mosaic mouse. *Cytogenetics*, 8, 87-96.

EVANS, E. P., LYON, M. F. and DAGLISH, M. (1967). A mouse translocation giving a metacentric marker chromosome. *Cytogenetics*, 6, 106-119.

FALCONER, D. S. (1947). Linkage of rex with shaker-z in the house mouse. *Heredity*, 1, 133-135.

FALCONER. D. S. (1951). Two new mutants, trembler and reeler, with neurological actions in the house mouse. *J. Genet.*, 50, 192-201.

FALCONER, D. S. (1952a). Location of reeler in linkage group III in the mouse. *Heredity*, 6, 255-257.

FALCONER, D. S. (1952b). A totally sex-linked gene in the house mouse. *Nature*, 169, 664-665.

FALCONER, D. S. (1953). Total sex-linkage in the house mouse. *Z. Ind. Abst. Vererbgslehre*, 85, 210-219.

FALCONER, D. S. (1954). Linkage in the mouse: the sex-linked genes and rough. *Z. Ind. Abst. Vererbgslehre*, 86, 263-268.

FALCONER, D. S. and ISAACSON, J. H. (1959). Adipose, a new inherited obesity of the mouse. *J. Hered.*, 50, 290-292.

FALCONER, D. S. and ISAACSON, J. H. (1962). The genetics of sex-linked anemia in the mouse. *Genet. Res.*, 3, 248-250.

FALCONER, D. S. and ISAACSON, J. H. (1966). Curly-whiskers and its linkage with tail-kinks in linkage group II of the mouse. *Genet. Res.*, 8, 111-113.

FALCONER, D. S. and SIERTS-ROTH, U. (1951). Dreher, ein neues gen der Tanzmausgruppe bei der Hausmaus. *Z. Ind. Abst. Vererbgslehre*, 84, 71-73.

FALCONER, D. S., SLIZYNSKI, B. M. and AUERBACH, C. (1952). Genetical effects of nitrogen mustard in the house mouse. *J. Genet.*, 51, 81-88.

FALCONER, D. S. and SNELL, G. D. (1952). Two new hair mutants, rough and frizzy in the mouse. *J. Hered.*, 43, 53-57.

FALCONER, D. S. and SOBEY, W. R. (1953). The location of trembler in linkage group VII of the house mouse. *J. Hered.*, 49, 159-160.

FELDMAN, H. W. (1924). Linkage of albino allelomorphs in rats and mice. *Genetics*, 9, 437-492.

FELDMAN, H. W. and PINCUS, G. (1926). On the inheritance of albinism and brown pigmentation in mice. *Amer. Nat.*, 60, 195-198.

FELDMAN, M. (1958). The antigen determined by a *Y* linked histocompatibility gene. *Transpl. Bull.*, 5, 15-16.

FIELDER, J. H. (1952). The taupe mouse. *J. Hered.*, 43, 74-76.

FINLAYSON, J. S., HUDSON, D. M. and ARMSTRONG, B. L. (1969). Location of the *Mup-a* locus on mouse linkage group VIII. *Genet. Res.*, 14, 329-331.

FISHER, R. A. (1946). A system of scoring linkage data, with special reference to the pied factors in mice. *Amer. Nat.*, 80, 568-578.

FISHER, R. A. (1949). A preliminary linkage test with agouti and undulated mice. I. The fifth linkage group. *Heredity*, 3, 229-241.

FISHER, R. A. (1950a). Polydactyly in mice. *Nature*, 165, 407.

FISHER, R. A. (1950b). Polydactyly in mice. *Nature*, 165, 796.

FISHER, R. A. (1953). The linkage of polydactyly with leaden in the house mouse. *Heredity*, 7, 91-95.

FISHER, R. A. and LANDAUER, W. (1953). Sex differences of crossingover in close linkage. *Amer. Nat.*, 87, 116.

FISHER, R. A., LYON, M. F. and OWEN, A. R. G. (1947). The sex chromosome in the house mouse. *Heredity*, 1, 355-365.

FISHER, R. A. and MATHER, K. (1936a). Verification in mice of the possibility of more than fifty per cent recombination. *Nature*, 137, 362-363.

FISHER, R. A. and MATHER, K. (1936b). A linkage test with mice. *Ann. Eugen.*, 7, 265-280.

FISHER, R. A. and SNELL, G. D. (1948). A twelfth linkage group in the mouse. *Heredity*, 2, 271-273.

FISHER, R. A. and YATES, F. (1953). *Statistical tables for biological, agricultural and medical research.* Edinburgh: Oliver and Boyd.

FLANAGAN, S. P. (1966). Nude, a new hairless gene with pleiotropic effects in the mouse. *Genet. Res.*, 8, 295-309.

FLANAGAN, S. P. and ISAACSON, J. H. (1967). Close linkage between genes which cause hairlessness in the mouse. *Genet. Res.*, 9, 99-110.

FORD, C. E. (1966). The murine *Y* chromosome as a marker. *Transplantation*, 4, 33-35.

FORD, C. E., CARTER, T. C. and HAMERTON, J. L. (1956). Cytogenetics of a mouse translocation. *Heredity*, 10, 284.

FORD, C. E. and EVANS, E. P. (1964). A reciprocal translocation in the mouse between the *X* chromosome and a short autosome. *Cytogenetics*, 3, 295-305.

FORD, C. E., HAMERTON, J. L., BARNES, W. H. and LOUTIT, J. F. (1956). Cytological identification of radiation chimaeras. *Nature*, 177, 452-454.

FORD, E. H. R. and WOLLAM, D. H. M. (1963). A study of the mitotic chromosomes of mice of the Strong A line. *Exp. Cell Res.*, 32, 320-326.

FORSTHOEFEL, P. S. and SHENK, T. E. (1970). Linkage relations of Strong's luxoid gene in the mouse. *J. Hered.*, 61, 64-66.

FOSTER, M., PETRAS, M. L. and GASSER, D. L. (1968). The *Ea-1* blood group locus of the house mouse: inheritance, linkage polymorphism and control of antibody synthesis. *Proc. XII Inter. Congr. Genet.*, 1, 245.

FOX, A. S. (1958). Genetics of tissue specificity. *Annals N.Y. Acad. Sci.*, 73, 611-634.

FRASER, A. S., SOBEY, S. and SPICER, C. C. (1953). Mottled, a sex modified lethal in the house mouse. *J. Genet.*, 51, 217-221.

FRASER, F. C. and HERER, M. L. (1950). The inheritance and expression of the lens rupture gene in the house mouse. *J. Hered.*, 41, 3-7.

FRASER, F. C. and SCHABTACH, G. (1962). Shrivelled: a heredity degeneration of the lens in the house mouse. *Genet. Res.*, 3, 383-387.

FREDGA, K. (1961). Studies of the chromosomes in American obese-hyperglycaemic mice. *Hereditas*, 47, 615-618.

GALTON, M. and HOLT, S. F. (1965). Asynchronous replication of the mouse sex chromosomes. *Exp. Cell Res.*, 37, 111-116.

GANSCHOW, R. and PAIGEN, K. (1967). Separate genes determining the structure and intracellular location of hepatic glucuronidase. *Proc. Nat. Acad. Sci. U.S.A.*, 58, 938-945.

GARBER, E. D. (1952). Bent tail, a dominant sex linked mutation in the mouse. *Proc. Nat. Acad. Sci. U.S.A.*, 38, 876-879.

GASSER, D. L. (1969). Genetic control of the immune response in mice. I. Segregation data and localization in the fifth linkage group of a gene affecting antibody production. *J. Immunol.*, 103, 66-70.

GATES, W. H. (1926). The Japanese waltzing mouse. *Car. Inst. Wash. Pub.*, 337, 83-138.

GATES, W. H. (1927a). Linkage of short ears and density in the house mouse. *Proc. Nat. Acad. Sci. U.S.A.*, 13, 575-578.

GATES, W. H. (1927b). A case of non-disjunction in the mouse. *Genetics*, 12, 295-306.

GATES, W. H. (1928a). Linkage of the factors for short-ear and density in the house mouse. *Genetics*, 13, 170-179.

GATES, W. H. (1928b). Linkage of characters albinism and shaker in the house mouse. *Anat. Rec.*, 41, 104.

GATES, W. H. (1929). Linkage of shaker and albinism and pink-eye in the house mouse. *Anat. Rec.*, 44, 287.

GATES, W. H. (1931). Linkage of the factor shaker with albinism and pink-eye in the house mouse. *Z. Ind. Abst. Vererbgslehre*, 59, 220-226.

GATES, W. H. (1934). Linkage tests of a new shaker mutation with other factors in the house mouse. *Amer. Nat.*, 68, 173-174.

GATES, W. H. and PULLIG, T. (1945a). Linkage of dominant white spotting with hairless in the house mouse. *Genetics*, 30, 4.

GATES, W. H. and PULLIG, T. (1945b). Linkage of dominant white spotting with hairless in the house mouse. *Proc. La. Acad. Sci.*, 9, 57-60.

GEYER-DUSZYNSKA, I. (1963). On the structure of the *XY* bivalent in *Mus musculus*. *Chromosoma*, 13, 521-525.

GMLM Pt B—7

GEYER-DUSZYNSKA, I. (1964). Cytological investigations on the *T* locus in *Mus musculus. Chromosoma,* 15, 478-502.

GILMAN, J. G. and SMITHIES, O. (1968). Fetal hemoglobin variants in mice. *Science,* 160, 885-886.

GITTES, R. F. and RUSSELL, P. S. (1961). Male histocompatibility antigens in mouse endoctrine tissues: functional and histologic evidence. *J. Nat. Cancer Inst.,* 26, 283-303.

GLUECKSOHN-WAELSCH, S. and CORI, C. F. (1970). Glucose-6-phosphatase deficiency: mechanisms of genetic control and biochemistry. *Biochem. Genet.,* 4, 195-201.

GOLDJMAN, I. L., SMERTENKO, I. I. and VILKINA, (1966). The normal karyotype in mice of a highly leukaemic AKR strain. *Byull. Eksp. Biol. Med.,* 61 (2), 75-78.

GOODWINS, I. R. and VINCENT, M. A. C. (1955). Further data on linkage between short-ear and maltese dilution in the house mouse. *Heredity,* 9, 413-414.

GORER, P. A., LYMAN, S. and SNELL, G. D. (1948). Studies on the genetic and antigenic basis of tumour transplantation. *Proc. Roy. Soc. B.,* 135, 499-505.

GORER, P. A. and MIKULSKA, Z. B. (1959). Some further data on the H-2 system of antigens. *Proc. Roy. Soc. B.,* 151, 57-69.

GRAHN, D., FRY, R. J. M. and HAMILTON, K. F. (1969). Genetic and pathologic analysis of the sex-linked allelic series, mottled, in the mouse. *Genetics,* 61, s22-s23.

GREEN, C. V. (1931). Size inheritance and growth in a mouse species cross. III. Inheritance of adult quantitative characters. *J. Exp. Zool.,* 59, 213-263.

GREEN, C. V. (1932). Genetic linkage in size inheritance. *Amer. Nat.,* 66, 87-91.

GREEN, E. L. (1954). Genetics of a new hair deficiency, furless, in the house mouse. *J. Hered.,* 45, 115-118.

GREEN, E. L. (1966). The Jackson laboratory: a center for mammalian genetics in the United States. *J. Hered.,* 57, 2-12.

GREEN, E. L. (1967). Shambling, a neurological mutant of the mouse. *J. Hered.,* 58, 65-68.

GREEN, E. L. (1968). Linkage of shambling with rex in linkage group VII of the mouse. *J. Hered.,* 59, 59.

GREEN, E. L. and MANN, S. J. (1961). Opossum, a semi-dominant lethal mutation affecting hair and other characteristics of mice. *J. Hered.,* 52, 223-227.

GREEN, M. C. (1955). Luxoid, a new hereditary leg and foot abnormality in the house mouse. *J. Hered.,* 46, 90-99.

GREEN, M. C. (1961). The position of luxoid in linkage group II of the mouse. *J. Hered.,* 52, 297-300.

GREEN, M. C. (1963). Methods for testing linkage. In Burdette, W. J. (Editor), *Methodology in Mammalian genetics.* San Francisco: Holden-Day.

GREEN, M. C. (1966). Mutant genes and linkages. In Green, E. L.

(Editor), *Biology of the laboratory mouse.* New York: McGraw-Hill.
GREEN, M. C. (1968). Linkage map of the mouse. In Sober, H. A. (Editor), *Handbook of Chemistry.* Cleveland: Chemical Rubber Co.
GREEN, M. C. and DICKIE, M. M. (1959). Linkage map of the mouse. *J. Hered.,* 50, 2-5.
GREEN, M. C. and LANE, P. W. (1967). Linkage group II of the house mouse. *J. Hered.,* 58, 225-228.
GREEN, M. C. and SIDMAN, R. L. (1962). Tottering, a neuromuscular mutation in the mouse and its linkage with oligosyndactylism. *J. Hered.,* 53, 233-237.
GREEN, M. C., SNELL, G. D. and LANE, P. W. (1963). Linkage group XVIII of the mouse. *J. Hered.,* 54, 245-247.
GRIFFEN, A. B. (1955). A late pachytene chromosome map of the male mouse. *J. Morph.,* 96, 123-136.
GRIFFEN, A. B. (1958a). Occurrence of chromosomal aberrations in presperematocytic cells of irradiated male mice. *Proc. Nat. Acad. Sci. U.S.A.,* 44, 691-649.
GRIFFEN, A. B. (1958b). Mammalian cytogenetics and the cancer problem. *Annals N.Y. Acad. Sci.,* 71, 1156-1162.
GRIFFEN, A. B. (1960). Mammalian pachytene chromosome mapping and somatic chromosome identification. *J. Cell. Comp. Physiol.,* 56, (Suppl. 1), 113-121.
GRIFFEN, A. B. (1966). Nuclear cytology. In Green, E. L. (Editor), *Biology of the laboratory mouse.* New York: McGraw-Hill.
GRIFFEN, A. B. (1967). A case of tertiary trisomy in the mouse, and its implications for the cytological classification of trisomics in other mammals. *Canad. J. Genet. Cytol.,* 9, 503-510.
GRIFFEN, A. B. and BUNKER, M. C. (1964). Three cases of trisomy in the mouse. *Proc. Nat. Acad. Sci. U.S.A.,* 52, 1194-1198.
GRIFFEN, A. B. and BUNKER, M. C. (1967). Four further cases of autosomal primary trisomy in the mouse. *Proc. Nat. Acad. Sci. U.S.A.,* 58, 1446-1452.
GRUNEBERG, H. (1935). A three factor linkage experiment in the house mouse. *J. Genet.,* 31, 157-172.
GRUNEBERG, H. (1936a). Some linkage tests with wavy mice. *J. Genet.,* 32, 1-3.
GRUNEBERG, H. (1936b). Further linkage data on the albino chromosome of the house mouse. *J. Genet.,* 33, 255-265.
GRUNEBERG, H. (1936c). Inheritance of tail tip pigmentation in the house mouse. *J. Genet.,* 33, 343-345.
GRUNEBERG, H. (1943). Two new mutant genes in the house mouse. *J. Genet.,* 45, 22-28.
GRUNEBERG, H. (1952). *The genetics of the mouse.* The Hague: Nijhoff.
GRUNEBERG, H. (1956). An annotated catalogue of the mutant genes of the house mouse. *Med. Res. Council Mem.,* 33: 28pp.
GRUNEBERG, H., BURNETT, J. B. and SNELL, G. D. (1941). The origin of jerker, a new gene mutation of the house mouse and linkage studies made with it. *Proc. Nat. Acad. Sci. U.S.A.,* 27, 562-565.

GRUNEBERG, H. and TRUSLOVE, G. M. (1960). Two closely linked genes in the mouse. *Genet., Res.,* 1, 69-90.

HAGEDOORN, A. L. (1914). Repulsion in mice (a reply). *Amer. Nat.,* 48, 699-700.

HALDANE, J. B. S. (1919). The combination of linkage values and the calculation of distance between the loci of linked factors. *J. Genet.,* 8, 299-309.

HALDANE, J. B. S., SPRUNT, A. D. and HALDANE, N. M. (1915). Reduplication in mice. *J. Genet.,* 5, 133-135.

HAUSCHKA, T. S. (1955). Probable *Y* linkage of a histocompatibility gene. *Transpl. Bull.,* 2, 154-155.

HAUSCHKA, T. S., GOODWIN, M. B. and BROWN, E. (1951). Evidence for a sex-linked lethal in the house mouse. *Genetics,* 36, 235-253.

HAUSCHKA, T. S., GRINNELL, S. T., MEAGHER, M. and AMOS, D. B. (1961). Sex-linked incompatibility of male skin and primary tumours transplanted to isologous female mice. *XIII Annual Symp. Fundamental Cancer Res.,* (1959), 271-294.

HAUSCHKA, T. S. and HOLDRIDGE, B. A. (1962). A cytogenetic approach to the Y-linked histocompatibility antigen of mice. *Ann. N.Y. Acad. Sci.,* 101, 12-22.

HAUSCHKA, T. S., JACOBS, B. B. and HOLDRIDGE, B. A. (1968). Recessive yellow and its interaction with belted in the mouse. *J. Hered.,* 59, 339-341.

HAUSCHKA, T. S., MEAGHER, M. A. and HOLDRIDGE, B. A. (1962). Autosomal translocation of the 'male antigen' of mice and immunoselection against the *Y* chromosome. *Inter. Symp. Tissue Transplantation Santiago,* 1961, 25-36.

HENDERSON, S. A. and EDWARDS, R. G. (1968). Chiasma frequency and material age in mammals. *Nature,* 218, 22-28.

HERTWIG, P. (1940). Vererbbare Semisteritat bei Mausen nach Rontgenbestrahlung verursacht durch reziproke chromosomentranslokationen. *Z. Ind. Abst. Vererbgslehre* 79, 1-29.

HERTWIG, P. (1942). Neue mutationen und Koppelungsgruppen bei der Hausmaus. *Z. Ind. Abst. Vererbgslehre,* 80, 220-246.

HERZENBERG, L.A. (1964). A chromosome region for gamma 2a and beta 2a globulin H chaim isoantigens in the mouse. *Cold Spring Harbor Symp. Quant. Biol.,* 29, 455-462.

HERZENBERG, L. A., McDEVITT, H. O. and HERZENBERG, L. A. (1968). Genetics of antibodies. *Ann. Rev. Genet.,* 2, 209-244.

HERZENBERG, L. A., MINNA, J. D. and HERZENBERG, L. A. (1967). The chromosome region for immunoglobulin heavy chains in the mouse. *Cold Spring Harbor Symp. Quant. Biol.,* 32, 181-186.

HERZENBERG, L. A., WARNER, N. J. and HERZENBERG, L. A. (1965). Immunoglobulin isoantigens (allotypes) in the mouse. I. Genetics and cross reactions of the 7S γ2A-isoantigen controlled by alleles at the I_{G-1} Locus. *J. Exp. Med.,* 121, 415-438.

HESTON, W. E. (1941). Relationship between susceptibility to induced pulmonary tumors and certain known genes in mice. *J. Nat. Cancer*

Inst., 2, 127-132.

HESTON, W. E. (1951). The vestigial tail mouse. *J. Hered.*, 42, 71-74.

HESTON, W. E. (1953). Linkage between pulmonary tumours and vestigial tail in the house mouse. *Nature*, 172, 1007-1008.

HESTON, W. E., DERINGER, M. K., HUGHES, I. R. and CORNFIELD, J. (1952). Interrelation of specific genes, body weight and development of tumors in mice. *J. Nat. Cancer Inst.*, 12, 1141-1157.

HIRST, B. B. (1957). The influence of sex on transplantability of isologous thymic tissue in normal C57BL mice. *Transpl. Bull.*, 4, 58.

HOECKER, G. (1956). Genetic mechanisms in tissue transplantation in the mouse. *Cold Spring Harbor Symp. Quant. Biol.*, 21, 355-361.

HOECKER, G., MARTINEZ, A., MARKOVIC, S. and PIZARRO, O. (1954). Agitans, a mutation with neurological effects in the house mouse. *J. Hered.*, 45, 10-14.

HOECKER, G. and PIZARRO, O. (1962). The histocompatibility antigens. *Inter. Symp. Tissue transplantation Santiago*, 1961, 54-71.

HOLLANDER, W. F. and GOWEN, J. W. (1959). A single gene antagonism between mother and foetus in the mouse. *Proc. Soc. Exp. Biol. Med.*, 101, 425-428.

HOLLANDER, W. F., GOWEN, J. W. and STADLER, J. (1956). A study of 25 gynandromorphic mice of the Bagg albino strain. *Anat. Rec.*, 124, 223-244.

HOLLANDER, W. F. and STRONG, L. C. (1951). Pintail, a dominant mutation linked with brown in the house mouse. *J. Hered.*, 42, 179-182.

HUDSON, D. M., FINLAYSON, J. S. and POTTER, M. (1967). Linkage of one component of the major urinary protein complex of mice to the brown coat colour locus. *Genet. Res.*, 10, 195-198.

HUNGERFORD, D. (1955). Chromosome number of 10-day foetal mouse cells. *J. Morph.*, 97, 497.

HUNT, H. R., MIXTER, R. and PERMER, D. (1933). Flexed tail in the mouse. *Genetics*, 18, 335-366.

HUSKINS, C. L. and HEARNE, E. M. (1934). Chromosome differences in mice susceptible and resistant to cancer. *Nature*, 133, 615.

HUSKINS, C. L. and HEARNE, E. M. (1936). Spermatocyte chiasma frequency in strains of mice differing in susceptibility or resistance to the spontaneous occurrence of malignant tumours. *Canad. J. Res. D.*, 14, 39-58.

HUTTON, J. J. (1969a). Linkage analysis using biochemical variants in mice. *Genetics*, 61, s28.

HUTTON, J. J. (1969b). Linkage analyses using biochemical variants in mice. I. Linkage of the hemoglobin beta-chain and glucosephosphate isomerase loci. *Biochem. Genet.*, 3, 507-515.

HUTTON, J. J., BISHOP, J., SCHWEET, R. and RUSSELL, E. S. (1962). Hemoglobin inheritance in inbred mouse strains. II. genetic studies. *Proc. Nat. Acad. Sci. U.S.A.*, 48, 1718-1724.

HUTTON, J. J. and COLEMAN, D. L. (1969). Linkage analyses using biochemical variants in mice. II. Levulinate dehydratase and autosomal glucose 6-phosphate dehydrogenase. *Biochem. Genet.*, 3, 517-523.

HUTTON, J. J. and RODERICK, T. H. (1970). Linkage analyses using biochemical variants in mice. III. Linkage relationships of eleven biochemical markers. *Biochem. Genet.*, 4, 339-350.

HUTTON, J. J., SCHWEET, R. S., WOLFE, H. G. and RUSSELL, E. S. (1964). Hemoglobin solubility and x-chain structure in crosses between two inbred mouse strains. *Science*, 143, 252-253.

JAFFE, J. (1952). Cytological observations concerning inversion and translocation in the house mouse. *Amer. Nat.*, 86, 101-107.

JOHNSON, D. R. (1969a). Brachyphalangy, an allele of extra-toes in the mouse. *Genet. Res.*, 13, 275-280.

JOHNSON, D. R. (1969b). Polysyndactyly, a new mutant gene in the mouse. *J. Embryol. Exp. Morph.*, 21, 285-294.

KEELER, C. E. (1927). Rodless retina, an ophthalmic mutation in the house mouse. *J. Exp. Zool.*, 46, 355-407.

KEELER, C. E. (1930). Hereditary blindness in the house mouse with special reference to the linkage relationship. *Harvard Med. School, Howe Lab. Ophthalmol. Bull.*, 3, 11 pp.

KEELER, C. E. (1931a). The independence of dominant spotting and recessive spotting (piebald) in the house mouse. *Proc. Nat. Acad. Sci. U.S.A.*, 17, 101-102.

KEELER, C. E. (1931b). A probable new mutation to white belly in the house mouse. *Proc. Nat. Acad. Sci. U.S.A.*, 17, 700-703.

KEELER, C. E. (1966). Retinal degeneration in the mouse is rodless retina. *J. Hered.*, 57, 47-50.

KELTON, D. E. and RAUCH, H. (1968). Linkage of open eyelids with linkage group VII of the mouse. *J. Hered.*, 59, 27-28.

KIDWELL, J. F., GOWEN, J. W. and STADLER, J. (1961). Pugnose, a recessive mutation in linkage group III of mice. *J. Hered.*, 52, 145-148.

KIDWELL, J. F., GOWEN, J. W. and STADLER, J. (1966). Pugnose linkage in the mouse. *J. Hered.*, 57, 229-230.

KIDWELL, J. H., NASH, G. D., STADLER, J. and GOWEN, J. W. (1961). An X-ray induced deficiency in linkage group VI of the mouse. *Amer. Zool.*, 1, 365.

KINDRED, B. M. (1961). Abnormal inheritance of the sex-linked Tabby gene. *Aust. J. Biol. Sci.*, 14, 415-418.

KING, J. W. B. (1956). Linkage group XIV of the house mouse. *Nature*, 178, 1126-1127.

KLEIN, E. and LINDER, O. (1961). Factorial analysis of the reactivity of C57BL females against isologous male skin grafts. *Transplantation*, 27, 457-459.

KLEIN, J., BEDNAROVA, D. and SCHREFFLER, D. C. (1969). Analysis of the mouse in two translocations stocks and an *H-2* crossover. *Genetics*, 61, s33.

KLEIN, J. and MARTINKOVA, J. (1968). A new non-*H-2* antigen of C57BL mice. *Folia Biol. (Praha)*, 14, 237-238.

KOLLER, P. C. (1944). Segmental interchange in mice. *Genetics*, 29, 247-263.

KOLLER, P. C., and AUERBACH, C. (1941). Chromosome breakage and sterility in the mouse. *Nature*, 148, 501.

KOLLER, P. C. and DARLINGTON, C. D. (1934). The genetical and mechanical properties of the sex chromosomes. I. *Rattus norvegicus*. *J. Genet.*, 29, 359-383.

KOSAMBI, D. D. (1944). The estimation of map distances from recombination values. *Ann. Eugen.*, 12, 172-175.

KREITNER, P. C. (1957). Linkage studies in a black-eyed white mutation in the house mouse. *J. Hered.*, 48, 300-304.

KROON, P. H. W. VAN DER and BUIS, A. J. M. (1970). Linkage of dwarf with obese in the mouse. *Genetica*, 41, 57-60.

KRZANOWSKA, H. (1966). Inheritance of reduced male fertility, connected with abnormal spermatozoa, in mice. *Acta Biol. Cracov., Zool. Ser.*, 9, 61-70.

KRZANOWSKA, H. (1969). Factor responsible for spermatozoan abnormality located on the chromosome in mice. *Genet. Res.*, 13, 17-24.

LAGNEAU, L. E. and MEWISSEN, D. J. (1964). Le caryotype de la souris C57BL/6. *C.R. Soc. Biol.*, 158, 1960-1963.

LANE, P. W. (1963). Whirler mice, a recessive behavior mutation in linkage group VIII. *J. Hered.*, 54, 263-266.

LANE, P. W. (1967). Linkage groups III and XVII in the mouse and the position of the light-ear locus. *J. Hered.*, 58, 21-24.

LANE, L. W. and DICKIE, M. M. (1961). Linkage of wabbler-lethal and hairless in the mouse. *J. Hered.*, 52, 159-160.

LANE, P. W. and DICKIE, M. M. (1968). Three recessive mutations producing disproportionate dwarfing in mice: achondroplasia brachymorphic and stubby. *J. Hered.*, 59, 300-308.

LANE, P. W. and GREEN, E. L. (1967). Pale ear and light ear in the house mouse. *J. Hered.*, 58, 17-20.

LANE, P. W. and GREEN, M. C. (1960). Mahogany, a recessive color mutation in linkage group V of the mouse. *J. Hered.*, 51, 228-230.

LAW, L. W. (1952). The flexed-tail gene and induced anaemia in mice. *J. Nat. Cancer Inst.*, 12, 1119-1126.

LAW, L. W., MORROW, A. G. and GREENSPAN, E. M. (1952). Inheritance of low liver glucuronidase activity in the mouse. *J. Nat. Cancer Inst.*, 12, 909-919.

LEONARD, A. and DEKUNDT, G. (1965). Le caryotype de la souris BALB/c. *C.R. Soc. Biol.*, 159, 2080-2083.

LEONARD, A. and DEKUNDT, G. (1967a). The value of chromosome examination in animal breeding. *Z. Vers. Tierk.*, 9, 109-114.

LEONARD, A. and DEKUNDT, G. (1967b). A new marker for chromosome studies in the mouse. *Nature*, 214, 504-505.

LEONARD, A. and DEKUNDT, G. (1969a). Etude cytologique d'une translocation chromosome *Y*-autosome chez la souris. *Experientia*, 25, 876-877.

LEONARD, A. and DEKUNDT, G. (1969b). Etude d'une translocation de type Robertsonien chez les souris de race AKR. *Acta Zool. Pathol.*

Antwerp., **48**, 43-57.

LEVAN, A. (1956). The significance of polyploidy for the evolution of mouse tumours. *Exp. Cell Res.*, 11, 613-629.

LEVAN, A., HSU, T. C. and STICH, (1962). The idiogram of the mouse. *Hereditas*, 48, 677-687.

LIEBERMANN, R., DRAY, S. and POTTER, M. (1965). Linkage in control of allotypic specificities on two different γG immunoglobulins. *Science*, 148, 640-642.

LIEBERMANN, R. and POTTER, M. (1966). Close linkage in genes controlling γ^A and γ^G heavy chain structure in BALB/C mice. *J. Mol. Biol.*, 18, 516-528.

LILLY, F. (1966). The inheritance of susceptibility to the Gross leukemia virus in mice. *Genetics*, 53, 529-539.

LILLY, F. (1967). The location of histocompatibility-6 in the mouse genome. *Transplantation*, 5, 83-85.

LITTLE, C. C. (1913). Experimental studies of the inheritance of color in mice. *Car. Inst. Wash. Pub.*, 179, 11-102.

LITTLE, C. C. (1915). The inheritance of black-eyed white spotting in mice. *Amer. Nat.*, 49, 727-740.

LITTLE, C. C. (1916). The occurrence of three recognized coat color mutations in mice. *Amer. Nat.*, 50, 335-349.

LITTLE, C. C. (1917). The relation of yellow coat colour and black eyed white spotting of mice in inheritance. *Genetics*, 2, 433-445.

LITTLE, C. C. (1920). Note on the occurrence of a probable sex-linked factor in mammals. *Amer. Nat.*, 54, 457-460.

LITTLE, C. C. (1927). Notes on a species cross in mice and on an hypothesis concerning the quantitative potentiality of genes. *Science*, 66, 542-543.

LITTLE, C. C. and PHILLIPS, J. C. (1913). A cross involving four pairs of Mendelian characters in mice. *Amer. Nat.*, 47, 760-762.

LYON, J. B. (1970). The X chromosome and the enzymes controlling muscle glycogen: phosphorylase kinase. *Biochem. Genet.*, 4, 169-185.

LYON, J. B., PORTER, J. and ROBERTSON, M. (1967). Phosphorylase b kinase inheritance in mice. *Science*, 155, 1550-1551.

LYON, M. F. (1955). Ataxia, a new recessive mutant of the house mouse. *J. Hered.*, 46, 77-80.

LYON, M. F. (1956). Hereditary hair loss in the tufted mutant of the house mouse. *J. Hered.*, 47, 101-103.

LYON, M. F. (1958). Twirler: a mutant affecting the inner ear of the house mouse. *J. Embryol. Exp. Morph.*, 6, 105-106.

LYON, M. F. (1959). A new dominant T allele in the house mouse. *J. Hered.*, 50, 140-142.

LYON, M. F. (1960a). A further mutation of the mottled type in the house mouse. *J. Hered.*, 51, 116-121.

LYON, M. F. (1960b). Effect of X-rays on the mutation of t alleles in the mouse. *Heredity*, 14, 247-252.

LYON, M. F. (1961a). Linkage relations and some pleiotropic effects of the dreher mutant of the house mouse. *Genet. Res.*, 2, 92-95.

LYON, M. F. (1961b). Gene action in the X chromosome of the mouse. *Nature*, 190, 372-373.

LYON, M. F. (1961c). The nature of t alleles in the mouse. *Ann. Hum. Genet.*, 25, 263.

LYON, M. F. (1962). Sex chromatin and gene action in the mammalian X chromosome. *Amer. J. Hum. Genet.*, 14, 135-148.

LYON, M. F. (1963). Attempts to test the inactive X theory of dosage compensation in mammals. *Genet. Res.*, 4, 93-103.

LYON, M. F. (1966a). Order of loci on the X chromosome of the mouse. *Genet. Res.*, 7, 130-133.

LYON, M. F. (1966b). Lack of evidence that inactivation of the mouse X chromosome is incomplete. *Genet. Res.*, 8, 197-203.

LYON, M. F. (1969). A true hermaphrodite mouse presumed to be an XO/XY mosaic. *Cytogenetics*, 8, 326-331.

LYON, M. F., BUTLER, J. M. and KEMP, R. (1968). The position of the centromeres in linkage groups II and IX of the mouse. *Genet. Res.*, 11, 193-199.

LYON, M. F. and HAWKES, S. G. (1970). An X linked gene for testicular feminization in the mouse. *Nature*, 227, 1217-1219.

LYON, M. F. and MEREDITH, R. (1964a). Investigations of the nature of t alleles in the mouse. I. Genetic analysis of a series of mutants derived from a lethal allele. *Heredity*, 19, 301-312.

LYON, M. F. and MEREDITH, R. (1964b). Investigations of the nature of t alleles in the mouse. II. Genetic analysis of an unusual mutant allele and its derivatives. *Heredity*, 19, 313-325.

LYON, M. F. and MEREDITH, R. (1966). Autosomal translocations causing male sterility and viable anenploidy in the mouse. *Cytogenetics*, 5, 335-354.

LYON, M. F. and MEREDITH, R. (1969). *Muted*, a new mutant affecting coat colour and otoliths of the mouse, and its position in linkage group XIV. *Genet. Res.*, 14, 163-166.

LYON, M. F., MORRIS, T., SEARLE, A. G. and BUTLER, J. (1967). Occurrences and linkage relations of the mutant 'extra-toes' in the mouse. *Genet. Res.*, 9, 383-385.

LYON, M. F. and PHILLIPS, R. J. S. (1959). Crossingover in mice heterozygous for t alleles. *Heredity*, 13, 23-32.

LYON, M. F., SEARLE, A. G., FORD, C. E. and OHNO, S. (1964). A mouse translocation suppressing sex-linked variegation. *Cytogenetics*, 3, 306-323.

MACDOWELL, E. C. (1950). Light, a new mouse colour. *J. Hered.*, 41, 35-36.

MacNEIL, M. (1956). A mutation to caracul in the house mouse. *Nature*, 178, 1242.

MAKINO, S. (1941). Studies on the murine chromosomes. I. Cytological investigations of mice included in the genus *Mus. J. Fac. Sci. Hokkaido Univ. Ser. VI, Zool.*, 7, 305-380.

MALLYON, S. A. (1951). A pronounced sex difference in recombination values in the sixth chromosome of the house mouse. *Nature*, 168,

118-119.

MATHER, K. and NORTH, S. B. (1940). Umbrous: a case of dominance modification in mice. *J. Genet.*, 40, 229-241.

MATTHEY, R. (1953). Les chromosomes des *Muridae*. *Rev. Suisse Zool.*, 60, 225-283.

McDEVITT, H. O. (1968). Genetic control of the antibody response. III. Qualitative and quantitative characterization of the antibody response to (T.C.)-A-L in CBA and C57 mice. *J. Immunol.*, 100, 485-492.

McDEVITT, H. O. and TYAN, M. L. (1968). Genetic control of the antibody response in inbred mice. *J. Exp. Med.*, 128, 1-11.

McLAREN, A. (1960). New evidence of imbalanced sex chromosome constitution in the mouse. *Genet. Res.*, 1, 253-261.

McNUTT, W. (1967). Porcine tail, a new mutation in the house mouse. *Anat. Rec.*, 157, 286.

MEDVEDEV, N. N. and EGOROV, I. K. (1966). [A histocompatibility gene located in the Y chromosome of CC57BR and CC57W mice.] *Genetika (Mosk.)*, 1966 (8), 86-88.

MICHIE, D. (1952). A new linkage in the mouse: vestigial and rex. *Nature*, 170, 585.

MICHIE, D. (1953). Affinity: a new genetic phenomenon in the house mouse. Evidence from distant crosses. *Nature*, 171, 26-27.

MICHIE, D. (1955a). Genetic studies with vestigial tail mice. I. The sex difference in crossingover between vestigial and rex. *J. Genet.*, 53, 270-279.

MICHIE, D. (1955b). Genetical studies with vestigial tail mice. II. The position of vestigial in the VII linkage group. *J. Genet.*, 53, 280-284.

MICHIE, D. (1955c). Genetical studies with vestigial tail mice. III. New independence data. *J. Genet.*, 53, 285-294.

MICHIE, D. (1955d). Affinity. *Proc. Roy. Soc.*, B, 144, 241-259.

MICHIE, D. and McLAREN, A. (1958). A proposed genetic analysis of the Eichwald-Silmser effect. *Transpl. Bull.*, 5, 17-18.

MINNA, J. D., IVERSON, G. M. and HERZENBERG, L. A. (1967). Identification of a gene locus for γ^{G1} immunoglobulin H chains and its linkage to the H chain chromosome region in the mouse. *Proc. Nat. Acad. Sci. U.S.A.*, 58, 188-194.

MORGAN, T. H. (1914). Multiple allelomorphs in mice. *Amer. Nat.*, 48, 449-458.

MORRIS, T. (1968). The XO and OY chromosome constitutions in the mouse. *Genet. Res.*, 12, 125-137.

MURRAY, J. M. (1933). Leaden, a recent color mutation in the house mouse. *Amer. Nat.*, 67, 278-283.

MURRAY, J. M. and SNELL, G; D. (1945). Belted, a new sixth chromosome mutation in the mouse. *J. Hered.*, 36, 266-268.

NAKAMURA, A. and TONOMURA, A. (1963). A characteristic chromosome found in the karyotype analysis of the mouse. *Ann. Rep. Nat. Inst. Genet. Jap.*, 13, 32-33.

NAKAMURA, A. and TONOMURA, A. (1964). An unusual Y chromosome found in a strain of mice. *Rep. Nat. Inst. Genet. Jap.*, 14,

47-48.

NASH, D. J. and VENIER, L. H. (1964). Genetics of grizzle-belly. *Proc. Pa. Acad. Sci.*, 38, 11.

NASRAT, G. S. (1956). Estimation of the recombination fraction between two linked genes *Re* and *sh-2* in the house mouse. *Proc. Zool. Soc. Bengal*, 9, 85-87.

OHNO, S. and CATTANACH, B. M. (1962). Cytological study of an *X* autosome translocation in *Mus musculus*. *Cytogenetics*, 1, 129-140.

OHNO, S., CHRISTIAN, C. and STENIUS, C. (1963). Significance in mammalian oogenesis of the non-homologous association of bivalents. *Exp. Cell Res.*, 32, 590-592.

OHNO, S., KAPLAN, W. D. and KINOSITA, R. (1957a). Note on non-chiasma type association between the *X* and *Y* chromosomes of *Drosophila melanogaster* and *Mus musculus*. *Exp. Cell Res.*, 13, 422-424.

OHNO, S., KAPLAN, W. D. and KINOSITA, R. (1957b). Heterochromatic regions and nucleolus organizers in chromosomes of the mouse. *Exp. Cell Res.*, 13, 358-364.

OHNO, S., KAPLAN, W. D. and KINOSITA, R. (1959a). On the end-to-end association of the *X* and *Y* chromosomes of *Mus musculus*. *Exp. Cell Res.*, 18, 282-290.

OHNO, S., KAPLAN, W. D. and KINOSITA, R. (1959b). Do *XY* and *O* sperm occur in *Mus musculus? Exp. Cell res.*, 18, 382-384.

OHNO, S. and LYON, M. F. (1965). Cytological study of Searle's *X* chromosome translocation in *Mus musculus*. *Chromosoma*, 16, 90-100.

OJIMA, Y., TAKAYAMA, S. and KATO, F. (1965). A study of the sex chromosomes in the mouse. *Japan J. Genet.*, 40, 319-324.

OMURA, T. (1968). Temporary centric connection of chromosomes in normal mice. *Proc. XII Inter. Gongr. Genet. Tokyo*, 1, 162.

OWEN, A. R. G. (1949). The theory of genetical recombination. I. Long chromosome arms. *Proc. Roy. Soc.* B, 136, 67-94.

OWEN, A. R. G. (1950). The theory of genetical recombination. *Advanc. Genet.*, 3, 117-157.

OWEN, A. R. G. (1953a). Super-recombination in the sex chromosome of the mouse. *Heredity*, 7, 103-110.

OWEN, A. R. G. (1953b). The analysis of multiple linkage data. *Heredity*, 7, 247-264.

PAIGEN, K. and NOELL, W. K. (1961). Two linked genes showing a similar timing of expression in mice. *Nature*, 190, 148-150.

PAINTER, T. S. (1927). The chromosome constitution of Gate's 'non-disjunction' mice. *Genetics*, 12, 379-392.

PAINTER, T. S. (1928a). A comparison of the chromosomes of the rat and mouse. *Genetics*, 13, 180-189.

PAINTER, T. S. (1928b). The chromosome constitution of the Little and Bagg abnormal eyed mice. *Amer. Nat.*, 62, 284-286.

PARSONS, P. A. (1958a). A balanced four-point linkage experiment for linkage group XIII of the house mouse. *Heredity*, 12, 77-95.

PARSONS, P. A. (1958b). Additional three-point data for linkage group V of the mouse. *Heredity*, 12, 357-362.

PARSONS, P. A. (1958c). On the phenomenon of genetic interference. *Proc. X Inter. Congr. Genet.*, 2, 213.

PARSONS, P. A. (1959). Possible affinity between linkage groups V and XIII of the house mouse. *Genetica*, 29, 304-311.

PASSMORE, H. C. and SHREFFLER, D. C. (1968). A sex-limited serum protein variant associated with the *H-2* region of the mouse. *Genetics*, 60, 210-211.

PASSMORE, H. C. and SHREFFLER, D. C. (1970). A sex-limited serum protein variant in the mouse: inheritance and association in the H-2 region. *Biochem. Genet.*, 4, 351-365.

PEARSON, P. L. and Bobrow, M. (1970). Definitive evidence for the short arm of the *Y* chromosome associating with the *X* chromosome during meiosis in the human male. *Nature*, 226, 959-961.

PELT, A. F. VAN, CAIN, K. and KNORR, R. (1969). A new hairless mutant in the house mouse. *J. Hered.*, 60, 75, 96.

PELZER, C. F. (1965). Genetic control of erythrocyte esterase forms in *Mus musculus. Genetics*, 52, 819-828.

PETRAS, M. L. and BIDDLE, F. G. (1967). Serum esterases in the house mouse. *Canad. J. Genet. Cytol.*, 9, 704-710.

PETRAS, M. L. and SINCLAIR, P. (1969). Another esterase variant in the kidney of the house mouse. *Canad. J. Genet. Cytol.*, 11, 97-102.

PHILLIPS, R. J. S. (1954). Jimpy, a new totally sex-linked gene in the house mouse. *Z. Ind. Abst. Vererbgslehre.*, 86, 322-326.

PHILLIPS, R. J. S. (1956). The linkages of congenital hydrocephalus in the house mouse. *J. Hered.*, 47, 302-304.

PHILLIPS, R. J. S. (1960). Lurcher, a new gene in linkage group XI of the house mouse. *J. Genet.*, 57, 35-42.

PHILLIPS, R. J. S. (1961). Dappled, a new allele at the mottled locus in the house mouse. *Genet. Res.*, 2, 290-295.

PHILLIPS, R. J. S. (1963). Striated, a new sex-linked gene in the house mouse. *Genet. Res.*, 4, 151-153.

PHILLIPS, R. J. S. (1966). A cis-trans position effect at the *A* locus of the house mouse. *Genetics*, 54, 485-495.

PIERRO, L. J. and CHASE, H. B. (1963). Slate, a new coat colour mutant in the mouse. *J. Hered.*, 54, 46-50.

PINCUS, G. (1929). A spontaneous mutation in the house mouse. *Proc. Nat. Acad. Sci. U.S.A.*, 15, 85-88.

PIZARRO, O. and DUNN, L. C. (1970). A study of recombination between the *H-2* locus and loci closely linked with it in the house mouse. *Transplantation*, 9, 207-218.

PIZARRO, O., HOECKER, G., RUBINSTEIN, P. and RAMOS, A. (1961). The distribution in the tissues and the development of H-2 antigens of the mouse. *Proc. Nat. Acad. Sci. U.S.A.*, 47, 1900-1907.

PLATE, L. (1910). Die Erbformeln der Farbenrassen von *Mus musculus. Zool. Anz.*, 35, 634-640.

PLATZ, R. D. and WOLFE, H. G. (1969). Mouse seminal vesicle proteins. *J. Hered.*, 60, 187-192.

POPP, D. M. (1967a). A description of *rho*: a non-H-2 isoantigen in RFM mice. *Transplantation*, 5, 290-299.

POPP, D. M. (1967b). A review of non-H-2 isoantigens in mice. *Transplantation*, 5, 300-309.

POPP, D. M. (1969). Histocompatibility - 14: correlation of the isoantigen *rho* and the *R - Z* locus. *Transplantation*, 7, 233-241.

POPP, R. A. (1962a). Studies on the mouse hemoglobin loci. II. Position of the hemoglobin locus with respect to albinism and shaker-1 loci. *J. Hered.*, 53, 73-75.

POPP, R. A. (1962b). Studies on the mouse hemoglobin loci. III. Heterogeneity of electrophoretically indistinguishable single type hemoglobin. *J. Hered.*, 53, 75-77.

POPP, R. A. (1962c). Studies on the mouse hemoglobin loci. IV. Independent segretation of *Hb* and *SOL*. *J. Hered.*, 53, 77-80.

POPP, R. A. (1962d). Studies on the mouse hemoglobin loci. VI. A third allele, SOL^3, at the *SOL* locus. *J. Hered.*, 53, 147-148.

POPP, R. A. (1965). Loci linkage of serum esterase patterns and oligosyndactylism. *J. Hered.*, 56, 107-108.

POPP, R. A. (1967). Linkage of *Es-1* and *Es-2* in the mouse. *J. Hered.*, 58, 186-188.

POPP, R. A. (1969a). Studies on the mouse hemoglobin loci. VIII. A fourth α-chain phenotype. *J. Hered.*, 60, 126-128.

POPP, R. A. (1969b). Studies on the mouse hemoglobin loci. IX. A fifth α-chain phenotype. *J. Hered.*, 60, 128-131.

POPP, R. A. (1969c). Studies on the mouse hemoglobin loci. X. Linkage of duplicate genes at the α-chain locus *Hba. J. Hered.*, 60, 131-133.

POPP, R. A. and AMAND, W. S. (1960). Studies on the mouse hemoglobin locus. I. Identification of hemoglobin types and linkage of hemoglobin with albinism. *J. Hered.*, 51, 141-144.

POPP, R. A. and AMAND, W. S. (1964). A sex difference in recombination frequency in the albinism-hemoglobin interval of linkage group I in the mouse. *J. Hered.*, 55, 101-103.

POPP, R. A. and HILSE, K. (1968). Alpha chains of hemoglobins among laboratory mice. *Genetics*, 60, 212.

POTTER, M. and LIEBERMAN, R. (1967). Genetics of immunoglobulins in the mouse. *Advanc. Immunol.*, 7, 92-145.

PREHN, R. T. and MAIN, J. M. (1956). The influence of sex on isologous skin grafting in the mouse. *J. Nat. Cancer Inst.*, 17, 35-36.

REED, S. C. (1937). The inheritance and expression of fused, a new mutation in the house mouse. *Genetics*, 22, 1-13.

REID, D. H. and PARSONS, P. A. (1963). Sex of parent and variation of recombination with age in the mouse. *Heredity*, 18, 107-108.

ROBERTS, E. and QUISENBERG, J. H. (1935). Linkage of the genes for non-yellow and pink-eyed-2 in the house mouse. *Amer. Nat.*, 69, 181-183.

ROBERTS, R. C. (1967). Small eyes, a new dominant eye mutant in the mouse. *Genet. Res.*, 9, 121-122.

ROBINS, M. W. (1959). A mutation causing congenital clubfoot in the house mouse. *J. Hered.*, 50, 188-192.

ROBINSON, R. (1959). Sable and umbrous mice. *Genetica*, 29, 319-326.

RODERICK, T. H., HUTTON, J. J. and RUDDLE, F. H. (1970). Linkage of esterase-3 (*Es-3*) and rex (*Re*) in linkage group VII of the mouse. *J. Hered.*, 61, 278-279.

RUDDLE, F. H. and RODERICK, T. H. (1965a). The genetic control of three kidney esterases in C57BL/6J and RF/J mice. *Genetics*, 51, 445-454.

RUDDLE, F. H. and RODERICK, T. H. (1965b). Allelically determined isozyme polymorphisms in laboratory populations of mice. *Annals N.Y. Acad. Sci.*, 151, 531-539.

RUDDLE, F. H. and RODERICK, T. H. (1966). The genetic control of two types of esterases in inbred strains of the mouse. *Genetics*, 54, 191-202.

RUDDLE, F. H., SHOWS, T. B. and RODERICK, T. H. (1968). Autosomal control of an electrophoretic variant of glucose-6-phosphate dehydrogenase in the mouse. *Genetics*, 58, 599-606.

RUDDLE, F. H., SHOWS, T. B. and RODERICK, T. H. (1969). Esterase genetics in *Mus musculus:* expression, linkage and polyporphism of locus Es-2. *Genetics*, 62, 393-399.

RUNNER, M. N. (1959). Linkage of brachypodism. *J. Hered.*, 50, 81-84.

RUSSELL, E. S. and McFARLAND, E. C. (1966). Analysis of pleiotropic effects of *W* and *f* genic substitutions in the mouse. *Genetics*, 53, 949-959.

RUSSELL, E. S., NASH, D. J., BERNSTEIN, S. E., KENT, E. L., MACFARLAND, E. C., MATTHEWS, S. M. and NORWOOD, M. S. (1970). Characterization and genetic studies of microcytic anemia in house mouse. *Blood*, 35, 838-850.

RUSSELL, L. B. (1961). Genetics of mammalian sex chromosomes. *Science*, 133, 1795-1803.

RUSSELL, L. B. (1962). Chromosome aberrations in experimental mammals. *Progress Med. Genetics*, 2, 230-294.

RUSSELL, L. B. (1963). Mammalian *X* chromosome action: inactivation limited in spread and in region of origin. *Science*, 31, 976-978.

RUSSELL, L. B. (1967). Complementation mapping of a small region of linkage group II in the mouse. *Genetics*, 56, 585.

RUSSELL, L. B. and BANGHAM, J. W. (1961). Variegated type position effects in the mouse. *Genetics*, 46, 509-525.

RUSSELL, L. B., BANGHAM, J. W. and SAYLORS, C. L. (1962). Delimitation of chromosomal regions involved in V-type position effects from *X*-autosome translocations in the mouse. *Genetics*, 47, 981-982.

RUSSELL, L. B. and CHU, E. H. Y. (1961). An *XXY* male in the mouse. *Proc. Nat. Acad. Sci. U.S.A.*, 47, 571-575.

RUSSELL, L. B., McDANIEL, M. N. C. and WOODIEL, F. N. (1963). Crossingover within the *a* locus of the mouse. *Genetics*, 48, 907.

RUSSELL, L. B. and MONTGOMERY, C. S. (1965). The use of *X*-autosome translocations in locating the *X* chromosome inactivation center. *Genetics*, 52, 470-471.

RUSSELL, L. B. and MONTGOMERY, C. S. (1969). Comparative studies on X-autosome translocations in the mouse. I. Origin viability, fertility and weight of five $(X,1)$'s. *Genetics*, 63, 103-120.

RUSSELL, L. B. and MONTGOMERY, C. S. (1970). Comparative studies on X-autosome translocations in the mouse. II. Inactivation of autosomal loci, segregation and mapping of autosomal breakpoints in five T(X;1)'s. *Genetics*, 64, 281-312.

RUSSELL, L. B. and RUSSELL, W. L. (1960). Genetic analysis of induced deletions and of spontaneous nondisjunction involving chromosome 2 of the mouse. *J. Cell. Comp. Physiol.*, 56 (Suppl. 1), 169-188.

RUSSELL, L. B. and SAYLORS, C. L. (1962). Induction of parternal sex chromosome losses by irradiation of mouse spermatozoa. *Genetics*, 47, 7-10.

RUSSELL, L. B., STEELE, M. S. and THOMPSON, H. M. (1966). Deficiencies in the mouse. *U.S. Atom. Energy Com. Rep., ORNL*, 3999, 90-92.

RUSSELL, W. L., RUSSELL, L. B. and GOWER, J. S. (1959). Exceptional inheritance of a sex-linked gene in the mouse explained on the basis that the XO sex chromosome constitution is female. *Proc. Nat. Acad. Sci. U.S.A.*, 45, 554-560.

SACHS, L. (1955). The possibilities of crossingover between the sex chromosomes of the house mouse. *Genetica*, 27, 309-322.

SACHS, L, and HELLER, E. (1958). The sex-linked histocompatibility antigens. *J. Nat. Cancer Inst.*, 20, 555-561.

SARVELLA, P. A. and RUSSELL, L. B. (1956). Steel, a new dominant gene in the house mouse with effects on coat pigment and blood. *J. Hered.*, 47, 123-218.

SCHAIBLE, R. and GOWEN, J. W. (1961). A new dwarf mouse. *Genetics*, 46, 896.

SCHLESINGER, K., ELSTON, R. C. and BOGGAN, W. (1966). The genetics of sound induced seizure in inbred mice. *Genetics*, 54, 95-103.

SEARLE, A. G. (1961). Tipsy, a new mutant in linkage group VII of the mouse. *Genet. Res.*, 2, 122-216.

SEARLE, A. G. (1964). The genetics and morphology of two luxoid mutants in the house mouse. *Genet. Res.*, 5, 171-197.

SEARLE, A. G. (1966). Curtailed, a new dominant T-allele in the house mouse. *Genet. Res.*, 7, 86-95.

SEARLE, A. G., BERRY, R. J. and BEECHEY, C. V. (1970). Cytogenetic radio-sensitivity and chiasma frequency in wild living male mice. *Mutation Res.*, 9, 137-140.

SEARLE, A. G. and TRUSLOVE, G. M. (1970). A gene triplet in the mouse. *Genet. Res.*, 15, 227-235.

SHORT, B. F. and SOBEY, W. R. (1957). The effect of sex on skin grafts within inbred lines of mice. *Transpl. Bull.*, 4, 110-112.

SHOWS, T. B. and RUDDLE, F. H. (1968). Function of the lactate dehydrogenase B gene in mouse erythrocytes: evidence for control by a regulatory gene. *Proc. Nat. Acad. Sci. U.S.A.*, 61, 574-581.

SHOWS, T. B., RUDDLE, F. H. and RODERICK, T. H. (1969).

Phosphoglucomutase electrophoretic variants in the mouse. *Biochem. Genet.*, **3**, 25-35.

SHREFFLER, D. C. (1960), Genetic control of serum transferrion type in mice. *Proc. Nat. Acad. Sci. U.S.A.*, **46**, 1378-1384.

SHREFFLER, D. C. (1963). Linkage of the mouse transferrion locus. *J. Hered.*, **54**, 127-129.

SHREFFLER, D. C. (1964a). Inheritance of a serum pre-albumin variant in the mouse. *Genetics*, **49**, 629-634.

SHREFFLER, D. C. (1964b). A seriologically detected variant in mouse serum: further evidence for genetic control by the histocompatibility-2 locus. *Genetics*, **49**, 973-978.

SHREFFLER, D. C. (1964c). A further subdivision of the complex *H-2* region of the mouse. *Genetics*, **50**, 285.

SHREFFLER, D. C. (1965). The *Ss* system of the mouse. *Wistar Inst. Symp. Monogr.*, **3**, 11-19.

SHREFFLER, D. C. (1966). A new erythrocytic antigen in the house mouse. *Genetics*, **54**, 362.

SHREFFLER, D. C. (1967). Genetic control of cellular antigens. *Proc. III Inter. Congr. Human Genet.*, 217-231.

SHREFFLER, D. C., AMOS, D. B. and MARK, R. (1966). Serological analysis of a recombination in the *H-2* region of the mouse. *Transplantations*, **4**, 300-322.

SHREFFLER, D. C. and OWEN, R. D. (1963). A serologically detected variant in mouse serum: inheritance and association with the histocompatibility-2 locus. *Genetics*, **48**, 9-25.

SHREFFLER, D. C. and SNELL, G. D. (1969). The distribution of thirteen *H-2* allo-antigenic specificities among the products of eighteen *H-2* alleles. *Transplantation*, **8**, 435-450.

SICK, K. and NIELSEN, J. T. (1964). Genetics of amylase isozymes in the mouse. *Hereditas*, **51**, 291-296.

SIDMAN, R. L. and GREEN, M. C. (1965). Retinal degeneration in the mouse. Location of the *rd* locus in linkage group XVII. *J. Hered.*, **56**, 23-29.

SIDMAN, R. L., GREEN, M. C. and APPEL, S. H. (1965). *Catalog of the neurological mutants of the mouse.* Cambridge, Mass.: Harvard University Press.

SIDMAN, R. L., LANE, P. W. and DICKIE, M. M. (1962). Staggerer, a new mutation in the mouse affecting the cerebellum. *Science*, **137**, 610-612.

SILAGI, S. (1962). A genetical and embryological study of partial complementation between lethal alleles at the *T* locus of the house mouse. *Devel. Biol.*, **5**, 35-67.

SILVERS, W. K. and BILLINGHAM, R. E. (1967). Genetic background and expressivity of histocompatibility genes. *Science*, **158**, 118-119.

SILVERS, W. K., BILLINGHAM, R. E. and SANFORD, B. H. (1968). The *H-Y* transplantation antigen: a *Y* linked or sex influenced factor? *Nature*, **220**, 401-403.

SIRLIN, J. L. (1956). Vacillans, a neurological mutant in the house mouse linked with brown. *J. Genet.*, **54**, 42-48.

SIRLIN, J. L. (1957). Location of vacillans in linkage group VIII of the house mouse. *Heredity*, 11, 259-260.

SLIZYNSKI, B. M. (1949). A preliminary pachytene chromosome map of the house mouse. *J. Genet.*, 49, 242-245.

SLIZYNSKI, B. M. (1952). Pachytene analysis of Snell's T(5:8)*a* translocation in the mouse. *J. Genet.*, 50, 507-510.

SLIZYNSKI, B. M. (1954). Partial sex-linkage in the mouse. *Nature*, 174, 310.

SLIZYNSKI, B. M. (1955a). Chiasmata in the male mouse. *J. Genet.*, 53, 597-605.

SLIZYNSKI, B. M. (1955b). The sex bivalent of *Mus musculus. J. Genet.*, 53, 591-596.

SLIZYNSKI, B. M. (1957b). Chromosomal mechanism in translocation infertility. *Proc. Roy. Phys. Soc. Edin.*, 26, 49-60.

SLIZYNSKI, B. M. (1957a). Cytological analysis of translocations in the mouse. *J. Genet.*, 55, 122-130.

SLIZYNSKI, B. M. (1958). Chiasmata in female mice. *Proc. X Inter. Congr. Genet.*, 2, 264.

SLIZYNSKI, B. M. (1960). Sexual dimorphism in mouse gametogenesis. *Genet. Res.*, 1, 477-486.

SLIZYNSKI, B. M. (1964). Cytology of the *XXY* mouse. *Genet. Res.*, 5, 328-329.

SLIZYNSKI, B. M. (1967). Oocyte pachytene analysis of Cattanach's *fd* translocation. *Genet. Res.*, 9, 17-22.

SNELL, G. D. (1928). A crossover between the genes for short-ear and density in the house mouse. *Proc. Nat. Acad. Sci. U.S.A.*, 14, 926-928.

SNELL, G. D. (1931). Inheritance in the house mouse; the linkage relations of short-ear, hairless and naked. *Genetics*, 16, 42-74.

SNELL, G. D. (1933). Genetic changes in mice induced by X-rays. *Amer. Nat.*, 67, 24-31.

SNELL, G. C. (1935). The induction by X-rays of hereditary changes in mice. *Genetics*, 20, 545-567.

SNELL, G. D. (1941). Gene and chromosome mutations. In Snell, G. D., (Editor). *Biology of the laboratory mouse.* New York: McGraw-Hill Book Co.

SNELL, G. D. (1945). Linkage of jittery and waltzing in the mouse. *J. Hered.*, 36, 279-280.

SNELL, G. D. (1946). An analysis of translocations in the mouse. *Genetics*, 31, 157-180.

SNELL, G. D. (1952). Preliminary data on crossingover between *H-2* and *Fu*, *Ki* and *T* in the house mouse. *Heredity*, 6, 247-254.

SNELL, G. D. (1955). Ducky, a new second chromosome mutation in the mouse. *J. Hered.*, 46, 27-29.

SNELL, G. D. (1956). A comment on Eichwald and Silmer's communication. *Transpl. Bull.*, 3, 29-31.

SNELL, G. D. (1958). Histocompatibility genes of the mouse. II. Production and analysis of isogenic resitant lines. *J. Nat. Cancer Inst.*, 21, 843-877.

SNELL, G. D. and AMES, F. B. (1939). Hereditary changes in the descendants of female mice exposed to Roentgen rays. *Amer. J. Roent. Rad. Therap.*, 41, 248-255.

SNELL, G. D., BODEMANN, E. and HOLLANDER, W. (1934). A translocation in the house mouse and its effects on development. *J. Exp. Zool.*, 67, 93-104.

SNELL, G. D. and BORGES, P. R. F. (1953). Determination of the histocompatibility locus involved in the resistance of mice of strains C57BL/10-x, C57BL/6-x and C57BL/6Ks to C57BL tumors. *J. Nat. Cancer Inst.*, 14, 481-484.

SNELL, G. D. and BUNKER, H. P. (1964). Histocompatibility genes of mice. IV. The position of H-3 in the fifth linkage group. *Transplantation*, 2, 743-751.

SNELL, G. D., GUDKOWICZ, G. and BUNKER, H. P. (1967). Histocompatibility genes of mice. VIII. *H-13*, a new histocompatibility locus in the fifth linkage group. *Transplantation*, 5, 492-503.

SNELL, G. D., DICKIE, M. M., SMITH, P. and KELTON, D. E. (1954). Linkage of loop-tail, leaden, splotch and fuzzy in the mouse. *Heredity*, 8, 271-273.

SNELL, G. D. and HIGGINS, G. F. (1951). Alleles at the histocompatibility-2 locus in the mouse as determined by tumor transplantation. *Genetics*, 36, 306-310.

SNELL, G. D., HOECKER, G., AMOS, D. B. and STIMPFLING, J. H. (1964). A revised nomenclature for the histocompatibility-2 locus of the mouse. *Transplantation*, 2, 777-784.

SNELL, G. D., HOECKER, G. and STIMPFLING, J. H. (1967). Evidence that the 'R' and 'Z' blood group specificities of mice are allelic and distinct from H-2. *Transplantation*, 5, 481-491.

SNELL, G. D. and LAW, L. W. (1939). Linkage between shaker-2 and wavy-2 in the house mouse. *J. Hered.*, 30, 447.

SNELL, G. D., SMITH, P. and GABRIELSON, F. (1953). Analysis of the histocompatibility-2 locus in the mouse. *J. Nat. Cancer Inst.*, 14, 457-480.

SNELL, G. D. and STEVENS, L. C. (1961). Histocompatibility genes of mice. III. *H-1* and *H-4*, two histocompatibility loci in the first linkage group. *Immunology*, 4, 366-379.

SOLARI, A. J. (1970). The spatial relationship of the X and Y chromosomes during meiotic prophase in mouse spermatocytes. *Chromosoma*, 29, 217-238.

SORSBY, A., KOLLER, P. C., ATTFIELD, M., DAVEY, J. B. and LUCAS, D. R. (1954). Retinal dystrophy in the mouse: histological and genetic aspects. *J. Exp. Zool.*, 125, 171-197.

STEINBERG, A. G. and FRASER, F. C. (1943). Marcelled, a new recessive mutation affecting the coat of the house mouse. *Genetics*, 28, 92.

STEVENS, L. C. and BUNKER, M. C. (1964). Karyotype and sex of primary testicular teratomes in mice. *J. Nat. Cancer Inst.*, 33, 65-78.

STICH, J. H. and HSU, T. C. (1960). Cytological identification of male and female somatic cells in the mouse. *Exp. Cell Res.*, 20, 248-249.

STIMPFLING, J. H. and RICHARDSON, A. (1965). Recombination within the histocompatibility-2 locus in the mouse. *Genetics*, 51, 831-846.

STIMPFLING, J. H. and SNELL, G. D. (1968). Detection of a non-H-2 blood group system with the aid of B10.129(5M) mice. *Transplantation*, 6, 468-475.

STRONG, L. C. (1946). Linkage and crossingover between black pigmentation and susceptibility to induced fibrosarcoma in mice. *Science*, 103, 554.

TATCHELL, J. A. H. (1961a). Pulmonary tumours, group VII and sex in the house mouse. *Nature*, 190, 837-838.

TATCHELL, J. A. H. (1961b). Pulmonary tumours and group VII in the house mouse. *Nature*, 190, 1125.

TSUJI, S. and MEIER, H. (1969). Linkage of serum esterase and tottering in the mouse. *J. Hered.*, 60, 221-222.

TUTIKAWA, K. (1955). Test for allelism of alopecia periodica and furless in the house mouse. *Ann. Rep. Nat. Inst. Genet., Japan*, 5, 16.

WACHTER, W. L. (1921). Data concerning linkage in mice. *Amer. Nat.*, 55, 412-420.

WACHTER, W. L. (1927). Linkage studies in mice. *Genetics*, 12, 108-114.

WALLACE, M. E. (1950). Locus of the fidget gene in the house mouse. *Nature*, 166, 407.

WALLACE, M. E. (1953). Affinity: a new genetic phenomenon in the house mouse. Evidence from within laboratory stocks. *Nature*, 171, 27-28.

WALLACE, M. E. (1954). A mutation or a crossover in the house mouse. *Heredity*, 8, 89-105.

WALLACE, M. E. (1957a). A balanced three-point experiment for linkage group V of the house mouse. *Heredity*, 11, 223-258.

WALLACE, M. E. (1957b). The use of affinity in chromosome mapping. *Biometrics*, 13, 98-110.

WALLACE, M. E. (1958a). New linkage and independence data for ruby and jerker in the mouse. *Heredity*, 12, 453-462.

WALLACE, M. E. (1958b). Experimental evidence for a new genetic phenomenon. *Phil. Trans. Roy. Soc.*, 241, 211-254.

WALLACE, M. E. (1959). An experimental test of the hypothesis of affinity. *Genetica*, 29, 243-255.

WALLACE, M. E. (1961). Affinity: evidence from crossing inbred lines of mice. *Heredity*, 16, 1-23.

WALLACE, M. E. (1965). Pseudoallelism at the agouti locus in the mouse. *J. Hered.*, 56, 267-271.

WALLACE, M. E. and HERBERTSON, B. M. (1969). Neonatal intestinal lipidosis in mice. *J. Med. Genet.*, 6, 361-375.

WATSON, M. L. and OLIVER, J. D. (1965). Chromosomes of the laboratory mouse. *Proc. Iowa Acad. Sci.*, 72, 537-538.

WELSHONS, W. J. and RUSSELL, L. B. (1959). The Y chromosome as the bearer of male determining factors in the mouse. *Proc. Nat. Acad.*

Sci. U.S.A., **45**, 560-566.

WHITE, B. J. and TJIO, J. H. (1967). A mouse translocation with 38 and 39 chromosomes but normal N.F. *Hereditas*, **58**, 284-296.

WOLFE, H. G. (1967). Mapping the hemoglobin locus in mice transmitting the flecked translocation. *Genetics*, **55**, 213-218.

WOLFE, H. G., RUSSELL, E. S. and PACKER, S. O. (1963). Hemoglobins in mice. Segregation of genetic factors affecting electrophoretic mobility and solubility. *J. Hered.*, **54**, 107-112.

WOOLLEY, G. W. (1945). Misty dilution in the mouse. *J. Hered.*, **36**, 269-270.

WORTIS, H. H. (1965). A gene locus concerned with an antigenic serum substance in *Mus musculus. Genetics*, **52**, 267-273.

WRIGHT, M. E. (1947a). Undulated: a new genetic factor in *Mus musculus* affecting the spine and tail. *Heredity*, **1**, 137-141.

WRIGHT, M. E. (1947b). Two sex-linkages in the house mouse with unusual recombination values. *Heredity*, **1**, 349-354.

YOON, C. H. (1959). Waddler, a new mutation, and its interaction with quivering. *J. Hered.*, **50**, 238-244.

YOON, C. H. (1961). Linkage relationship of the waddler gene in mice with evidence for temperature effect in crossingover. *J. Hered.*, **52**, 279-281.

YOON, C. H. (1969). Disturbances in developmental pathways leading to a neurological disorder of genetic origin, leaner, in mice. *Develop. Biol.*, **20**, 158-181.

YOON, C. H. and LES, E. P. (1957). Quivering, a new first chromosome mutation in mice. *J. Hered.*, **48**, 176-180.

YOSIDA, T. H. (1960a). Study on the new mutant 'falter' found in the house mouse. *Ann. Rep. Nat. Inst. Genet. Japan*, **10**, 30-31.

YOSIDA, T. H. (1960b). Genetical study on the new mutant 'falter' found in the house mouse. *Bull. Exp. Anim.*, **9**, 179-182.

ZAALBERG, O. B. (1959). An analysis of the Eichwald-Silmser effect. *Transpl. Bull.*, **6**, 433-435.

ZELENY, V. (1967). Morphological identification of the murine *Y* chromosome. *Folia Biol. (Praha)*, **13**, 158-159.

APPENDIX

An unusually large amount of unpublished data exists for the mouse. Much of the data is of a negative nature and the mouse has passed the stage where these would be regarded as of great importance. The exception, of course, is where a new mutant has failed to show linkage with any member of the known groups and is currently regarded as a candidate for the unknown group XIX. Only a few genes have been sufficiently tested at this time to warrant this distinction.

However, a fair amount of positive linkage has not reached the stage of formal publication. The reason for this is the existence of *Mouse News Letter*, an informal magazine circulating among mammalian geneticists. Among the news items carried, are brief notifications of new mutants and linkages. In keeping with its newsletter status and somewhat restricted circulation, reports appearing in *Mouse News Letter* cannot be cited without permission of the authors. Ordinarily, this is a small matter, because much of the research is eventually reported elsewhere. However, over the years, a fair number of items has not attained formal publication. In addition, a large amount of work is too recent to have seen publication other than in *Mouse News Letter*.

It became obvious that any review with a claim to comprehensiveness could not afford to ignore this material. Accordingly, permission was sought to include details of results from *Mouse News Letter*. A circulated appeal brought forth an enthusiastic response. In many cases, moreover, the data were re-evaluated and up-dated while, in others, new data were submitted. My most cordial thanks are extended to those people listed in the following tabulation and who have given so generously of their time and research material. The unpublished data of M. M. Dickie are due to the kindness of J. L. Southard.

TABLE 11.19. Summary of informally published and personally
communicated crossover values in the house mouse.

Group	Loci	Sex	Mating type	Crossover value	Reference
I	I c - da	-	CII	23±3.9	D. S. Falconer, Per. Com. 1970
	c - ex	?	?	37±5	J. G. M. Shire. Per. Com. 1970
	c - Nil	♂	?BB	3.7±1.4	Wallace and Herbertson (1969), M. E. Wallace, Per. Com. 1970
		♀	?BB	8.62±1.7	Wallace and Herbertson (1969), M. E. Wallace, Per. Com. 1970
	c - p	♂	?	14.1±1.3	W. St. Amand, Per Com. 1970
		♀	?	16.3±1.5	W. St. Amand, Per. Com. 1970
	c - pu	♂	?	30.4±1.8	W. St. Amand, Per Com. 1970
		♀	?	32.2±1.8	W. St. Amand, Per. Com. 1970
	c - tp	♀	RBB	2.1±0.5	L. B. Russell (1963) MNL, **29**, 73
	da - p	-	RII	15.4±7.5	D. S. Falconer, Per. Com. 1970
	Nil - p	♂	?BB	3.2±1.3	Wallace and Herbertson, (1970), M. E. Wallace, Per. Com. 1970
		♀	?BB	2.5±1.3	Wallace and Herbertson (1970), M. E. Wallace, Per. Com. 1970
	nv - p	?	?	34.8±2.8	J. H. F. van Abeelen, P. H. W. van der Kroon and A. J. M. Buis, Per. Com. 1970
	p - pu	♂	?	16.3±1.4	W. St. Amand, Per. Com. 1970
		♀	?	16.0±1.4	,, ,, ,, ,, ,, ,,
	p - ru-2	♀	RBB	3.7±1.2	F. Lilly, Per. Com. 1970
		-	RII	2.2±2.1	,, ,, ,, ,,
II	d - Fv-2	?	-BB	16.0±7.3	,, ,, ,, ,,
		-	RII	15.5±4.8	,, ,, ,, ,,
	d - wy	-	CII	25.3±1.9	R. J. Burns and J. A. Burns, Per. Com. 1970
	du - tk	-	RII	23.8±4.7	M. M. Dickie (1969) MNL, **41**, 31

Group	Loci	Sex	Mating type	Crossover value	References
III	Ds - hr	?	RBB	5.6±2.7	K. P. Hummel and D. B. Chapman (1966) MNL, **34**, 31
	Fkl - hr	?	CBB	36.1±4.4	M. S. Lyon, Per. Com. 1970
	Fkl - s	♂	CBB	23.3±3.9	,, ,, ,, ,, ,,
			CBB	9.1±3.1	,, ,, ,, ,, ,,
	hr - s	♀	CBB	18.3±1.2	P. W. Lane, Per. Com. 1970
IV	av - dl	?	CBB	11.9±1.2	,, ,, ,, ,, ,,
	av - dy	?	C?	28.2±2.1	E. S. Russell and J. Southard (1966) MNL, **35**, 32
		?	R?	23.4±3.3	E. S. Russell and J. Southard (1966) MNL, **35**, 32
	av - eb	-	RII	28.8±3.1	K. P. Hummel and D. B. Chapman (1966) MNL, **34**, 31
	av - gl	-	RII	35.4±5.5	P. W. Lane, Per. Com. 1970
	av - si	♂	?BB	35.0±3.8	R. H. Schaible (1961) MNL, **24**, 38
		♀	?BB	32.1±3.1	R. H. Schaible (1961) MNL, **24**, 38
		?	CBB	30.6±4.7	R. H. Schaible, Per. Com. 1970
	av - Sl	?	?BB	14.9±1.6	P. W. Lane, Per. Com. 1970
		?	?BB	12.4±2.5	R. H. Schaible (1963) MNL, **28**, 39, Per. Com. 1970
	dl - gl	?	RII	6.5±2.6	P. W. Lane, Per. Com. 1970
	dl - Sl	?	CBB	22.8±1.0	,, ,, ,, ,, ,,
	dy - Sl	♂	CBB	31.9±2.2	H. G. Wolfe (1969) MNL, **29**, 40
	eb - Sl	?	CBB	20.2±2.0	K. P. Hummel and D. B. Chapman (1966) MNL, **34**, 31
	gl - Sl	?	CM	30.0±4.6	P. W. Lane, Per. Com. 1970
	jc - Sl	?	?	28.7±3.1	J. L. Southard, Per. Com. 1970
	pg - si	?	RII	20.1±5.9	D. S. Falconer and J. H. Isaacson (1965) MNL, **32**, 30
	py - Sl	?	CBB	14.2±4.0	D. S. Falconer and J. H. Isaacson (1965) MNL, **32**, 30
	si - Sl	?	CBB	18.2±4.4	M. C. Green, Per. Com. 1970
		?	CBB	20.4±4.1	R. H. Schaible, Per. Com. 1970
V	a - Cm	?	CBB	12.9±2.4	H. D. Bunker (1967) MNL, **37**, 34

Group	Loci	Sex	Mating type	Crossover value	References
	a - dm	?	CBB	3.7 ± 1.9	M. C. Green, Per. Com. 1970
		-	CII	3.3 ± 1.3	,, ,, ,, ,,
		?	?	10.0 ± ?	J. M. Mackensen (1962) MNL, **27**, 38
	a - H-6	♀	?BB	26.2 ± 2.0	F. Lilly, Per. Com. 1970
	a - kr	-	CII	0.4 ± 0.6	P. W. Lane (1959) MNL, **21,** 47
	a - ld	?	CBB	28.0 ± 5.2	M. C. Green, Per. Com. 1970
	a - ls	?	?	25.7 ± ?	R. J. S. Phillips (1966) MNL, **34**, 27
	a - lst	?	CBB	26.8 ± 5.4	P. F. Forsthoefel, Per. Com. 1970
		?	RBB	15.1 ± 6.9	P. F. Forsthoefel, Per. Com. 1970
	a - mg	?	CBB	13.0 ± 2.4	M. C. Green, Per. Com. 1970
	a - pa		CBB	18.6 ± 1.8	F. Lilly, Per. Com. 1970
	a - rh	?	?	35.0 ± ?	D. Varnum and L. C. Stevens (1970) MNL, **43**, 34
	a - Sut	♂	RBB	6.3 ± 1.7	I. K. Egorov and Z. K. Blandova (1968) MNL, **39,** 41, Per. Com. 1970
		♀	RBB	3.1 ± 1.4	I. K. Egorov and Z. K. Blandova (1968) MNL, **39,** 41, Per. Com. 1970
	dm - mg	?	CBB	6.5 ± 2.3	M. C. Green, Per. Com. 1970
		-	CII	6.2 ± 1.9	,, ,, ,, ,,
	dm - Ra	?	?	29.0 ± ?	J. M. Mackensen (1962) MNL, **27**, 38
	H-6 - pa	♂	-BB	8.0 ± 1.2	J. M. Mackensen (1962) MNL, **27**, 38
	ld - lst	?	?	Closely linked	P. F. Forsthoefel, Per. Com. 1970
	ld - mg	?	CBB	4.0 ± 2.3	M. C. Green, Per. Com. 1970
	ls - Ra	?	?	3.1 ± ?	R. J. S. Phillips (1965) MNL, **34,** 27
	lst - Ra	?	CBB	39.5 ± 5.7	P. F. Forsthoefel, Per. Com. 1970
VI	bt - Ca	♂	?BB	11.1 ± 0·7	S. A. Mallyon and M. E. Wallace, Per. Com. 1970
		♀	?BB	3.8 ± 0·4	S. A. Mallyon and M. E. Wallace, Per. Com. 1970
		♂	?	12.7 ± 0·9	W. St. Amand, Per. Com. 1970
		♀	?	7.1 ± 0.7	,, ,, ,,
	bt - hl	♂	CBB	9.3 ± 0.9	W. F. Hollander (1959) MNL, **20,** 34, Per. Com. 1970

Group	Loci	Sex	Mating type	Crossover value	References
		♂	RBB	7.3±1.1	W. F. Hollander (1959) *MNL*, **20**, 34, Per. Com. 1970
		♀	CBB	7.1±0.6	W. F. Hollander (1959) *MNL*, **20**, 34, Per. Com. 1970
		♀	RBB	8.4±0.9	W. F. Hollander (1959) *MNL*, **20**, 34, Per. Com. 1970
	bt - Ht	♂	?	7.8±0.7	W. St. Amand Per. Com. 1970
		♀	?	5.4±0.6	,, ,, ,,
	bt - N	♂	?BB	13.0±0.7	S. A. Mallyon and M. E. Wallace, Per. Com. 1970
		♀	?BB	4.3±0.5	S. A. Mallyon and M. E. Wallace, Per. Com. 1970
		♂	RBB	11.5±1.8	C. Stieler Per. Com. 1970
		♀	RBB	7.3±3.5	,, ,, ,,
	bt - uw	♂	CBB	38.3±2.8	M. M. Dickie (1963) *MNL*, **39**, 27
		♀	CBB	43.7±2.6	M. M. Dickie (1963) *MNL*, **39**, 27
		-	RII	38.4±3.6	M. M. Dickie (1963) *MNL*, **39**, 27
	bt - Ve	♂	CBB	9.8±1.2	G. Stieler, Per. Com. 1970
		♀	CBB	7.3±1.5	,, ,, ,,
	Ca - hl	♂	CBB	5.8±0.7	W. F. Hollander (1959) *MNL*, **20**, 34; Per. Com. 1970
		♂	RBB	3.5±0.7	W. F. Hollander (1959) *MNL*, **20**, 84; Per. Com. 1970
		♀	CBB	1.2±0.3	W. F. Hollander (1959) *MNL*, **20**, 34; Per. Com. 1970
		♀	RBB	1.7±0.3	W. F. Hollander (1959) *MNL*, **20**, 34; Per. Com. 1970
	Ca - Ht	♂	?	4.9±0.6	W. St. Amand, Per. Com. 1970
		♀	?	1.7±0.3	,, ,, ,,
	Ca - med	?	Close linkage		A. G. Searle, Per. Com. 1970
	Ca - mn	?	About 20 per cent		S. A. Mallyon and M. E. Wallace, Per. Com. 1970
	Ca - N	♂	?BB	2.3±0.3	S. A. Mallyon and M. E. Wallace, Per. Com. 1970
		♀	?BB	4.5±1.6	S. A. Mallyon and M. E. Wallace, Per. Com. 1970
		♂	CBB	1.0±0.2	W. F. Hollander (1959) *MNL*, **20**, 34, Per. Com. 1970
		♂	RBB	3.1±0.9	W. F. Hollander (1959) *MNL*, **20**, 34, Per. Com. 1970
		♀	CBB	0.3±0.1	W. F. Hollander (1959) *MNL*, **20**, 34, Per. Com. 1970

Group	Loci	Sex	Mating type	Crossover value	References
		♀	RBB	0.3±0.2	W. F. Hollander (1959)
					MNL, **20**, 34, Per. Com. 1970
	Ca - sw	?	Closely linked		P. W. Lane (1970) *MNL*, **42**, 30
	Ca - uw	♂	CBB	48.3±4·6	M. M. Dickie (1968)
					MNL, **39**, 27
		♀	CBB	40.5±4.1	M. M. Dickie (1968)
					MNL, **39**, 27
	mn - N	?	About 20 per cent		S. A. Mallyon and M. E. Wallace, Per. Com. 1970
	N - Ve	♂	RBB	1.8±0.7	C. Stieler, Per. Com. 1970
		♀	RBB	1.8±5.8	,, ,, ,,
VII	Al - Re	?	RBB	6.7±2.5	P. W. Lane, Per. Com. 1970
	Al - sh-2	?	CBB	23.0±2.9	,, ,, ,,
		?	RBB	24.7±3.7	,, ,, ,,
	df - Re	♀	CBB	34.4±4.4	A. Bartke (1965) *MNL*, **32**, 52
		♀	RBB	18.3±2.6	,, ,, ,,
	lt - Re	?	CBI	4.0±1.7	M. M. Dickie, Per. Com. 1970
	Re - tn	♂	CM	15.7±6.2	P. W. Lane, Per. Com. 1970
		♀	CM	24.6±3.4	,, ,, ,,
		♀	RM	19.8±4.4	,, ,, ,,
	Re - vb	♂	CM	17.6±7.6	,, ,, ,,
		♀	CM	17.6±3.7	,, ,, ,,
		♀	RM	18.5±4.0	,, ,, ,,
	sh-2 - tu	-	CII	44.9±7.4	,, ,, ,,
		-	RII	44.0±6.6	,, ,, ,,
	sh-2 - vb	-	RII	35.0±6.2	,, ,, ,,
VIII	an - b	?	RIB	14.5±2.6	E. C. McFarland, Per. Com. 1970
	an - Pt	?	CIB	20.6±2.9	E. C. McFarland, Per. Com. 1970
	b - db	-	CII	9.9±2.5	K. P. Hummel, Per. Com. 1970
		-	CII	8.7±9.7	P. W. Lane, Per. Com. 1970
	b - pf	♂	?BB	21.3±4.2	R. Meredith, Per. Com. 1970
	·b - Pt	♂	?BB	5.6±2.4	,, ,, ,,
		?	CBB	4.0±8.0	W. F. Hollander, Per. Com. 1970
		?	RBB	4.9±0.9	E. C. McFarland, Per. Com. 1970
	b - sno	♂	CBB	15.0±1.2	W. F. Hollander (1966)
					MNL, **34**, 29, Per. Com. 1970
		♀	CBB	18.3±1.2	W. F. Hollander (1966)
					MNL, **34**, 29, Per. Com. 1970

Group	Loci	Sex	Mating type	Crossover value	References
	db - m	-	CII	1.0±0.3	K. P. Hummel, Per. Com. 1970
	b - wi	-	CII	3.5±6.1	P. W. Lane, Per. Com. 1970
	db - wi	-	CII	12.0±9.7	,, ,, ,,
	dep - Pt	?	CBB	6.5±2.4	M. C. Green (1970) *MNL*, **43**, 31
	m - Pt	?	RBB	4.0±0.8	W. F. Hollander, Per. Com. 1970
	m - sno	?	RBB	11.0±1.3	W. F. Hollander, Per. Com. 1970
	pf - Pt	♂	?BB	15.7±3.8	R. Meredith, Per. Com. 1970
IX	*H-2 - T*	♂	-BB	18.8±2.3	M. C. Green and J. H. Stimpfling, Per. Com. 1970
		♀	-BB	16.2±3.3	M. C. Green and J. H. Stimpfling, Per. Com. 1970
	H-2 - tf	♂	-BB	5.6±1.9	M. C. Green and J. H. Stimpfling, (1966) *MNL*, **35**, 32
		♀	-BB	7.9±1.7	M. C. Green and J. H. Stimpfling, (1966) *MNL*, **35**, 32
	Low - T	♂	CBB	2.9± ?	L. C. Dunn and D. Bennett, Per. Com. 1970
	Low - tf	♂	CBB	0.6± ?	L. C. Dunn and D. Bennett, Per. Com. 1970
	qk - T	♂	CBB	6.3±2.5	M. M. Dickie and J. L. Southard, Per. Com. 1970
	T - tf	♂	RBB	13.2±2.8	M. C. Green and J. H. Stimpfling (1966), *MNL*, **35**, 32
		♀	RBB	8.3±1.7	M. C. Green and J. H. Stimpfling (1966), *MNL*, **35**, 32
		♂	CBB	2.5±0.8	L. C. Dunn and D. Bennett, Per. Com. 1970
		♂	RBB	5.3±1.0	L. C. Dunn and D. Bennett, Per. Com. 1970
		♀	CBB	12.4±4.1	L. C. Dunn and D. Bennett, Per. Com. 1970
X	*gr - ji*	-	RII	8.1±5.2	M. C. Green and E. F. Woodworth, Per. Com. 1970
	gr - kd	?	CBB	16.0±7.3	E. V. Hulse and M. F. Lyon, Per. Com. 1970

Group	Loci	Sex	Mating type	Crossover value	References
		?	RBB	16.7±10.8	E. V. Hulse and M. F. Lyon, Per. Com. 1970
		-	CII	37.6±23.9	E. V. Hulse and M. F. Lyon, Per. Com. 1970
		-	RII	16.8±8.7	E. V. Hulse and M. F. Lyon, Per. Com. 1970
XI	Cd - Mi	?	RBB	11.5±4.4	M. C. Green and E. F. Woodworth (1970), *MNL*, **43,** 32
	Hd - Ob	?	RBB	16.7±4.4	K. P. Hummel, Per. Com. 1970
	Hd - wa-1	?	RBB	14.7±4.0	„ „ „
	Lc - Mi	?	CBB	12.7±2.7	A. G. Searle, Per. Com. 1970
	Lc - Sig	?	RBB	34.1±3.6	„ „ „
	Mi - ob	♂	CM	25.2±5.6	P. W. Lane, Per. Com. 1970
		♀	CM	25.2±6.8	„ „ „
	Mi - Sig	?	RBB	42.2±3.8	A. G. Searle (1970), *MNL*, **43,** 29
	Mi - tc	♂	CBB	4.5±1.2	P. W. Lane, Per. Com. 1970
		♂	RBB	6.7±2.3	„ „ „
		♀	CBB	10.5±2.1	„ „ „
		♀	RBB	5.1±1.5	„ „ „
	ob - wa-1	-	RII	23.5±3.2	M. M. Dickie and P. W. Lane, Per. Com. 1970
	tc - wa-1	?	Closely linked		P. W. Lane, Per. Com. 1970
XIII	dt - fz	-	RII	15.6±5.6	D. S. Falconer and J. H. Isaacson (1965) *MNL*, **32,** 31
		-	RII	35.6±7.8	D. E. Kelton (1965) *MNL*, **32,** 60
	dt.- ln	-	RII	28.2±4.8	D. S. Falconer and J. H. Isaacson (1965), *MNL*, **32,** 31
		-	RII	36.4±7.8	D. C. Kelton (1965) *MNL*, **32,** 60
		-	RII	38.2±6.8	P. W. Lane, Per. Com. 1970
	fz - ln	-	CII	36.5±4.4	D. S. Falconer and J. H. Isaacson (1965) *MNL*, **32,** 31
		-	CII	41.3±8.2	D. E. Kelton (1965) *MNL*, **32,** 60
		-	CII	34.7±3.0	M. M. Dickie, Per. Com. 1970
		-	CII	33.8±5.0	P. W. Lane, Per. Com. 1970
	fz - tb	-	RII	8.6±3.8	M. M. Dickie and J. L. Southard, Per. Com. 1970

Group	Loci	Sex	Mating type	Crossover value	References
	fz - vl	-	RII	51.0±3.6	M. M. Dickie, Per. Com. 1970
	ln - Sp	♂	CBB	3.3±1.0	M. M. Larsen (1965) *MNL*, **33,** 68
		♀	CBB	4.0±1.0	M. M. Larsen (1965) *MNL*, **33,** 68
	ln - tb	-	RII	26.9±3.6	M. M. Dickie and J. L. Southard, Per. Com. 1970
	ln - th	♂	CBB	1.3±0.7	M. M. Larsen (1965) *MNL*, **33,** 68
		♀	CBB	2.5±0.7	M. M. Larsen (1965) *MNL*, **33,** 68
		-	RII	21.2±8.5	M. F. Lyon (1967) *MNL*, **36,** 34
	ln - vl	-	RII	15.5±4.4	M. M. Dickie, Per. Com. 1970
	Lp - tb	?	CBI	52.4±3.6	M. M. Dickie and J. L. Southard, Per. Com. 1970
	Sp - tb	♂	CBB	27.1±1.4	M. M. Dickie and J. L. Southard, Per. Com. 1970
	Sp - th	♂	CBB	4.6±1.2	M. M. Larsen (1965) *MNL*, **33,** 68
		♀	CBB	6.5±1.1	M. M. Larsen (1965) *MNL*, **33,** 68
XIV	*bg - cr*	-	RII	0.3±0.3	P. W. Lane. Per. Com. 1970
	bg - f	-	CII	26.8±2.9	E. L. Green, Per. Com. 1970
		-	RII	26.8±2.4	,, ,, ,,
	bg - fs	-	CII	29.4±2.6	,, ,, ,,
		-	RII	35.5±3.1	,, ,, ,,
	bg - sa	♂	CBB	6.7±1.4	W. St. Amand, Per. Com. 1970
		♂	RBB	7.5±1.5	,, ,, ,, ,,
		♀	CBB	8.7±1.6	,, ,, ,, ,,
		♀	RBB	10.3±1.7	,, ,, ,, ,,
		-	RII	8.4±1.9	,, ,, ,, ,,
	f - fs	♀	CBB	12.0±3.1	E. L. Green, Per. Com. 1970
		-	CII	8.3±1.3	,, ,, ,,
		-	RII	12.4±4.5	,, ,, ,,
	f - pe	♀	CBB	23.0±2.1	,, ,, ,,
		-	CII	23.8±3.1	,, ,, ,,
		-	RII	19.6±4.1	,, ,, ,,
	fs - pe	♀	CBB	25.6±3.1	,, ,, ,,
		♀	RBB	12.0±6.5	E. L. Green, Per. Com. 1970
		-	CII	19.5±6.4	,, ,, ,,
		-	RII	12.2±1.1	,, ,, ,,
	pe - sa	-	RII	42.9±5.8	W. St. Amand (1970), Per. Com. 1970

Group	Loci	Sex	Mating type	Crossover value	References
XVI	ft - ma	♂	CBB	4.8±1.0	P. W. Lane, Per. Com. 1970
		♀	CBB	3.9±1.2	,, ,, ,,
	ft - Va	♂	CBB	22.1±3.6	,, ,, ,,
		♀	CBB	28.5±3.4	,, ,, ,,
	ma - Va	♂	CBB	28.6±3.8	,, ,, ,,
		♀	CBB	22.6±3.8	,, ,, ,,
	Spa - Va	♂	CBB	25.6±3.0	,, ,, ,,
		♂	RBB	25.7±7.3	,, ,, ,,
XVII	bf - Hm	?	CBB	38.6±3.6	M. M. Dickie, Per. Com. 1970
	bf - W	?	CBB	21.7±3.1	,, ,, ,,
	bl - W	?	CBB	11.5±2.5	R. J. S. Phillips (1970) MNL, **42**, 26
	go - Hm	♂	CBB	28.4±3.7	W.F.Hollander,Per.Com.1970
		♀	CBB	31.2±3.4	,, ,, ,,
		?	CBB	25.4±2.6	M. M. Dickie, Per. Com. 1970
	go - W	♀	CBB	6.5±2.4	W.F.Hollander,Per.Com.1970
		?	CBB	8.0±1.6	M. M. Dickie, Per. Com. 1970
	Hm - jg	♂	CBB	44.7±7.3	M. C. Green, Per. Com. 1970
		-	CM	37.1±6.2	,, ,, ,, ,,
	Hm - lx	?	CBB	4.6±2.0	,, ,, ,, ,,
	Hm - mc	♀	CBB	25.5±3.4	M. C. Green and E. F. Woodworth, Per. Com. 1970
	Hm - W	?	CBB	19.5±1.8	M. M. Dickie, Per. Com. 1970
		♂	CBB	22.1±3.2	M. C. Green, Per. Com. 1970
		♂	RBB	30.0±10.3	,, ,, ,, ,,
		♀	CBB	20.9±1.7	,, ,, ,, ,,
		♀	RBB	22.4±3.2	,, ,, ,, ,,
		?	CBB	20.4±3.9	,, ,, ,, ,,
		♀	CBB	21.7±3.3	M. C. Green and E F Woodworth, Per. Com. 1970
	jg - le	♂	CBB	4.9±2.8	M. C. Green, Per. Com. 1970
		?	CM	4.7±2.9	,, ,, ,, ,,
		-	CII	4.4±2.9	,, ,, ,, ,,
	jg - W	♂	CBB	16.1±3.9	,, ,, ,, ,,
		?	CM	17.8±3.7	,, ,, ,, ,,
		?	RM	16.6±5.8	,, ,, ,, ,,
		-	CII	24.0±3.3	,, ,, ,, ,,
	le - rd	?	CBB	2.0±1.0	R. L. Sidman (1968) MNL, **39**, 24
	le - W	♂	CBB	14.5±3.2	R. L. Sidman (1968) MNL, **39**, 34
		♀	CBB	9.4±3.4	R. L. Sidman (1968) MNL,

Group	Loci	Sex	Mating type	Crossover value	References
					39, 34
	lx - W	♀	CBB	17.6±3.7	M. C. Green, Per. Com. 1970
	mc - W	♀	CBB	6.2±1.9	M. C. Green and E. F. Woodworth, Per. Com. 1970
	rd - W	♂	CBB	12.9±3.0	R. L. Sidman (1968) *MNL,* **39,** 24
		♀	CBB	9.4±3.4	R. L. Sidman (1968) *MNL,* **39,** 24
XVIII	*e - Hk*	?	RBB	34.8±5.7	D. S. Falconer and J. H. Isaacson (1965) *MNL,* **32,** 31
	e - hy-3	?	CM	11.4±3.2	M. C. Green, Per. Com. 1970
	e - nr	♀	CBB	50.8±2.6	,, ,, ,, ,,
	e - Os	?	CBB	38.1±3.4	D. S. Falconer and J. H. Isaacson (1962) *MNL,* **27,** 30, (1965) *MNL,* **32,** 31
		?	RBB	41.4±3.9	D. S. Falconer and J. H. Isaacson (1962) *MNL,* **27,** 30, (1965) *MNL,* **32,** 31
		♂	?BB	28.0±3.0	W. F. Hollander (1966) *MNL,* **35,** 30
		♀	?BB	28.0±7.0	W. F. Hollander (1966) *MNL,* **35,** 30
		?	CBB	25.6±2.5	A. G. Searle (1970) *MNL,* **42,** 27
		?	RBB	25.3±3.6	A. G. Searle (1970) *MNL,* **42,** 27
		?	CBB	29.9±1.8	M. C. Green, Per. Com. 1970
	e - Q	♂	?BB	30.0±3.0	W. F. Hollander, Per. Com. 1970
		♀	?BB	30.0±7.0	W. F. Hollander, Per. Com. 1970
	Ea-1 - Es-1	?	-BI	5.3±3.6	M. Foster, M. L. Petras, and P. Tomlin, Per Com. 1970
	Ea-1 - Es-5	?	-BI	9.9±6.3	M. Foster, M. L. Petras, and P. Tomlin, Per Com. 1970
	hy-3 - Os	?	CM	16.8±2.7	M. C. Green, Per. Com. 1970
	la - Os	?	CIB	linked	J. L. Southard (1970) *MNL,* **43,** 33
	nr - Os	♀	CBB	26.6±2.3	M. C. Green, Per. Com. 1970
		?	CM	22.2±2.8	,, ,, ,, ,,
		?	RM	23.9±2.9	,, ,, ,, ,,

Group	Loci	Sex	Mating type	Crossover value	References
	Os - Q	♂	CBB	5.8±1.2	W. F. Hollander (1968) MNL, **38,** 23, Per. Com. 1970
		♀	CBB	2.8±1.9	W. F. Hollander (1968) MNL, **38,** 23, Per. Com. 1970
	Hk - Os	?	RBB	15.9±4.4	D. S. Falconer and J. H. Isaacson (1965) MNL, **32,** 31
	Hk - Q	?	Close linkage		W. F. Hollander, Per. Com. 1970
XX	Bn - Gy	♀	?	36.7±3.1	M. F. Lyon (1966) MNL, **35,** 28
	Bn - Mo	♀	CBB	15.6±4.2	P. W. Lane, Per. Com. 1970
		♀	RBB	13.5±3.9	,, ,, ,, ,, ,,
	Cg - Gs	♀	RBB	7.5±1.5	D. Grahn, R. A. Lea, and J. Hulesch, Per. Com. 1970
	Gg - Mo	♀	CBB	13.1±1.4	D. Grahn, R. A. Lea, and J. Hulesch, Per. Com. 1970
		♀	RBB	16.0±4.0	D. Grahn, R. A. Lea, and J. Hulesch, Per. Com. 1970
	Gs - Mo	♀	RBB	4.9±0.9	D. Grahn, R. A. Lea, and J. Hulesch, Per. Com. 1970
	Gy - Mo	♀	?	24.6±4.0	M. S. Lyon (1966) MNL, **35,** 28
	Gy - Ta	♀	?	20.9±2.4	M. S. Lyon (1966) MNL, **35,** 28
	Mo - Ta	♀	RBB	2.8±1.9	L. B. Russell (1960) MNL, **23,** 59
	Mo - Ym	♀	RBB	2.6±0.1	P. R. Hunsicker (1929) MNL, **40,** 46

Note: ? in the sex column indicates either that sex was not stated or data from both sexes were pooled; ? in the phase column indicates the phase was not stated or that data from both phases were pooled; M in the phase column indicates either IB or BI, the order not being explicit in original data; Per. Com. means personal communication; *MNL* means *Mouse News Letter*.

CHAPTER 12

Norway Rat

Rattus norvegicus $(n = 20 + X + Y)$

After the house mouse, the rat has the largest number of known linked pairs of genes. Curiously, not the largest number of linkage groups (this falls to the rabbit) because most of the linked genes belong to one system. Some 40 to 45 mutant gene/loci have been described and 35 have featured in studies of linkage and independent assortment. A detailed discussion of the early work in linkage may be found in Robinson (1960); while a general description of most of the mutant genes is provided by Robinson (1965).

TABLE 12.1 Genes/loci of the rat which have featured in studies on linkage; assigned linkage groups, symbol, and conventional designation.

Linkage Group	Symbol	Designation	Prime Characteristics
IV	*a*	Non-agouti	Coat colour
	Ag-A	Antigen	Immunogenetics
	Ag-B	Antigen	Immunogenetics
	Ag-C	Antigen	Immunogenetics
II	*an*	Anaemia	Blood
II	*b*	Brown	Coat colour
I	*c*	Albinism	Coat colour
	Ca	Cataract	Eye
II	*Cu-1*	Curly-1	Hair texture
	Cu-2	Curly-2	Hair texture
	cw	Cow-lick	Hair texture
	d	Dilute	Coat colour
	dx	Anaphylactoid reaction	Physiology
IV	*f*	Fawn	Coat colour
I	*fz*	Fuzzy	Hypotrichosis
	h	Hooded	White spotting
	Ha	Antigen	Immunogenetics
I	*Hb-1*	Antigen	Immunogenetics
I	*Hbb*	Haemoglobin β chain	Electrophoretic variant
I	*he*	Haemotomas	Physiology
	Hm	Antigen	Immunogenetics
	hr	Hairless	Hypotrichosis

Linkage Group	Symbol	Designation	Prime Characteristics
II	*in*	Incisorless	Teeth
	j	Jaundice	Enzyme system
III	*k*	Kinky	Hair texture
I	*lg*	Gruneberg's lethal	Skeleton
	n	Naked	Hypotrichosis
	Nė	Hydronephrosis	Physiology
I	*p*	Pink eye	Coat colour
I	*r*	Red eye	Coat colour
	R-1	Antigen	Immunogenetics
I	*Rw*	Warfarin resistance	Physiology
II	*s*	Silvering	Coat colour
II	*Sh*	Shaggy	Hair texture
	sr	Shaker	Behaviour
III	*st*	Stub	Tail
I	*w*	Waltzing	Behaviour
	Wo	Wobbly	Behaviour

LINKAGE GROUPS

Twenty-three pairs of genes have been observed to display linked segregation and these have been placed into four groups. A summary of the data is given by Table 12.2. A sex difference in crossingover is revealed for some pairs as shown by Table 12.3. A general trend for crossingover to occur more frequently in the female is evident although, statistically, the differences are barely significant.

Group I (c, fz, Hb-1, Hbb, he, lg, p, r, Rw, w)
This is the most important group. Not only was it the first to be discovered but it also contains the largest number of mutant loci. The mean crossover values for the intercepts c - w, lg - p, p - r, and p - w are 42.5 ± 1.5, 22.1 ± 1.6, 19.0 ± 1.1, and 52.0 ± 6.2 respectively, estimation resting upon the analysis of more than one segregation.

Marked heterogeneity is apparent for the c - p gene interval. Even without the two deviating crossover values reported by Harris *et al.* (1963) and French *et al.* (1971), there is heterogeneity due to two divergent crossover values for females in backcrosses (16.0 ± 1.1 and 21.9 ± 0.6, respectively). It is possible that the divergence is due to chance since, taken together, their effects on the mean roughly cancel. If they are included, the mean value is 19.2 ± 0.3; if they are excluded, the mean value is 18.4 ± 0.4. The greater deviation is that of 21.9 and it may be noted that this particular sample provides part of the evidence for a sex difference in crossingover. Taking the data as a whole, the sex difference for the c - p intercept cannot be held to be firmly established.

Genes c and r have been the subject of many studies, with little inconsistency among the observed crossover values. The mean value is 0.34 ± 0.06. An exception are the data of Dunn (1920), where certain animals from two classes of a repulsion double intercross were individually analysed for the presence of crossover gametes. It is impossible to partition Dunn's figures (as published) for an exact analysis but a rough assessment showed no disagreement with Dunn's own conclusions. However, the derived value of 1.87 ± 0.83 is significantly higher than the mean above but not tremendously so. If Dunn's result is excluded, the mean crossover value is 0.28 ± 0.06.

Identification of w is complicated by partial penetrance of the abnormal behaviour. When due allowance is made for this, estimates of the crossover value differ from those of King and Castle (1937) and of Castle and King (1941). The greatest discrepancies are for the p - w interval: Castle and King proposing a value of 35, with a subsequent correction to 45 (Castle 1944, 1946,), whereas a more realistic value is 52 ± 6.2. This fact, incidentally, clarifies the dispute on the precise order of the three genes c, p, and w (Whittinghill, 1944; Castle, 1946).

The gene referred to by Greaves and Ayres (1969) as Rw shows a similar amount of crossingover to p as does he. These authors suggest that the two genes may in fact be identical or be alleles. The two genes apparently affect a common feature, blood clotting (or the underlying physiology) which manifests as warfarin resistance in the observations of Greaves and Ayres but as haemorrhagic lesions in the observations of Dunning and Curtis (1939).

The fuzzy type of hairlessness found by Palm (1971) is linked to c with a mean crossover value of 17.7 ± 3.2. No other information is available as yet, hence the fz gene cannot be precisely positioned in the group.

The trigon of Fig. 12.1 probably fixes the linear order of the various loci of the group. The arrangement is arbitrary in the sense that the one chosen is that which minimizes total map length. Until the genes *fz, he, lg, Rw,* and *w* are tested among themselves, the order must be regarded as subject to revision.

Yosida (1960) and Yosida, Kurita and Taneda (1961) have described a new ruby-eyed mutant which they ascribe to an allele of the *p* locus. Part of the evidence for this is phenotypic and part is based on linkage with *c*. Allowing for the fact that *c* is epistatic to the ruby-eyed gene, the RII joint segregation differs significantly from random assortment. Unfortunately, the number of animals is small and it is impossible to obtain a sensible estimate of the crossover value. On the other hand, the observed frequencies are not at variance with a crossover value of 19 or slightly less (the mean crossover value for the *c - p* intercept).

Brdicka (1966, 1967, 1968) has found linkage between a codominant blood antigenic difference (the locus for which he denotes as *Hb-1*) and genes *c* and *p*. In earlier results the assortment data were badly distorted from expectation but the 1968 data was not affected. The estimates of crossingover are based upon the latter.

It may be remarked that the main reason why this linkage group contains so many genes is that many of the more widely used stocks of laboratory rats are homozygous for *a, c,* and *h* and frequently for *p*. Should outcrosses be necessary for any purpose, these genes would segregate and associations may be

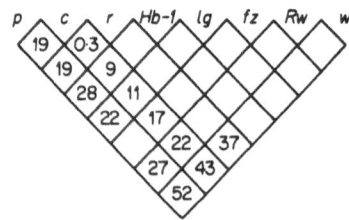

Fig. 12.1. Linkage group I of the rat.

observed. Since two of the genes belong to the first group, this has also had the advantage of precisely locating the new mutant in a few cases.

Group II (an, b, Cu-1, in, s, Sh)
The mean crossover values for the intercepts *an - b*, *b - Cu-1*, *b - s*, and *Cu-1 - Sh* are 45.2 ± 2.3, 45.2 ±b 1.0, 7.6 ± 1.2, and 3.9 ± 1.3, respectively, all of which are derived from more than one group of segregations. A hint of heterogeneity is shown by the data for *Cu-1 - Sh* but scarcely large enough to warrant detailed consideration. A sex difference of crossingover may occur for the *b - Cu-1* intercept but in the opposite direction to the general tendency of greater crossingover in the female.

Significant heterogeneity is apparent between the three groups of data for the *an* and *Cu-1* genes. The discordant segregation is the backcross which is characterized by a crossover value of 10.3 ± 2.3, in contrast to a mean value of 2.3 ± 0.8 for the other two. The latter value is probably the more accurate estimate, as judged by the almost identical crossover values of 14 and 15 for the *an - in* and *Cu-1 - in* intervals. These results imply that *an* and *Cu-1* must be closely linked. The value of 2.3 has been entered in Table 12.2 and Fig. 12.2.

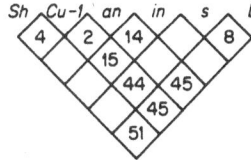

Fig. 12.2. Linkage group II of the rat.

The order of the loci is almost certainly that of the trigon of Fig. 12.2. The inconsistencies are very minor but the number of blanks in the diagram is an indication that further studies are desirable. In particular, both *s* and *Sh* could be tested against *an* and *in* to improve the overall picture.

Group III (k, st)
The crossover value for *k* and *st* was originally assessed by the

method of Castle (1939) because of the severe inviability of *st* homozygotes. However, this method may not always lead to detection of all +*st* animals, depending on the number of young bred per individual and the relative inviability of *stst*. This is equivalent to partial penetrance of *st* and an attempt has been made to allow for this – probably not wholly successfully. The upshot is that the estimated crossover value of 25.9 ± 5.2 is less than the value reported by Castle and King (1944).

Group IV (a, f)

The estimation of amount of crossingover for these two genes was straightforward.

SEX-LINKAGE

No sex-linked genes on the *X* chromosomes have been reported at present date. Billingham *et al.* (1962) give results, however, which suggest the presence of a locus on the *Y* chromosome capable of producing a weak histoincompatibility reaction. The expression of the reaction is irregular, hence the suggestion of a locus on the *Y* may still invite caution.

It is a fact that few of the known genes have been specifically tested for partial sex-linkage, in spite of an appeal by Darlington, Haldane, and Koller as early as 1932. The possibility of partial sex-linkage may be especially relevant for the rat because of the conclusion of Darlington and Koller (1934) that the *XY* bivalent have a mutual pairing region apart from the differential segments. On the other hand, the pairing region may be small and thinly populated by gene loci, hence the chances of discovery may be small.

AGE INFLUENCE

Castle and Wachter (1924) have specifically examined data on crossingover between *c* - *p* for an effect of age. Diheterozygous animals of both sexes were investigated over the range of three to 18-20 months. No consistent trend was evident for either sex. Castle (1919) refers to a similar study on the *p* - *r* interval which also produced negative results.

However, the observations of Bryden (see later), which indicate a decrease in mean frequency of chiasma per bivalent with age, would imply that age differences of crossingover may exist. It is not to be presumed, on the other hand, that such differences may be large enough for easy detection or that they may apply to all regions of a chromosome.

STRAIN AND INDIVIDUAL INFLUENCE

No evidence could be found by Castle (1919) for the p-r intercept or by Castle and Wachter (1924) for the c-p intercept, that individual diheterozygous males were producing exceptionally high or low frequencies of crossingover. Castle and Wachter did credit one male with an unusually high rate of crossingover but this peculiarity was not transmitted to his sons; hence the high rate may have been the odd deviant to be expected in a series of repetitive tests. Feldman (1924) has briefly considered the possible effects of individual or parentage upon the amount of crossingover for c-p but without uncovering any positive tendency.

There are reasons for believing that crossingover is an adaptive feature of the genotype and, as such, would be expected to be under the influence of heredity. This does not mean that the differences may be great and it may require large numbers of offspring under propitious circumstances before any tendency can be revealed. Inter-strain comparison would seem to be the most promising source of significant variation. The differences mooted in this section are visualized as part of the normal variability and distinct from the differences which could arise from major changes in the chromosome constitution due to inversion or to translocation.

AFFINITY

No signs of the affinity phenomenon have been observed in the rat. Robinson (1960) made a search among the data on independent assortment but with negative results. The method employed was not very sensitive and possibly would only have yielded positive results had the phenomenon been marked.

TABLE 12.2 Established linkage groups in the rat

Group	Loci	Sex	Mating type	Crossover value	References
I	c - fz	♂	CBB	16.0±4.2	Palm (1971)
		♀	CBB	19.7±5.9	„ „
	c - Hb-1	?	?BB	9.8±2.8	Brdicka (1968)
	c - Hbb	-	II	4.9±2.6	French et al. (1971)
	c - lg	-	RII	10.6±5.2	Gruneberg (1939)
	c - p	♂	RBB	18.5±1.1	Feldman (1924)
		♂	RBB	18.5±0.4	Castle and Wachter (1924)
		♀	RBB	16.0±1.1	Feldman (1924)
		♀	RBB	21.9±0.6	Castle and King (1941)
		?	RBB	21.3±2.4	„ „ „ „
		?	CBB	16.7±3.0	Greaves and Ayres (1969)
		-	RII	20.7±4.2	Castle (1916, 1919)
		-	RII	41.7±19.0	Harris et al. (1963)
	Hbb - p	-	II	19.7±6.2	French et al. (1971)
	c - r	♂	RBB	0.18±0.06	Castle and Wachter (1924)
		♀	RBB	0.53±0.17	„ „ „ „
		-	RII	0.19±0.23	Castle (1916, 1919), Whiting and King (1918), Dunn (1920), Ibsen (1920)
		-	RII	1.87±0.83	Dunn (1920)
	c - Rw	?	CBB	21.8±2.5	Greaves and Ayres (1969)
	c - w	?	CBB	43.5±1.6	King and Castle (1937), Castle and King (1941)
		-	CII	35.3±4.5	King and Castle (1937)
	Hb-1 - p	?	?BB	28.1±4.0	Brdicka (1968)
		-	RII	54.2±10.3	French et al. (1971)
	he - p	-	CII	29.2±5.6	Dunning and Curtis (1939)
	lg - p	♂	CIB	24.5±6.0	Gruneberg (1939)
		♀	CIB	24.7±3.4	„ „
		-	CII	20.9±2.0	„ „
	p - r	♂	CBB	13.6±2.7	Castle (1916, 1919)
		♂	RBB	18.8±2.1	„ „ „
		♀	CBB	19.9±2.4	„ „ „
		♀	RBB	21.5±1.9	„ „ „
		-	CII	18.2±3.3	„ „ „
	p - Rw	?	CBB	26.7±3.6	Greaves and Ayres (1969)
	p - w	?	RBB	50.8±2.4	King and Castle (1937), Castle and King (1941)
			RII	54.4±2.4	King and Castle (1937), Castle and King (1941)
	r - w	?	CBB	36.8±5.9	Castle and King (1949)
II	an - b	♂	CIB	44.6±2.6	Castle and King (1941)

Group	Loci	Sex	Mating type	Crossover value	References
		♀	CIB	48.9±5.7	„ „ „ „
	an - Cu-1	♂	RBI	2.2±0.9	„ „ „ „
		♀	RBI	2.8±0.2	„ „ „ „
		?	RBB	10.3±2.3	Castle and King (1944)
	an - in	?	RBB	13.7±2.6	„ „ „ „
	b - Cu-1	♂	CBB	43.3±3.1	King and Castle (1935)
		♂	RBB	47.0±1.5	Castle and King (1941)
		♀	CBB	38.5±2.6	King and Castle (1935)
		♀	RBB	39.5±3.2	Castle and King (1941)
		?	RBB	47.8±2.6	„ „ „ „
	b - s	?	CBB	6.7±1.3	Castle (1953)
		-	RII	10.4±2.3	„ „
	b - Sh	?	CBB	51.2±4.5	Castle and King (1947a)
	Cu-1 - in	?	CBB	14.9±2.7	Castle and King (1944)
	Cu-1 - s	?	RBB	43.8±2.6	Castle (1953)
	Cu-1 - Sh	♂	RBB	3.2±1.6	Castle and King (1947a)
		♀	RBB	4.8±2.1	„ „ „ „
III	k - st	♂	RBB	25.9±5.2	Castle and King (1944)
IV	a - f	?	CBB	44.7±1.4	Castle and King (1949)

INDEPENDENT ASSORTMENT

Prima facie evidence is shown by Table 12.4 for independent inheritance of 131 pairs of gene combinations. The majority of genes of the table doubtless segregate independently although a few conceivably could be loosely linked. Some idea of the magnitude of this is outlined by Fig. 12.3. The figure serves the twin purposes of (a) showing the extent of those tests completed to date and (b) the value of linkage which would be compatible with the available data. The latter is approximated by multiplying the standard error by 196 and subtracting the product from the observed recombination value. It is obvious that loose linkage cannot be ruled out for a large number of gene pairs.

A number of statements occur in the literature that certain genes are inherited independently but are unsupported by numerical data. This procedure is clearly regrettable at the present level of knowledge. The following instances may be recorded. Wilder, *et al.* (1932) stated that *c* is recombined freely

TABLE 12.3 Sex differences in crossover values for certain chromosome
intercepts in the rat

Group	Loci	Sex	Crossover value	References
I	c - p	♂	18.5±1.1	Feldman (1924)
		♀	16.0±1.0	,, ,,
		Diff.	2.5±1.5	
		♂	18.5±0.4	Castle and Wachter (1924)
		♀	21.9±0.6	,, ,, ,, ,,
		Diff.*	3.4±0.6	
	c - r	♂	0.18±0.06	,, ,, ,, ,,
		♀	0.53±0.17	,, ,, ,, ,,
		Diff.	0.35±0.18	
	lg - p	♂	24.5±6.0	Gruneberg (1939)
		♀	24.7±3.5	,, ,,
		Diff.	0.2±6.9	
	p - r	♂	17.0±1.6	Castle (1919)
		♀	20.9±1.5	,, ,,
		Diff.	3.9±2.2	
II	an - b	♂	44.6±2.6	Castle and King (1941)
		♀	48.9±5.7	,, ,, ,, ,,
		Diff.	4.4±6.2	
	an - Cu-1	♂	2.2±0.9	,, ,, ,, ,,
		♀	2.8±2.1	,, ,, ,, ,,
		Diff.	0.6±2.3	
	b - Cu-1	♂	46.3±1.3	King and Castle (1935), Castle and King (1941)
		♀	40.9±2.0	King and Castle (1935), Castle and King (1941)
		Diff.*	5.6±2.4	
	Cu-1 - Sh	♂	3.2±1.6	Castle and King (1947a)
		♀	4.8±2.1	,, ,, ,, ,,
		Diff.	1.6±2.6	

* Significant difference

with *hr*. King and Castle (1935) noted that *a*, *c* and *h* are apparently independent of each other. King and Castle (1937) said that *w* assorts freely with *d*, *hr*, and *k*. Castle and King (1940) quoted J. W. MacArthur that *j* segregates freely with *a*, *c* and *h*. Castle and King (1944) reported negative results for linkage tests of *st* with *Cu-2*, *d* and *wo*. Castle (1953) commented that no sign of linkage assortment could be seen between *s* and *a*, *Cu-2*, *k* or *wo*. In a few instances, data have been gleaned from other sources to verify these statements. The situation for the blood group loci is somewhat confusing. The original stock carrying *Ha* and *Hm*, defined by Burhoe (1947), has been lost. The fact that neither were found to be linked with other known loci is not reassuring that they will ever be positively rediscovered. Owen (1962) thought that he may have found *Ha* but, if so, it has been lost again. Out of the several blood group loci mentioned by Owen, one (*Ho* or the *C-D* system) is currently available. No genetic data has been given but it is stated that *Ho* is not sex-linked and is inherited independently of *a*, *c*, *h*, and *p*. Some of the confusion has been straightened out by Palm and Black (1971). These authors have introduced a new system of terminology to cover both existing loci and those which may be discovered in the future. The basic symbol is *Ag* with qualifying letters for loci and numerical superscripts to denote alleles.

Among the loci recognized by Palm and Black, *Ha* and *Ho* or *C - D* have been designated under the new system as *Ag-A* and *Ag-C*, respectively. The *R-1* locus of Bogden and Aptekman (1960, 1962) is redesignated as *Ag-B*. This locus is probably identical to that symbolized as *RtH-1* by Stark *et al.* (1968). *Ag-B* or *RtH-1* is being extensively studied and the indications are that it may be the rat homologue of the important complex locus *H-2* of the mouse (Palm, 1964; Stark *et al.*, 1968). Some eight alleles have been identified and up to 17 antigenic specificities have been defined (Stark *et al.*, 1968). Here, *RtH-1* will be shortened to *R-1*.

Lozzio, Chernoff, Machado, and Lozzio (1967) have presented details of the joint segregation of *j* and a dominant gene causing a congenital hydronephrosis. The authors did not symbolize the latter and here the gene will be referred to as *Ne*. Only a small portion of their published pedigree furnished critical evidence for independent inheritance of the two genes.

With so many tests for linkage, it is inevitable that a few will produce ambiguous results. In the usual course of events, these would seemingly be further investigated. However, a number of ambiguous situations still exist and these will be briefly discussed.

The genes *c - Cu-2* show a significant recombination value of 40.6 ± 4.4, according to the data of Blunn and Gregory (1937).

Inspection of the records for individual letters revealed, however, that one particular female was responsible for about half the observed deviation and that her litters were unusually small. The authors concluded that linkage cannot be entertained under these circumstances. The possibility of a unique translocation in the female might be the explanation for the curious result.

The data of Castle, King, and Daniels (1941) is suggestive of linkage for the genes *hr* and *wo* (crossover value, 40.3 ± 4.6). However, subsequent results (Castle 1955) have failed to confirm the linkage and the estimated recombination from the pooled data is 44.3 ± 2.9.

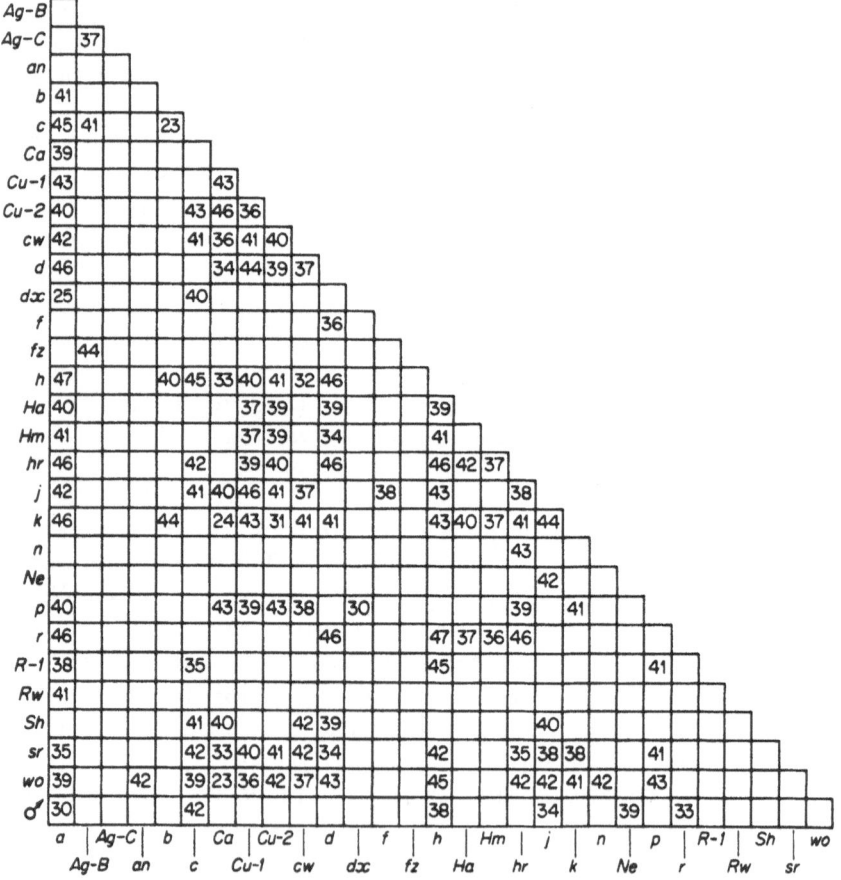

Fig. 12.3. Summary of the extent of independent assortment in the rat.

In 1955, Castle, Dempter, and Shurrager reported possible linkage between *hr* and *n*. The recombination value is 34.7 ± 7.1, formally indicative, but certainly not conclusively so. Subsequent data of Castle (1955) did not confirm the above result and the recombination value for the whole data is 45.3 ± 3.4. Note that a suggestion of linkage had been found between *hr* and *wo* (preceding paragraph) but no hint of an association has been detected between *n* and *wo*. Hence, the possibility of linkage for any of these loci appears *prima facie* remote.

Castle, Dempter, and Shurrager (1955) found a recombination value of 30.8 ± 8.0 for the genes *d* and *sr*. The value is formally significant but the sample is small and the deviation was due to a deficiency in numbers for one of the two recombination phenotypes. The other classes are to expectation and an accident of sampling is probably the cause of fluctuation of frequencies observed.

Two instances of excessive recombination may be briefly noted. Feldman's (1935) observations imply a recombination value of 75.5 ± 6.1 for *Cu-1* and *k*. However, King and Castle (1935) failed to duplicate the results and it seems that reduced viability of *k* is the main reason for the anomalous value.

Castle and King (1940) reported an F_2 repulsion generation for ·the genes *d* and *j* which yielded a value of 61.5 ± 6.0. Statistically, the results are merely on the borderline of significance and, in the absence of other confirming data, may be attributed to chance. An overabundance of the double recessive class is the tangible reason for the high value.

CHROMOSOME MORPHOLOGY

The haploid number of elements is 21. The early investigations on the number and form of the chromosomes have been discussed in detail by Robinson (1965). Most of this early work lacked the refinement of current research. For instance, it was thought that all of the chromosomes were telocentric or acrocentric, as for the house mouse. However, this is no longer accepted. The most informative reports on rat karyology are: Tobias (1947), Makino and Hsu (1954), Yosida (1955a, b),

TABLE 12.4 Quasi-independent gene assortment in the rat

Loci	Recombination value	Phase balance	References
a - b	50±4	0	King and Castle (1935)
a - c	50±2	95	Whiting and King (1918), Hanson and Stewart (1923), King and Castle (1935), Roberts and Quisenberg (1936), Harris et al. (1963), Greaves and Ayres (1969)
a - Ca	47±6	60	Castle and King (1948)
a - Cu-1	49±3	100	King and Castle (1935)
a - Cu-2	47±5	0	Blunn and Gregory (1937)
a - cw	48±4	0	Castle, Dempster and Shurrager (1955)
a - d	51±2	100	Roberts and Quisenberg (1936)
a - dx	40±8	0	Harris et al. (1963)
a - h	49±1	100	Whiting and King (1918), Ibsen (1920), Hanson and Stewart (1923), King and Castle (1935), Roberts and Quisenberg (1936)
a - Ha	48±5	0	Burhoe (1947)
a - Hm	55±4	100	,, ,,
a - hr	50±2	100	Roberts et al. (1940)
a - j	52±4	?	Castle and King (1940)
a - k	50±2	54	King and Castle (1935), Mitchell (1935)
a - p	47±3	63	Yosida (1960), Yosida et al. (1961), Greaves and Ayres (1969)
a - r	50±2	75	Ibsen (1920), Roberts and Quisenberg (1936)
a - R-1	48±5	-	Stark et al. (1969)
a - Rw	48±4	100	Greaves and Ayres (1969)
a - sr	47±7	0	Castle et al. (1955)
a - wo	45±6	0	Castle et al. (1941)
a - ♂	51±9	86	Whiting and King (1918)
Ag-B - Ag-C	47±5	-	Palm (1971)
Ag-B - c	49±4	-	,, ,,
Ag-B - fz	53±4	-	,, ,,
an - wo	49±4	0	Castle et al. (1941)
b - c	42±14	0	King and Castle (1935)
b - h	51±5	0	,, ,, ,, ,,
b - k	46±3	0	,, ,, ,, ,,
c - Cu-2	41±4	100	Blunn and Gregory (1937)
c - cw	55±4	0	Castle et al. (1955)
c - dx	49±5	0	Harris et al. (1963)
c - h	49±2	95	Whiting and King (1918), Hanson

Loci	Recombination value	Phase balance	References
			and Stewart (1923), King and Castle (1935)
c - hr	51±3	0	Roberts (1926)
c - j	50±5	?	Castle and King (1940)
c - R-1	43±4	-	Stark et al. (1969)
c - Sh	46±5	100	Castle and King (1947a)
c - sr	47±4	0	Castle et al. (1955)
c - wo	43±6	0	Castle et al. (1941)
c - ♂	46±4	51	Whiting and King (1918)
ca - Cu-1	49±4	50	Castle and King (1948)
Ca - Cu-2	48+2	0	,, ,, ,, ,,
Ca - cw	48±7	100	Castle et al. (1955)
Ca - d	43±8	100	Castle and King (1948)
Ca - h	62±9	35	,, ,, ,, ,,
Ca - j	53±5	31	,, ,, ,, ,,
Ca - k	53±13	100	,, ,, ,, ,,
Ca - p	48±4	100	,, ,, ,, ,,
Ca - Sh	51±5	0	Castle and King (1947a)
Ca - sr	51±9	100	Castle et al. (1955)
Ca - wo	50±14	100	Castle and King (1948)
Cu-1 - Cu-2	54±7	0	Gregory and Blunn (1936)
Cu-1 - cw	46±4	100	Castle et al. (1955)
Cu-1 - d	50±3	100	King and Castle (1935)
Cu-1 - h	59±10	100	,, ,, ,, ,,
Cu-1 - Ha	49±7	100	Burhoe (1947)
Cu-1 - Hm	48±7	100	,, ,,
Cu-1 - hr	49±5	100	King and Castle (1935)
Cu-1 - j	49±2	?	Castle and King (1940)
Cu-1 - k	58±3	100	Feldman (1935), King and Castle (1935)
Cu-1 - p	57±5	100	King and Castle (1935)
Cu-1 - sr	43±5	100	Castle et al. (1955)
Cu-1 - wo	51±7	?	Castle et al. (1941)
Cu-2 - cw	48±5	100	Castle et al. (1955)
Cu-2 - d	54±6	100	Blunn and Gregory (1937)
Cu-2 - h	47±5	0	,, ,, ,, ,,
Cu-2 - Ha	48±6	100	Burhoe (1947)
Cu-2 - Hm	54±5	100	,, ,,
Cu-2 - hr	49±5	100	Blunn and Gregory (1937)
Cu-2 - j	46±4	?	Castle and King (1940)
Cu-2 - k	66±9	100	Blunn and Gregory (1937)
Cu-2 - p	48±4	100	,, ,, ,, ,,
Cu-2 - sr	53±5	100	Castle et al. (1955)
Cu-2 - wo	50±4	0	Castle et al. (1941)
cw - d	46±7	0	,, ,, ,, ,,

Loci	Recombination value	Phase balance	References
cw - h	55±9	0	Castle et al. (1941)
cw - j	46±7	0	,, ,, ,, ,,
cw - k	46±5	0	,, ,, ,, ,,
cw - p	59±6	0	,, ,, ,, ,,
cw - Sh	50±4	100	,, ,, ,, ,,
cw - sr	46±3	0	,, ,, ,, ,,
cw - wo	53±6	0	,, ,, ,, ,,
d - f	45±7	100	Castle and King (1947b)
d - h	47±2	100	Roberts and Quisenberg (1936)
d - Ha	47±6	100	Burhoe (1947)
d - Hm	54±8	100	,, ,,
d - hr	51±2	90	Roberts and Quisenberg (1936)
d - j	62±6	0	Castle and King (1940)
d - k	50±5	0	Mitchell (1935)
d - r	51±2	84	Roberts and Quisenberg (1935)
d - Sh	52±6	0	Castle and King (1947a)
d - sr	31±8	0	Castle et al. (1955)
d - wo	51±4	?	Castle et al. (1941)
dx - p	48±9	0	Harris et al. (1963)
h - Ha	50±6	100	Burhoe (1947)
h - Hm	50±5	0	,, ,,
h - hr	50±2	100	Roberts et al. (1940)
h - j	47±3	?	Castle and King (1940)
h - k	57±4	100	King and Castle (1935), Mitchell (1935)
h - r	50±2	61	Ibsen (1920), Roberts and Quisenberg (1936)
h - R-1	55±5	-	Stark et al. (1970)
h - sr	47±4	0	Castle et al. (1955)
h - wo	47±3	?	Castle et al. (1941)
h - ♂	47±6	80	Whiting and King (1918)
Ha - hr	47±4	100	Burhoe (1947)
Ha - k	50±5	0	,, ,,
Ha - r	54±7	0	,, ,,
Hm - hr	53±7	0	,, ,,
Hm - k	48±7	0	,, ,,
Hm - r	47±7	0	,, ,,
hr - j	52±6	0	Castle and King (1940)
hr - k	51±5	0	Mitchell (1935)
hr - n	45±4	33	Castle (1955), Castle et al. (1955)
hr - p	52±5	100	King and Castle (1935)
hr - r	50±2	100	Roberts et al. (1940)
hr - sr	60±8	0	Castle (1955)
hr - wo	44±3	0	Castle (1955), Castle et al. (1941)

principal objections have been detailed by Makino (1943) and, particularly, by Matthey (1957). The main argument hinges on the alleged end-to-end association of the X and Y in such a manner to exclude the formation of chiasmata. Tijo and Levan (1956b) could not be definite that chiasmata did not occur for the sex bivalent but, on the other hand, these authors were not prepared to discount the possibility that the end-to-end association was the result of a preceding terminal chiasma.

A problem in the study of the sex bivalent, is the presence of a sex vesicle during early prophase, which obscures the movement of the chromosomes. However, by means of special pre-treatments, Ohno et al. (1957, 1958), have concluded that the X chromosome is initially bent into a V and that the Y lies in a side-by-side association with the arm of the X containing the centromere. This alignment implies the homologous nature of the adjacent portions. By early diakinesis, the X had straightened and the bivalent had changed to the well-known end-to-end configuration. This description, of course, suggests that chiasmata can occur and probably occur regularly. The essence of the matter is whether both sex chromosomes possess a minute second arm. The whole question receives an exhaustive discussion in Robinson (1965).

A spontaneous reciprocal translocation is described by Bouricius (1948). The first evidence was partial infertility (of the order of 40 per cent of ova) and the transmission of the partial infertility from parent to offspring (of the order of 50 per cent); see Tyler and Chapman (1948). Some difficulty was experienced in obtaining good diplotene figures but Bouricius concluded that the translocation involved a large and a small chromosome.

CHIASMA FREQUENCY

The frequency of chiasmata has been briefly noted by several observers but only one has been willing to devote time to collect quantitative data. Before these are outlined, the more fragmentary observations will be described. This seems worthwhile because of the importance of chiasma frequency in cytogenetic theory.

TABLE 12.5. Summary of karyological investigations of the Norway rat.

Defined autosomal groups	No. of metacentrics			No. of acrocentrics			Sex chromosomes		References
	Large	Medium	Small	Large	Medium	Small	X	Y	
6	0	4	3	2	10	1	Ma	Sa	Tobias (1947)
—	0	1	5	2	5	7	Ma	Sa	Makino and Hsu (1954)
3	0	0	5	2	6	7	Ma	Sa	Yosida (1955a, b)
3	0	0	7	2	8	3	Ma	Sa	Ohno and Kinosita (1955)
3	0	0	7	2	5	6	Ma	Sa	Tijo and Levan (1956a)
3	0	2	6	2	8	2	Ma	Sa	Tanaka and Kano (1957)
3	0	0	5	2	8	5	Ma	Sa	Makino (1957)
3	0	0	7	2	8	3	Ma	Sa	Makino and Sasaki (1958)
3	0	0	5	2	9	4	Ma	Sa	Ohno et al. (1959)
2	0	0	9	2	0	9	?	Sa	Yosida et al. (1960)
2	0	0	10	2	3	5	Ma	Sa	Fitzgerald (1961)
9	0	0	5	2	8	5	Ma	Sa	Hungerford and Nowell (1963); Nowell et al. (1968)
5	0	0	7	2	8	3	Ma	Sa	Honda (1964)
3	0	0	7	2	5	6	Ma	Sa	Vrba (1964)
3	0	0	7	2	8	3	Ma	Sa	Yosida and Amano (1965)
10	0	0	7	2	8	3	Ma	Sa	Zieverink and Moloney (1965)
8	0	0	7	2	8	3	Ma	Sa	Tagaki and Makino (1966)
5	0	0	5	2	8	5	Ma	Sa	Bianchi and Molina (1966)
10	0	0	7	2	8	3	Ma	Sa	Dyban and Udalova (1967)
3	0	0	7	2	5	6	Ma	Sa	Zidek (1968)
4	0	0	7	2	9	2	Ma	Sa	Rees et al. (1968)
—	0	0	8	2	9	1	Ma	Sa	Mittwoch (1970)

M = medium; S = small; a = acrocentric; acrocentric embraces subtelocentric and telocentric elements

TABLE 12.6. Classification of chromosomes by centromere position.

Autosomes			Sex chromosomes		References
Metacentric	Subtelocentric	Telocentric	X	Y	
6	13	0	st	st	Tobias (1947)
5	3	12	st	t	Makino and Hsu (1954)
5	7	8	st	st	Yosida (1955a, b)
7	5	8	st	t	Ohno and Kinosita (1955)
7	5	8	t	t	Tijo and Levan (1956a)
7	4	9	st	m	Tanaka and Kano (1956)
5	9	6	st	t	Makino (1957)
7	5	8	st	t	Makino and Sasaki (1958)
5	4	11	st	t	Ohno et al. (1959)
7	5	8	t	t	Fitzgerald (1961)
7	5	8	t	t	Hungerford and Nowell (1963), Nowell et al. (1963)
7	5	8	t	t	Honda (1964)
7	5	8	t	t	Vrba (1964)
7	5	8	t	t	Yosida and Amano (1965)
7	4	9	t	t	Yosida and Amano (1965)
7	5	8	t	t	Zieverink and Moloney (1965)
7	5	8	t	t	Tagaki and Makino (1966)
7	4	9	t	t	Bianchi and Molina (1966)
7	3	10	t	t	Dyban and Udalova (1967)
7	5	8	t	t	Zidek (1968)
8	4	8	t	t	Rees et al. (1968)

m = metacentric, st = subtelocentric, t = telocentric

Pincus (1927) did not examine in detail the occurrence of chiasmata in his material but Bryden (1932) made a belated examination in his published figures of metaphased plates. In Bryden's opinion, the bivalents in Pincus' plates can be interpreted as displaying a percentage distribution of 49, 48, and 3 for 1, 2, and 3 chiasmata per bivalent. The mean frequency is 1.58. These figures agree closely with those found by Bryden. The percentage distribution for Bryden's observations is 48, 47, and 5 for the same number of chiasmata at metaphase (Table 12.7, entries 3 and 5). The mean frequencies are correspondingly consistent.

Koller and Darlington (1934) are primarily concerned with the sex chromosomes but comment pertinently on chiasmata among the autosomes. A length ratio of roughly 1 : 6 was reported between the larger and shortest chromosomes. At diplotene, chiasmata could be regularly observed, the number per bivalent varying from 1 to 4, with a mean of 2. The shortest bivalents always had one chiasma while the longest has a mean of 3. There is, therefore, a lack of simply proportionality between number of chiasmata and chromosome length. No obvious reduction of mean chiasmata was observed to occur between diplotene and metaphase. Koller and Darlington's depiction of chiasmata in the sex chromosomes has been disputed but, if their interpretation is accepted, the sex chromosomes regularly have one chiasma but may have as many as two or three at low frequencies.

The most extensive observations on chiasmata frequency are those of Bryden (1932, 33a, b, 1935, 1936a, b, 1937). It was found impossible to enumerate the frequency over all bivalents of a nucleus, so recourse was made to sampling of clear bivalent configurations over a large number of cells. In most nuclei, only about 16 to 17 bivalents could be expected to yield good interpretations of chiasma formation. It is obvious that more than one sort of bias may result from this method; bias which may not be corrected by simply sampling from a large number of cells.

Bryden's results are summarized by Tables 12.7 to 12.9. For technical reasons, the majority of observations are on spermatogenesis. It was thought at one time that chiasmata in the longer chromosomes might be localized but further observation failed to substantiate the belief. There was a tendency for the frequency to be related to the length of chromosome for the longer elements but not for the shorter. It is as if the length must exceed a threshold before a length/frequency relationship becomes apparent. The figures of Table 12.7 show that chiasmata frequency declines as meiosis proceeds, probably because of movement of the chiasmata towards the free ends of the chromosome and their eventual 'slipping-off'. It may be noticed that the means for the observations of entries 3 and 5 of Table 12.7 are significantly different in a formal statistical sense. However, the difference is not great and could be due to causes other than a change in chiasmata frequency.

A sex difference is brought out by Table 12.8. The mean chiasma frequency is significantly greater in the female compared with that of the male.

Table 12.9 indicates that the mean frequency of chiasma declines with age, the effect being most marked for old males. The decline is probably an aspect of a decrease in spermatogenesis. The 17 month old animal was found to possess about 65 per cent of spermic tubules devoid of meiotic activity. This compares with a value of 13 per cent for males of the other two age groups.

All four of Bryden's papers from 1935 to 1937 are indirectly concerned with chiasmata. Primarily, the effects of certain experimental treatments (e.g. artificial cryptorchidism) on mitosis and meiosis was investigated and chiasma formation was one of the variables examined. In the present context, the results for control animals are reproduced since these may be held to be free of possible experimental bias. However, the results obtained by Bryden imply that the experimental treatments had little effect. The distribution of frequencies and mean values do not differ greatly from those of the controls. Attention is drawn to the availability of this supplementary data should the occasion warrant.

The chiasmata data have several implications for genetic mapping. The first is a rough estimate of the average length of genetic chromosome, if the assumption is made that one chiasma per bivalent corresponds to 50 crossover units. Summing entries 1 and 4 of Table 12.7 gives a mean frequency

TABLE 12.7. Frequency distribution of chiasmata per autosomal bivalent at different phases of meiosis in the male Norway rat.

Phase	No. of bivalents	Chiasmata frequency				Mean	Reference
		1	2	3	4		
Diplotene	320	90	175	50	5	1.91 ± 0.04	Bryden (1932)
Diakinesis	320	141	156	23	0	1.63 ± 0.04	Bryden (1932)
Metaphase	320	150	156	14	0	1.58 ± 0.03	Bryden (1932)
Diplotene	1500	480	594	331	95	2.03 ± 0.02	Bryden (1936b, 1937)
Metaphase	1200	582	552	66	0	1.57 ± 0.02	Bryden (1936b, 1937)

TABLE 12.8. Frequency distribution of chiasmata per bivalent according to sex of the rat.

Sex and phase	No. of bivalents	Chiasmata frequency					Mean	Reference
		1	2	3	4	5		
Male								
Diplotene	200	55	85	47	13	0	2.09 ± 0.06	Bryden (1933a)
Metaphase	100	49	43	8	0	0	1.59 ± 0.06	Bryden (1933a)
Female								
Diplotene	100	19	24	30	22	5	2.70 ± 0.11	Bryden (1933a)
Metaphase	100	31	50	11	8	0	1.96 ± 0.13	Bryden (1933a)

TABLE 12.9. Variation of chiasmata with age in the male rat.

Age and phase	No. of bivalents	Chiasmata frequency				Mean	Reference
		1	2	3	4		
Two months							
Diplotene	900	288	378	180	54	2.00 ± 0.03	Bryden (1933b)
Metaphase	900	468	306	126	0	1.62 ± 0.02	Bryden (1933b)
Eight months							
Diplotene	900	279	399	186	36	1.98 ± 0.03	Bryden (1933b)
Metaphase	900	458	369	73	0	1.58 ± 0.02	Bryden (1933b)
17 months							
Diplotene	900	363	351	156	30	1.84 ± 0.03	Bryden (1933b)
Metaphase	900	548	342	10	0	1.40 ± 0.02	Bryden (1933b)

of 2.00 ± 0.02 chiasmata per bivalent. This would correspond to a mean genetic map of 100.00 ± 1.00 crossover units. Some chromosomes are known to be longer than others (two, at least, much longer), hence the range is much greater than the standard error would indicate. This estimate must be regarded as very tentative, of course, and may have to be modified pending: (a) more accurate assessment of the mean chiasma frequency either over the whole karyotype, (b) for peculiar chromosomes, and (c) development of the functional relationship between chiasmata frequency and variation of chromosome length. A theoretical mean genetic map of some 100 units would be capable of engendering a large number of loose linkages, particularly while the number of known genes is few.

Two further implications may be deduced, both of which appear to be reasonably established. The first is a decline in the rate of crossingover with age. The decline may be slight for middle aged animals but may become pronounced in the old. A decline is almost certain for the male but, since spermatogenesis and oogenesis differ rather fundamentally, it may be wise to be cautious of a similar decline for the female. The second is a higher rate of crossingover in the female compared with that of the male. This expectation need not apply to all chromosomes, or to all regions of a particular chromosome, but would be in keeping with what appears to be a general tendency in this direction for most mammalian species.

BIBLIOGRAPHY

BIANCHI, N. O. and MOLINA, O. (1966). Autosomal polymorphism in a laboratory strain of rat. *J. Hered.*, 57, 231-232.

BILLINGHAM, R. E., HODGE, B. A. and SILVERS, W. (1962). An estimate of the number of histocompatibility loci in the rat. *Proc. Nat. Acad. Sci.*, 48, 138-147.

BLUNN, C.T. and GREGORY, P. W. (1937). Linkage studies with curly$_2$ in the rat. *J. Hered.*, 28, 43-44.

BODGEN, A. E. and APTEKMAN, P. M. (1960). The R-1 factor, a histocompatibility antigen in the rat. *Cancer Res.*, 20, 1372-1382.

BOGDEN, A. E. and APTEKMAN, P. M. (1962). Histocompatibility antigens and hemagglutinogens in the rat. *Annal. N.Y. Acad. Sci.*, 97, 43-56.

BOURICIUS, J. K. (1948). Embryological and cytological studies in rats heterozygous for a probable reciprocal translocation. *Genetics*, 33, 577-587.

BRDICKA, R. (1966). Evidence for linkage between haemoglobin and chromogen loci. *Folia Biol.* (Praha), 12, 305-306.

BRDICKA, R. (1967). Genetics of rat haemoglobin. *Proc. X. Europ. Confer. Anim. Blood Biochem. Polymorph.*, Paris, 1966, 407-411.

BRDICKA, R. (1968). The chromosome I of the laboratory rat. *Acta Univ. Carol. Med.*, 14, 93-98.

BRYDEN, W. (1932). Cytogenetic studies on the rat; somatic chromosome complex, meiosis and chiasma frequency. *J. Genet.*, 26, 395-415.

BRYDEN, W. (1933a). The effect of sex on the frequency of chiasma formation and its relation to crossingover in the Wistar rat. *Cytologia*, 4, 241-247.

BRYDEN, W. (1933b). The relation of age to the frequency of chiasma formation in the Wistar rat. *J. Genet.*, 27, 415-420.

BRYDEN, W. (1935). Some observations upon the mitotic and meiotic divisions in the Wistar rat. I. The effect of changes in temperature. *Cytologia*, 6, 300-307.

BRYDEN, W. (1936a). Some observations upon the mitotic and meiotic divisions in the Wistar rat. II. Changes in temperature over localised areas. *Cytologia*, 7, 389-395.

BRYDEN, W. (1936b). Some observations upon the mitotic and meiotic divisions in the Wistar rat. III. Effects produced by experimental cryptorchidism. *Cytologia*, 7, 499-503.

BRYDEN, W. (1937). Some observations upon the mitotic and meiotic divisions in the Wistar rat. IV. Effect of factors influencing the functional development of the male gonad. *Cytologia*, Fujii jub. vol., 627-632.

BURHOE, S. O. (1947). Blood groups of the rat. (*Rattus norvegicus*) and their inheritance. *Proc. Nat. Acad. Sci. U.S.A.*, 33, 102-109.

CASTLE, W. E. (1916). Further studies of piebald rats and selection, with observations on gametic coupling. *Car. Inst. Wash. Pub.*, 241.

CASTLE, W. E. (1919). Observations on the occurrence of linkage in rats and mice. *Car. Inst. Wash. Pub.*, 288, 29-36.

CASTLE, W. E. (1920). The measurement of linkage. *Amer. Nat.*, 264-267.

CASTLE, W. E. (1926). A sex difference in linkage in rats and mice. *Genetics*, 10, 580-582.

CASTLE, W. E. (1939). On a method for testing for linkage between lethal genes. *Proc. Nat. Acad. Sci. U.S.A.*, 25, 593-594.

CASTLE, W. E. (1942). Experimental studies of heredity in small mammals. *Car. Inst. Yearb.*, no. 41, 225-226.

CASTLE, W. E. (1944). Linkage of waltzing in the rat. *Proc. Nat. Acad. Sci. U.S.A.*, 30, 226-230.

CASTLE, W. E. (1946). Linkage in the albino chromosome of the rat. *Proc. Nat. Acad. Sci. U.S.A.*, 32, 33-36.

CASTLE, W. E. (1947). The domestication of the rat. *Proc. Nat. Acad.*

Sci. U.S.A., 33, 109-117.

CASTLE, W. E. (1952). Genetic linkage in the common rat, *Rattus norvegicus. Virginia J. Sci.*, 3, 95-100.

CASTLE, W. E. (1953). Silver, a new mutation of the rat. *J. Hered.*, 44, 205-206.

CASTLE, W. E. (1955). Further studies of linkage in the third chromosome of the rat. *J. Hered.*, 46, 84-86.

CASTLE, W. E., DEMPSTER, E. R. and SHURRAGER, W. C. (1955). Three new mutations of the rat. *J. Hered.*, 46, 9-14.

CASTLE, W. E. and KING, H. D. (1940). Linkage studies of the rat (*Rattus norvegicus*) III. *Proc. Nat. Acad. Sci. U.S.A.*, 26, 578-580.

CASTLE, W. E. and KING, H. D. (1941). Linkage studies of the rat (*Rattus norvegicus*) V. *Proc. Nat. Acad. Sci. U.S.A.*, 27, 394-398.

CASTLE, W. E. and KING, H. D. (1944). Linkage studies of the rat. (*Rattus norvegicus*). VI. *Proc. Nat. Acad. Sci. U.S.A.*, 30, 79-82.

CASTLE, W. E. and KING, H. D. (1947a). Linkage studies of the rat. VII. Shaggy, a new dominant. *J. Hered.*, 38, 341-343.

CASTLE, W. E. and KING, H. D. (1947b). Linkage studies of the rat. VIII. Fawn, a new color dilution gene. *J. Hered.*, 38, 343-344.

CASTLE, W. E. and KING, H. D. (1948). Linkage studies of the rat. IX. Cataract. *Proc. Nat. Acad. Sci. U.S.A.*, 35, 135-136.

CASTLE, W. E. and KING, H. D. (1949). Linkage studies of the rat. X. *Proc. Nat. Acad. Sci. U.S.A.*, 35, 545-546.

CASTLE, W. E., KING, H. D. and DANIELS, A. L. (1941). Linkage studies of the rat (*Rattus norvegicus*). IV. *Proc. Nat. Acad. Sci. U.S.A.*, 27, 250-253.

CASTLE, W. E. and WACHTER, W. L. (1924). Variation of linkage in rats and mice. *Genetics*, 9, 1-12.

CASTLE, W. E. and WRIGHT, S. (1915). Two colour mutations of rats which show coupling. *Science*, 42, 193-195.

DARLINGTON, C. D., HALDANE, J. B. S. and KOLLER, P. C. (1934). Possibility of incomplete sex linkage in mammals. *Nature*, 133, 417.

DETLEFSEN, J. A. (1925). The linkage of dark-eye and color in mice. *Genetics*, 10, 17-32.

DUNN, L. C. (1920). Linkage in rats and mice. *Genetics*, 5, 325-343.

DUNNING, W. F. and CURTIS, M. R. (1939). Linkage in rats between factors determining a pathological condition and a coat colour. *Genetics*, 24, 70.

DYBAN, A. P. and UDALOVA, L. D. (1967). [The characteristic morphological features of the X chromosome and the third pair of autosomes in different strains of rats.] *Genetika (Mosk.)*, 1967 (2), 125-135.

FELDMAN, H. W. (1924). Linkage of albino allelomorphs in rats and mice. *Genetics*, 9, 487-492.

FELDMAN, H. W. (1935). A recessive curly-haired character of the Norway rat. *J. Hered.*, 26, 252-254.

FITZGERALD, P. H. (1961). Cytological identification of sex in somatic cells of the rat. *Exp. Cell Res.*, 25, 191-193.

FRENCH, E. A., ROBERTS, K. B. and SEARLE, A. G. (1971). Linkage between a haemoglobin locus and albinism in the Norway rat. *Biochem. Genet.*, 5, 597-404.

GREAVES, J. H. and AYRES, P. (1969). Linkages between genes for coat colour and resistance to Warfarin in *Rattus norvegicus. Nature*, 224, 284-285.

GREGORY, P. W. and BLUNN, C. T. (1936). Curly$_2$, a recent dominant mutation in the Norway rat. *J. Hered.*, 27, 38-40.

GRUNEBERG, H. (1939). The linkage relations of a new lethal gene in the rat (*Rattus norvegicus). Genetics*, 24, 732-741.

HANSON, F. B. and STEWART, D. R. (1923). A study of albino rats carrying factors for agouti and the hooded pattern. *Wash. Univ. Stud.*, 11, 71-88.

HARRIS, J. M., KALMUS, H. and WEST, C. B. (1963). Genetical control of the anaphylactoid reaction in rats. *Genet. Res.*, 4, 346-355.

HONDA, T. (1964). Chromosomes of the regenerating rat liver after partial hepatectomy. *Jap. J. Genet.*, 36, 69-73.

HUNGERFORD, D. A. and NOWELL, P. C. (1963). Sex chromosome polymorphism and the normal karyotype in three strains of the laboratory rat. *J. Morph.*, 113, 275-286.

IBSEN, H. L. (1920). Linkage in rats. *Amer. Nat.*, 54, 61-67.

KING, H. D. and CASTLE, W. E. (1935). Linkage studies of the rat (*Rattus norvegicus). Proc. Nat. Acad. Sci. U.S.A.*, 21, 390-399.

KING, H. D. and CASTLE, W. E. (1937). Linkage studies of the rat (*Rattus norvegicus). II. Proc. Nat. Acad. Sci. U.S.A.*, 23, 56-60.

KOLLER, P. C. and DARLINGTON, C. D. (1934). The genetical and mechanical properties of the sex chromosomes. I. *Rattus norvegicus. J. Genet.*, 29, 159-173.

LOZZIO, B. B., CHERNOFF, A. I., MACHADO, E. R. and LOZZIO, C. B. (1967). Hereditary renal disease in a mutant strain of rats. *Science*, 156, 1742-1744.

MAKINO, S. (1943). Studies on the murine chromosomes. III. A comparative study of chromosomes in five species of *Rattus. J. Fac. Sci., Hokkaido Univ.*, VI, 9, 19-58.

MAKINO, S. (1957). The chromosome cytology of the ascites tumours of rats, with special reference to the concept of the stemline cell. *Inter. Rev. Cytol.*, 6, 25-84.

MAKINO, S. and HSU, T. C. (1954). Mammalian chromosomes *in vitro*. V. The somatic complement of the Norway rat, *Rattus norvegicus. Cytologia*, 19, 23-28.

MAKINO, S. and SASAKI, M. (1958). Cytological studies of tumors. XXI. A comparative ideogram study of the Yoshida sarcoma and its sublime derivatives. *J. Nat. Cancer Inst.*, 20, 465-488.

MATTHEY, R. (1957). Les bases cytologiques de l'hérédité "relativement" liée au sexe chez les mammifères. *Experientia*, 13, 341-346.

MITCHELL, A. L. (1935). Inheritance and linkage relations of kinky coat, a new mutation in the Norway Rat. *Proc. Nat. Acad. Sci. U.S.A.*, 21,

453-456.

MITTWOCH, U. (1970). How does the Y chromosome affect gonadal differentiation? *Phil. Trans. Roy. Soc., B*, **259**, 113-117.

NOWELL, P. C., FERRY, F. and HUNGERFORD, D. A. (1963). The chromosomes of primary granulocytic leukemia (chloroleukemia) in the rat. *J. Nat. Cancer Inst.*, **30**, 687-703.

OHNO, S., KAPLAN, W. D. and KINOSITA, E. (1957). Conjugation of the heteropycnotic *X* and *Y* chromosomes of the rat spermatocyte. *Exp. Cell Res.*, **12**, 395-397.

OHNO, S., KAPLAN, W. D. and KINOSITA, R. (1958). A photographic representation of mitosis and meiosis in the male of *Rattus norvegicus*. *Cytologia*, **23**, 422-428.

OHNO, S., KAPLAN, W. D. and KINOSITA, R. (1959). The centromeric and nucleolus-associated heterochromatin of *Rattus norvegicus*. *Exp. Cell Res.*, **16**, 348-357.

OHNO, S. and KINOSITA, R. (1955). The primary and secondary constrictions on the chromosomes of the rat lymphoblast. *Exp. Cell Res.*, **8**, 558-562.

OWEN, R. D. (1962). Earlier studies of blood groups in the rat. *Ann. N.Y. Acad. Sci.*, **97**, 37-42.

PALM, J. (1962). Current status of blood groups in rats. *Annals N.Y. Acad. Sci.*, **97**, 57-68.

PALM, J. (1964). Serological detection of histocompatibility antigens in two strains of rats. *Transplantation*, **2**, 603-612.

PALM, J. (1971). (Personal Communication).

PALM, J. and BLACK, G. (1971). Interrelations of inbred rat strains with respect to *Ag-B* and non-*Ag-B* antigens. *Transplantation*, **11**, 184-189.

PINCUS, G. (1927). A comparative study of the chromosomes of the Norway rat (*Rattus norvegicus*) and the black rat (*Rattus rattus*). *J. Morph. Physiol.*, **44**, 515-536.

REES, E. D., SHUCK, A. E., CHRISTIAN, J. C. and PUGH, J. R. (1968). Karyotypes of rats from strains of different susceptibility to mammary cancer induction. *Cancer Res.*, **28**, 823-830.

ROBERTS, E. (1926). Further data on inheritance of hypotrichosis in rats. *Anat. Rec.*, **14**, 172.

ROBERTS, E. and QUISENBERG, J. H. (1936). Linkage studies in the rat. *Amer. Nat.*, **70**, 395-399.

ROBERTS, E., QUISENBERG, J. H. and THOMAS, L. C. (1940). Hereditary hypotrichosis in the rat. *J. Invest. Dermatol.*, **3**, 1-29.

ROBINSON, R. (1960). A review of independent and linked segregation in the Norway rat. *J. Genet.*, **57**, 173-192.

ROBINSON, R. (1965). *Genetics of the Norway rat*. Oxford: Pergamon Press.

STARK, O., KREN, V., FRENZL, B. and BRDICKA, R. (1968). The main histocompatibility system of the rat. In Dausset, J., Hamburger, J., and Mathe, G. (Editors). *Advances in transplantation*. Baltimore: Williams and Wilkins.

STARK, O., KREN, V., FRENZL, B. and KRSIAKOVA, M. (1969).

Independent segregation of *RtH-1* alleles and coat colour genes in the rat. *Folia Biol.* *(Praha)*, 15, 470-473.

TAGAKI, N. and MAKINO, S. (1966). An autoradiographic study of the chromosomes of the rat, with special regard to the sex chromosomes. *Chromosoma*, 18, 359-370.

TANAKA, T. and KANO, K. (1956). On the somatic chromosomes of rats. *Proc. Inter. Genet. Symp.*, 1956, 196-201.

TIJO, J. H. and LEVAN, A. (1956a). Comparative idiogram analysis of the rat and the Yosida rat sarcoma. *Hereditas*, 42, 218-234.

TIJO, J. H. and LEVAN, A. (1956b). Note on the sex chromosomes of the rat during meiosis. *Anal. Estac. Exp. Aula Dei*, 4, 173-184.

TOBIAS, P. V. (1947). The characterisation of the spermatogonial chromosomes of the albino rat. (*Rattus norvegicus albinus*). *S. Afr. J. Sci.*, 43, 312-319.

TYLER, W. J. and CHAPMAN, A. B. (1948). Genetically reduced prolificacy in rats. *Genetics*, 33, 565-576.

VRBA, M. (1964). Idiogram of the rat and reliability in identification of individual chromosomes. *Folia Biol. (Praha)*, 10, 75-80.

WHITING, P. W. and KING, H. D. (1918). Ruby-eyed dilute grey, a third allelomorph in the albino series of the rat. *J. Exp. Zool.*, 26, 55-64.

WHITTINGHILL, M. (1944). Concerning linkage of waltzing in rats. *Proc. Nat. Acad. Sci. U.S.A.*, 30, 221-226.

WILDER, W., BETHKE, R. M., KICK, C. H. and SPENCER, W. P. (1932). A hairless mutation in the rat. *J. Hered.*, 23, 480-484.

YOSIDA, T. H. (1955a). Chromosome constitutions in male germ cells of the rat. *Ann. Rep. Nat. Inst. Genet. Jap.*, No. 5, 18-19.

YOSIDA, T. H. (1955b). Origin of V-shaped chromosomes occurring in tumor cells of some ascites tumors in the rat. *Proc. Jap. Acad.*, 31, 237-242.

YOSIDA, T. H. (1960). Genetic aspects of two color mutants of rats caught in the suburbs of Misima. *Ann. Rep. Nat. Inst. Jap.*, 10, 29-30.

YOSIDA, T. H. and AMANO, K. (1965). Autosomal polymorphism in laboratory bred and wild Norway rats. *Chromosoma*, 16, 658-667.

YOSIDA, T. H., ISHIHARA, T. and ODASHIRO, T. (1960). Chromosome attraction and development of tumours. IV. Comparative idiogram analysis in cells of normal rat liver and rat ascites hepatoma No. 7974. *Jap. J. Genet.*, 35, 35-40.

YOSIDA, T. H., KURITA, Y. and TANEDA, S. (1961). Genetic study of two wild mutants of rats. *Bull. Exper. Anim.*, 10, 20-22.

ZIDEK, Z. (1968). Karyotypes of four inbred strains of rats: AVN, BP, LEW, WP *Folia Biol. (Praha)*, 14, 74-79.

ZIEVERINK, W. D. and MOLONEY, W. C. (1965). Use of the *Y* chromosome in the Wistar/Furth rat as a cellular marker. *Proc. Soc. Exp. Biol. Med.*, 119, 370-373.

CHAPTER 13

Rabbit

Oryctolagus cuniculus $(n = 21 + X + Y)$

The domestic rabbit featured prominently in the early days of mammalian genetics but this popularity has waned in recent decades. This has reflected upon the rate of discovery of mutants for the species and upon the number of known linkage groups. Despite this, the rabbit still remains a useful research animal for a variety of topics. Some 45-50 mutant gene/loci have been reported, of which 34 have been employed in studies of independent assortment and linkage. A detailed collation of early work is presented by Robinson (1956).

Three variable systems of esterase content of the red blood cells have been described by Schiff and Stormont (1970). All three systems are controlled by codominant genes. As yet, these are unsymbolized. Preliminary results indicate that two of the systems (nos. 1 and 2) are closely linked (zero recombination among 51 backcross progeny). Ancillary data of an unsystematic nature suggest that crossingover may occur but at a low frequency. At this stage, it is not possible to assign genes to a definite linkage group.

TABLE 13.1. Genes/loci of the rabbit which have featured in studies on linkage: assigned linkage group, symbol, and conventional designation

Linkage Group	Symbol	Designation	Prime Characteristic
IV	*a*	Non-agouti	Coat colour
	Aa	Immunoglobin	Immunogenetics
	Ab	Immunoglobin	Immunogenetics
	Ac	Immunoglobin	Immunogenetics
	Ae	Immunoglobin	Immunogenetics
	Af	Immunoglobin	Immunogenetics
V	*An*	Antigen A	Immunogenetics
VI	*As*	Atropinesterase	Physiology
	ax	Ataxia	Behaviour
I	*b*	Brown	Coat colour
V	*br*	Brachydactyly	Feet

Linkage Group	Symbol	Designation	Prime Characteristics
I	c	Albinism	Coat colour
	d	Dilute	Coat colour
II	du	Dutch	White spotting
IV	dw	Dwarf	Body size
VI	e	Yellow	Coat colour
II	En	English	White spotting
	Ess	α_1 aryl esterase	Immunogenetics
V	f	Furless	Hypotrichosis
	H	Blood antigen	Immunogenetics
	H-6	Blood antigen	Immunogenetics
	Hph	Haptoglobin	Immunogenetics
II	l	Long hair	Hair texture
	Lpq	Lipoprotein	Immunogenetics
	Mtz	α_2 macroglobulin	Immunogenetics
III	r-1	French rex	Hair texture
	r-2	German rex	Hair texture
	r-3	Normandy rex	Hair texture
	s	White marked	White spotting
	sa	Satin	Hair texture
	v	Viennese white	White spotting
IV	w	Wide-band	Coat colour
	wa	Waved	Hair texture
I	y	Yellow fat	Physiology

LINKAGE GROUPS

Twelve pairs of linked loci have so far been discovered, forming the basis for six linkage groups. A summary of the data is presented by Table 13.2. A sex difference in the rate of crossingover has been established for at least two pairs of genes as revealed by Table 13.3.

An association between the immunoglobin locus Aa and a newly postulated locus is outlined by Hamers, Hamers-Casterman, and Lagnaux (1966). Two systems of red-cell esterase activity (one of aliesterases and the other probably of acetylcholinesterases) are reported to be linked by Schiff and Stormont (1968). Linkage is imputed by Dubiski (1969) to explain an association between Aa and another immunoglobin locus Ae. In none of these instances has the situation been described in detail with adequate supporting data.

Group I (b, c, y)

The mean crossover values for the *b - c* and *b - y* intercepts are 34.6 ± 1.1 and 28.2 ± 1.4, respectively, based upon consistent data. The backcross data of Castle (1936) for the genes *c* and *y* produce a mean crossover value of 13.7 ± 1.1. Pease (1928) also contributes data for the *c - y* interval but his intercross segregation yields the suspiciously high value of 33.1 ± 6.8. However, the data do not permit of straightforward estimation and the value of 33.1 is seemingly falsely inflated.

Obvious differences occurred in the amount of crossingover between the sexes, especially for the *b - c* and *c - y* intercepts; these two attaining statistical significance (Table 13.3). The linear order of the genes is not in doubt and, allowing for the sex difference in crossingover, may be portrayed as:

$$\frac{c \qquad y \qquad b}{8 \qquad 27}$$

$$\frac{c \qquad y \qquad b}{14 \qquad 28}$$

Group II (du, En, l)

This group is interesting because it contains one of the closest linkages known in mammalian genetics. This is between the genes *du* and *En*, with a mean crossover value of 0.17 ± 0.12. In fact, the actual data are composed of two crossover individuals among the equivalency of 1178 backcrossed progeny. Both genes produce white spotting but it is a moot point whether this has special significance. The two types of spotting are very different phenotypically, hence true allelism was always doubtful and pseudo-allelism, while possible, could not be automatically assumed.

The mean crossover values for the *du - l* and *En - l* intercepts are 13.5 ± 0.8 and 13.2 ± 0.7, respectively. Similar values are to be expected because of the exceedingly close linkage of *du* and *En*.

For many years, the accepted order of the loci has been *du - En - l* on the basis of slightly greater crossingover between *du - l* compared with *En* and *l* (e.g., Sawin, 1955). However, Rifaat (1954) has proposed that the order may be *En - du - l.*

He makes the point that the solitary crossover animal reported by Castle (1926) in his tri-coupling experiment must have arisen by a double crossover by the former arrangement but by a single crossover by the latter. Rifaat's intricate analysis favours the former but the data are insufficient to establish this with complete certainty. The question of precise order is one which may not often arise in practice, because the closeness of the linkage is such that the two genes segregate as a unit in crosses (Castle, 1932).

Group III (r-1, r-2)

The phenotypic resemblance between these two genes is a complication but the mean crossover value is 16.3 ± 2.4.

Group IV (a, dw, w)

The mean crossover value for the gene pair *a* and *dw* is 14.7 ± 3.8. The mean crossover value for the *a - w* intercept is 31.1 ± 3.2, based upon two reliable backcross segregations. Sawin (1934) presents details of two further segregations. In one of these, the number of crossovers is in excess of independent assortment and, in the other, there is no numerical evidence that crossover occurred at all. The sample sizes are small and this probably accounts for the curious results. If these samples are included, the mean crossover value becomes 28.9 ± 3.1.

The linear order of these genes has yet to be determined. The order *a - dw - w* gives the shortest map length and a theoretical *dw - w* intercept of approximately 14.2 ± 4.9. The alternative order of *dw - a - w* implies a *dw - w* intercept in the region of 43.6 ± 4.9.

Group V (An, br, f)

The mean crossover value for the *An - br* intercept is 36.8 ± 4.4. The precise alignment of the three genes of this group is unknown because the amount of crossingover between the *An* and *f* genes has yet to be ascertained.

The shorter total map length is obtained by the order *An - f - br* which results in a predicted *f - An* intercept of about 8.6 ± 6.3. The alternative order of *An - br - f* is not impossible although the *An - f* intercept would be as large as 65.1 ± 6.3, should simple addition of the intervening intercepts hold true. It is impossible to exclude the latter order solely on the magnitude of predicted crossover values.

Group VI (As, e)
The mean crossover value for the intercept is 26.2 ± 6.8.

SEX-LINKAGE

No mutants residing in the sex chromosomes have so far been reported.

MULTI-POINT CROSSES

A three-point cross has been reported by Castle (1933b, 1936) for the triad b - c - y of linkage group I. Some interesting data emerged on interference. Very different results were obtained for each sex. In the male, $C = 2.09 ± 0.64$ and $K = 4.07 ± 1.04$ and in the female, $C = 0.551 ± 0.082$ and $K = 0.719 ± 0.167$. These figures imply negative interference for the male but positive interference for the female. However, the coincidence does not differ significantly from unity for the male although the Kosambi coefficient does so. On the other hand, the coincidence differs significantly from unity for the female but the Kosambi coefficient does not.

The data for the female consisted of 799 individuals and appear to be internally consistent. There appears to be no doubt that interference occurs for the b - c intercept in this sex. The male data, however, consisted of 109 animals and the double crossover class is over-represented. In fact, there are more double crossovers than those recorded for one of the single crossover classes. The male data, as these stand at present, either indicate a marked difference of crossover behaviour between the sexes or the sample is aberrant.

INDEPENDENT ASSORTMENT

Those studies which have failed to detect linkage between loci

TABLE 13.2. Established linkage groups in the rabbit

Group	Loci	Sex	Mating type	Crossover value	References
I	b - c	♂	CBB	29.4 ± 2.9	Castle (1926)
		♂	RBB	25.7 ± 4.2	Castle (1936)
		♀	CBB	36.2 ± 1.9	Castle (1926)
		♀	RBB	38.3 ± 1.7	Castle (1926, 1936)
		-	RII	38.8 ± 16	Zelnik (1965)
	b - y	♂	CBB	26.6 ± 4.2	Castle (1936)
		♀	CBB	28.4 ± 1.6	,, ,,
	c - y	♂	RBB	8.3 ± 2.6	,, ,,
		♀	RBB	14.4 ± 1.2	,, ,,
		-	RII	33.1 ± 6.9	Pease (1928)
	du - En	♂	CBB	0.19 ± 0.19	Castle (1926)
		♀	CBB	0	Castle (1926), Punnett (1928)
		?	CBB	0	Castle (1919)
		?	RBB	0.26 ± 0.56	Castle (1934)
		-	CII	0	Castle (1919)
	du - l	♂	CBB	9.2 ± 1.2	Castle (1926)
		♂	RBB	17.0 ± 5.2	,, ,,
		♀	CBB	10.6 ± 2.6	,, ,,
		♀	RBB	13.7 ± 1.2	,, ,,
II	du - En	♂	CBB	13.5 ± 0.9	,, ,,
		♂	RBB	18.2 ± 4.8	,, ,,
		♀	CBB	11.7 ± 3.6	,, ,,
		♀	RBB	17.9 ± 4.7	,, ,,
III	r-1 - r-2	♀	RBB	17.1 ± 2.8	Castle and Nachtsheim (1933)
		-	RII	12.9 ± 5.2	,, ,,
IV	a - dw	?	CBI	17.6 ± 4.5	Castle and Sawin (1941)
		?	RBI	8.2 ± 6.8	,, ,, ,, ,,
	a - w	♀	CBB	29.2 ± 4.0	Sawin (1934)
		♀	RBB	34.2 ± 5.4	,, ,,
		-	RII	0	,, ,,
V	An - br	♀	CBB	40.3 ± 5.3	Sawin et al. (1944)
		♀	RBB	29.1 ± 8.1	,, ,, ,, ,,
	Br - f	♂	RBB	28.3 ± 4.5	Castle and Sawin (1941)
VI	As - e	?	CBB	26.7 ± 8.0	Sawin and Glick (1943)
		?	RBB	25.1 ± 12.6	,, ,, ,, ,,

TABLE 13.3 Sex differences in crossover values for certain chromosome intercepts in the rabbit

Group Loci		Sex	Crossover value	References
I	b - c	♂	28.3 ± 2.6	Castle (1926, 1936)
		♀	37.4 ± 1.2	,, ,, ,,
		Diff.*	9.1 ± 2.7	
	b - y	♂	26.6 ± 4.2	Castle (1936)
		♀	28.4 ± 1.6	,, ,,
		Diff.	1.8 ± 4.5	
	c - y	♂	8.3 ± 2.6	,, ,,
		♀	14.4 ± 1.2	,, ,,
		Diff.*	6.1 ± 2.9	
II	du - l	♂	13.4 ± 1.3	Castle (1926)
		♀	13.6 ± 1.1	,, ,,
		Diff.	0.2 ± 1.7	
	En - l	♂	15.9 ± 1.0	,, ,,
		♀	13.6 ± 1.4	,, ,,
		Diff.	2.3 ± 1.7	

* Significant difference

are summarized by Table 13.4 and Fig. 13.1. The great majority of pairs of tested loci undoubtedly reside in different chromosomes although it is conceivable that some could be loosely linked. An idea of the strength of linkage which would be compatible with the observed joint assortment is shown by the entries of the figure. This value is approximated by multiplying the standard error by 196 and subtracting from the observed crossover value. It is clear from the distribution of entries that, except for the early known genes, there has been little systematic testing. This implies that a few linkages may yet be discovered even among those genes currently available. It is also clear that the amount of information for a number of gene pairs is inadequate to exclude linkages of less than 40.

Castle (1926) mentions that the linkage phase for the test of c - En was unknown. The recombination value, therefore, is 47 ± 5 or 53 ± 5 according to the phase assumed. However, the lower value has been adopted for Table 13.4 and for Fig. 13.1.

Castle and Law (1936) state that recombination occurs with such frequency as to indicate free assortment of sa with the ten

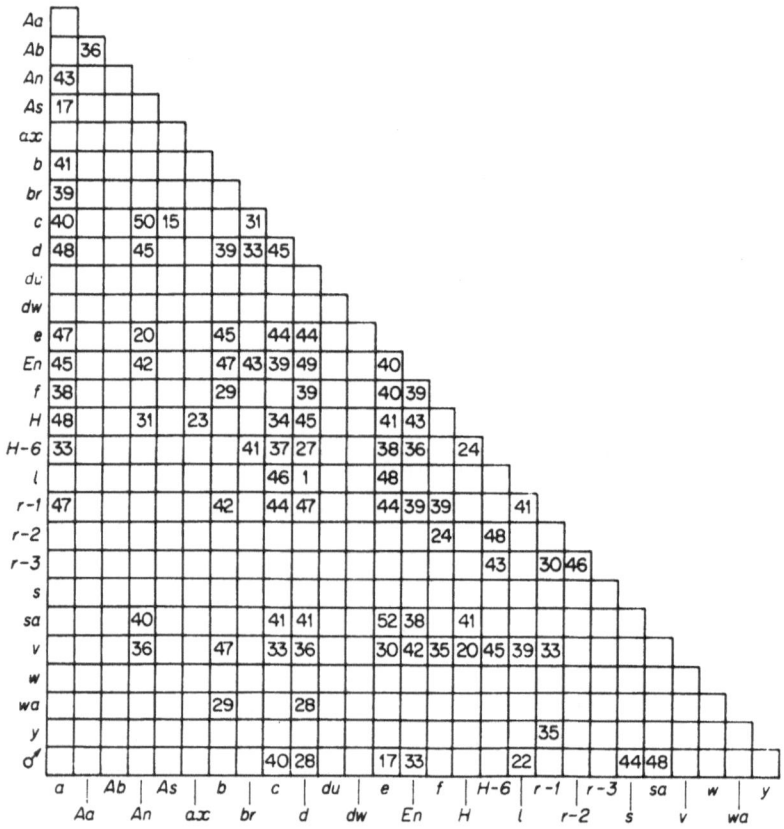

Fig. 13.1. Summary of the extent of independent assortment in the rabbit.

genes *a, b, c, e, En, f, H, l, r-1,* and *r-3,* without giving details of the data upon which this conclusion is based.

It is probable that the minor spotting gene symbolized by *s* is an allele of *du* but direct evidence of this is lacking (Robinson 1951).

Within recent years, a number of loci governing immunological and electrophoretic variables have been discovered. Several statements occur in the literature that the loci concerned are inherited independently. Specifically, Dubiski, Rapacz, and Dubiska (1962), Dray, Young, and Gerald (1963), Kelus and Gell (1963, 1967), and Dray (1963) state that the *Aa* and *Ab* loci (heavy and light chain immunoglobulin) assort independently or are not closely linked. Similarly, Albers and Dray (1968) noted that the low density lipo-protein locus *Lpq* is not closely linked with either *Aa* or *Ab*. The α_2 macroglobulin locus *Mtz*, isolated by Knight and Dray (1968), is said to be independent of *Aa* and *Ab* and of *Lpq* (Albers and Dray 1969). The haptoglobin locus *Hph* described by Chiaso and Dray (1969) is represented as not closely linked to either *Aa*, *Ab*, *Lpq*, or *Mtz*. The *Af* immunoglobin locus is said by Conway, Dray, and Lichter (1969) not to be closely linked to either *Aa*, *Ab*, *Lpq*, *Mtz*, or *Hph*. The α_1 aryl esterase locus *Ess* is reported by Albers, Dray, and Knight (1969) not to be closely linked with either *Aa*, *Ab*, *Lpq*, *Mtz*, *Hph*, or *Af*. Similarly, the light chain immunoglobin locus *Ac* does not show close linkage with either *Aa*, *Ab*, *Af*, *Lpq*, or *Mtz*. (Gilman-Sachs, Mage, Young, Alexander and Dray, 1969). System 3 of the three unsymbolized erythocytic esterase systems described by Schiff and Stormont (1970) is said to assort independently of the other two.

All of these statements are true so far as they go but the statistical evidence is inadequate (at this time) to exclude anything but tight linkage. An exception is the small amount of data contributed by Dray (1964) on *Aa* and *Ab*, and Gilman-Sachs *et al.* (1969) on *Aa* and *Ac*.

As in other species, the borderline cases deserve to be carefully examined. Some may attain formal significance by chance alone — an event which should occur if the distribution of test results is truly at random. Seemingly these are not, for there is evidence for the rabbit (in common with other species) that a suspicious deviation from random recombination will be followed up to ascertain if the deviation is real.

Castle (1926) has tabulated a series of six groups of data for the genes l and v which together produce a recombination value of 44.4 ± 2.4. The deviation from random assortment is significant ($\chi^2 = 5.47$). The heterogeneity between groups is negligible ($\chi^2 = 2.78$ for 5 degrees of freedom) — the data are thus consistent. In spite of this, Castle is cautious in accepting linkage. He makes the point that v might be expected to exhibit linkage with *En* (since *En* and l are known to be linked) if v and l are in the same chromosome segment. The data on the *En* and v pair would tend to invalidate close linkage (say for a loci order of v - En - l) but not a loose linkage (as would be expected if the order is v - l - En). Rather, the observations would harmonize with the order v - l - En since the observed recombination value of 53.1 ± 5.4 does not differ significantly from the sum of the two smaller recombination values (57.6 ± 2.5). In fact, it is on the right side of expectation. This particular case clearly merits further investigation.

The data of Pickard (1941) on segregation of the b and wa genes could be held to be suggestive of linkage. However, linkage seems improbable because the positive evidence stems from one mating in which the number of wa differs sharply from expectation. Three other groups of data, in which wa occurs to expectation, give no indication of linkage (Robinson, 1956).

Knopfmacher (Robinson, 1956) records 21 crossover young in a backcross sample of 57, yielding a recombination value of 36.9 ± 6.6, for the genes H and H-6. The deviation from random assortment is on the borderline of significance (χ^2 = 3.95). The sample number is small, however, and this fact alone throws doubt on the result.

Two instances of excess recombination occur in the data. The affected comparisons are e - sa and e - δ (χ^2 = 5.69 and 6.60, respectively). No obvious explanation can be offered for these peculiar results, unless they arise from sampling variation. Both occurred in the same experiment (Spendlove and Robinson, 1970) which could argue for an affinity effect of dissociation.

CHROMOSOME MORPHOLOGY

The haploid number of chromosomes is 22. The early observations on rabbit chromosomes have been discussed by Robinson (1958). Much of this work is only of historic interest since most lack the refinement now possible with the aid of modern techniques. Only recent studies have attempted to characterize the karyotype in some detail and to establish criteria for the recognition of individual chromosomes or groups of chromosomes. The following papers are pertinent: Melander (1956), Carvalho and Malaquias (1961), Chiarelli, Carli and Nuzzo (1962), Clausen and Syverton (1962), Teplitz and Ohno

TABLE 13.4. Quasi-independent gene assortment in the rabbit.

Loci	Recombination value	Phase balance	References
a - An	58±8	100	Sleeper (1940)
a - As	50±17	?	Sawin and Glick (1943)
a - b	46±3	100	Punnett (1915, 1930), Pap (1921), Parkhurst and Wilson (1933)
a - br	51±6	?	Castle and Sawin (1941)
a - c	47±4	100	Hurst (1905), Castle (1905, 1922), Castle et al. (1909), Punnett (1912)
a - d	51±2	83	Castle (1926, 1934a), Parkhurst and Wilson (1933)
a - e	50±4	60	Punnett (1912, 1915, 1924, 1930), Castle (1926, 1934a), Castle et al. (1909)
a - En	49±2	73	Castle (1926, 1934a)
a - f	47±5	100	Castle (1933a)
a - H	55±4	95	Castle and Keeler (1933), Knopfmacher (1942)
a - H-6	49±8	100	Sawin et al. (1944)
a - r-1	51±2	0	Castle (1929a, b), Nachtsheim (1929), Parkhurst and Wilson (1933)
Aa - Ab	49±7	?	Dray (1963)
An - c	59±5	100	Sleeper (1940)
An - d	61±8	14	,, ,,
An - e	57±19	100	,, ,,
An - En	58±8	100	,, ,,
An - H	43±6	?	Knopfmacher (1942)
An - sa	52±6	100	Sawin et al (1944)
An - v	45±5	100	,, ,, ,, ,,
As - c	50±18	?	Sawin and Glick (1943)
ax - H	33±13	?	Knopfmacher (1942)
b - d	45±3	78	Parkhurst and Wilson (1933), Pickard (1941)
b - e	52±4	3	Punnett (1915, 1924, 1930), Castle (1926)
b - En	54±4	100	Castle (1926)
b - f	61±16	0	Castle (1933a)
b - r-1	47±3	13	Nachtsheim (1929), Parkhurst and Wilson (1933), Pickard (1941)
b - v	55±4	0	Castle (1926)
b - wa	42±7	100	Pickard (1941)
br - c	52±11	?	Castle and Sawin (1941)
br - d	43±5	?	,, ,, ,, ,,
br - En	51±3	?	,, ,, ,, ,,

Loci	Recombination value	Phase balance	References
br - H-6	47±3	100	Sawin et al. (1944)
c - d	53±4	35	Castle (1922, 1926), Robinson (1953)
c - e	50±3	13	Punnett (1912, 1926, 1930), Pap (1921), Castle (1922, 1926), Castle et al. (1909), Spendlove and Robinson (1970)
c - En	47±4	?	Castle (1926), Spendlove and Robinson (1970)
c - H	46±6	?	Knopfmacher (1942)
c - H-6	52±7	100	Sawin et al. (1944)
c - l	53±4	25	Assheton (1910), Castle (1922, 1926),
c - r-1	51±4	0	Castle (1929a, b), Nachtsheim (1929), Pickard (1941)
c - sa	51±5	100	Robinson (1953), Spendlove and Robinson (1970)
c - v	53±11	0	Pap (1921), Castle (1926)
c - ♂	48±4	61	Assheton (1910), Castle (1922), Robinson (1953), Spendlove and Robinson (1970)
d - e	48±2	99	Castle (1922, 1926, 1934a), Punnett (1924)
d - En	52±2	100	Castle (1926, 1934a)
d - f	55±8	0	Castle (1933a)
d - H	55±5	91	Castle and Keeler (1933)
d - H-6	42±8	100	Sawin et al. (1944)
d - l	32±16	0	Castle (1922)
d - r-1	52±3	0	Castle (1929a, b), Nachtsheim (1929), Parkhurst and Wilson (1933), Pickard (1941)
d - sa	57±8	0	Robinson (1953)
d - v	43±4	100	Castle (1926)
d - wa	44±8	0	Pickard (1941)
d - ♂	41±7	75	Castle (1922), Robinson (1953)
e - En	51±2	92	Castle (1926, 1934a), Spendlove and Robinson (1970)
e - f	48±4	0	Castle (1933a)
e - H	49±4	91	Castle and Keeler (1933), Knopfmacher (1942)
e - H-6	50±6	100	Sawin, et al. (1944)
e - l	54±3	58	Castle (1922, 1926)
e - r-1	51±3	0	Castle (1929a, b), Nachtsheim (1929), Spendlove and Robinson (1970)
e - sa	64±6	0	Spendlove and Robinson (1970)
e - v	46±8	10	Pap (1921), Castle (1926)

Loci	Recombination value	Phase balance	References
e - ♂	65±6	0	Castle (1922), Spendlove and Robinson (1970)
En - f	49±5	100	Castle (1933a)
En - H	48±2	100	Castle and Keeler (1933)
En - H-6	50±7	0	Sawin, et al. (1944)
En - r-1	45±3	100	Castle (1929a)
En - sa	50±6	100	Spendlove and Robinson (1970)
En - v	53±5	100	Castle (1926)
En - ♂	45±6	100	Spendlove and Robinson (1970)
f - r-1	47±4	0	Castle (1933a)
f - r-2	38±7	0	,, ,,
f - v	53±9	0	,, ,,
H - H-6	37±6	?	Knopfmacher (1942)
H - sa	57±8	?	,, ,,
H - v	42±12	?	,, ,,
H-6 - r-2	58±5	100	Sawin et al. (1944)
H-6 - r-3	52±5	100	,, ,, ,, ,,
H-6 - v	55±5	100	,, ,, ,, ,,
l - r-1	50±5	0	Castle (1929b), Nachtsheim (1929), Pickard (1941), Zelnik (1965)
l - v	44±3	0	Castle (1926)
l - ♂	50±14	100	Castle (1922)
r-1 - r-3	53±12	0	Castle and Nachtsheim (1933)
r-2 - r-3	76±15	0	,, ,, ,, ,,
r-1 - v	44±5	0	Castle (1929a, b), Nachtsheim (1929)
r-1 - wa	70±39	100	Pickard (1941)
r-1 - y	62±14	100	Castle (1929a)
s - ♂	51±4	100	Robinson (1951)
sa - ♂	60±6	78	Robinson (1953), Spendlove and Robinson (1970)

(1963), Myers, O'Leary and Fox (1964), Dave, Tagagi, Oishi, and Kikuchi (1965), Nichols, Levan, Hansen-Melander, and Melander (1965), Prunieras, Jacquemont, and Mathivon (1965), Ray and Williams (1966) Issa, Atherton and Blank (1968), and Shaver and Carr (1969).

The classification of chromosomes in the manner of Table 13.5 cannot avoid being arbitrary to some extent. For this reason, the discrepancies between investigations in the main are probably more apparent than real. However, the comparison reveals a steady trend towards more detail scrutiny of individual chromosomes and more exacting karyotypic analysis. The observations of Issa, Atherton and Blank (1968) are particularly noteworthy in this respect. The larger chromosomes are clearly those most easily characterized and the successive entries of the table reveal that four large metacentrics and four large acrocentrics could be consistently identified for the karyotype of the rabbit. After this, differences between medium and small elements and between metacentrics and those grouped as acrocentrics become increasingly blurred.

CHIASMA FREQUENCY

No detailed analysis has yet been made on the distribution of chiasmata per bivalent. S. Muldal (see Robinson, 1958) has noted that the mean chiasma frequency is in the region of 1.3 per bivalent, with never more than one in the XY. In this connection, Sharma, Parshad, and Ghuman (1963) state that the autosomal bivalents usually have between one and two chiasmata each.

BIBLIOGRAPHY

ALBERS, J. J. and DRAY, S. (1968). Identification and genetic control of two rabbit low density lipoprotein allotypes. *Biochem. Genet.*, 2, 25-35.

ALBERS, J. J. and DRAY, S. (1969). Identification of genetic control of two new low-density lipoprotein allotypes: phenogroups of the *Lpq* locus. *J. Immunol.*, 103, 155-162.

ALBERS, L. V., DRAY, S. and KNIGHT, K. L. (1969). Allotypes and isozymes of rabbit alpha aryl esterase. Allelic products with different enzymatic activities for the same substrates. *Biochemistry*, 8, 4416-4424.

ASSHETON, R. (1910). Observations on crossing wild rabbits with certain tame breeds. *Guy's Hospital Rep.*, 64, 313-342.

CARVALHO, A. A. M. S., de and MALAQUIAS, M. I. C. O. (1961). Contribucias para o estudio do cariotipo do *Lepus cuniculus*. *Folia Anat. Univ. Conimbrigensis*, 35 (9), 6 pp.

CASTLE, W. E. (1905). Heredity of coat characters in guinea-pigs and rabbits. *Car. Inst. Wash. Pub.*, 23.

CASTLE, W. E. (1919). Studies of heredity in rabbits, rats and mice. *Car. Inst. Wash. Pub.*, 288, 4-28.

CASTLE, W. E. (1920). Linked genes in rabbits. *Science*, 52, 156-157.

TABLE 13.5. Summary of karyological investigations of the rabbit.

Defined autosomal groups	No. of metacentrics			No. of acrocentrics			Sex chromosomes		References
	Large	Medium	Small	Large	Medium	Small	X	Y	
—	4	5	3	4	3	2	Ma	Sa	Melander (1956)
10	4	3	2	5	2	5	Mm	Sa	Carhalho and Malaquias (1961)
5	3	4	5	4	3	2	Lm	Sm	Chiarelli et al. (1962)
—	4	3	3	2	6	3	Sm	Sa	Clausen and Syveton (1962)
—	4	4	5	4	1	3	Ma	Sa	Teplitz and Ohno (1963)
7	3	6	3	4	2	2	Mm	Sa	Myers et al. (1964)
—	4	5	4	4	2	2	Mm	Sa	Dave et al. (1965)
4	4	3	4	4	4	2	Mm	Sm	Nichols, et al. (1965)
6	4	3	6	4	2	2	Mm	Sa	Prunerias et al. (1965)
7	4	5	2	4	4	2	Mm	Sa	Ray and Williams (1966)
11	4	5	4	4	2	2	Mm	Sm	Issa et al. (1968)
—	4	3	4	4	4	2	Mm	Sm	Shaver and Carr (1969)

L = large, M = medium, S = small, a = acrocentric, m = metacentric; acrocentric embraces subtelocentric and telocentric elements.

CASTLE, W. E. (1921). More linked genes in rabbits. *Science*, 54, 634-636.

CASTLE, W. E. (1922). Genetic studies of rabbits and rats. *Car. Inst. Wash. Pub.*, 320, 1-50.

CASTLE, W. E. (1924a). Linkage of Dutch, English and Angora in rabbits. *Proc. Nat. Acad. Sci. U.S.A.*, 10, 107-108.

CASTLE, W. E. (1924b). Occurrence in rabbits of linkage in inheritance between albinism and brown pigmentation. *Proc. Nat. Acad. Sci. U.S.A.*, 10, 486-488.

CASTLE, W. E. (1926). Studies of colour inheritance and of linkage in rabbits. *Car. Inst. Wash. Pub.*, 337, 3-47.

CASTLE, W. E. (1929a). The rex rabbit. *J. Hered.*, 20, 193-199.

CASTLE, W. E. (1929b). Linkage studies in Castorrex rabbits. *Z. Ind. Abst. Vererbgsl.*, 52, 53-60.

CASTLE, W. E. (1932). English and Dutch spotting and the genetics of the Hotot rabbit. *Car. Inst. Wash. Pub.*, 427, 3-13.

CASTLE, W. E. (1933a). The Furless rabbit. *J. Hered.*, 24, 81-86.

CASTLE, W. E. (1933b). The linkage relations of yellow fat in rabbits. *Proc. Nat. Acad. Sci. U.S.A.*, 19, 947-950.

CASTLE, W. E. (1934a). Size inheritance in rabbits; further data on the backcross to the small race. *J. Exp. Zool.*, 67, 105-114.

CASTLE, W. E. (1934b). Genetics of the Dutch pattern. *J. Exp. Zool.*, 68, 377-391.

CASTLE, W. E. (1936). Further data on linkage in rabbits. *Proc. Nat. Acad. Sci. U.S.A.*, 22, 222-225.

CASTLE, W. E. and KEELER, C. F. (1933). Tests for linkage between the blood group genes and other known genes in the rabbit. *Proc. Nat. Acad. Sci. U.S.A.*, 19, 98-100.

CASTLE, W. E. and LAW, L. W. (1936). Satin, a new hair mutation in the rabbit. *J. Hered.*, 27, 235-240.

CASTLE, W. E. and NACHTSHEIM, H. (1933). Linkage interrelations of three genes for rex (short) coat in the rabbit. *Proc. Nat. Acad. Sci. U.S.A.*, 19, 1006-1011.

CASTLE, W. E. and SAWIN, P. B. (1941). Genetic linkage in the rabbit. *Proc. Nat. Acad. Sci. U.S.A.*, 27, 519-523.

CASTLE, W. E., WALTER, H. E., MULLENIX, R. C. and COBB, S. (1909). Studies of inheritance in rabbits. *Car. Inst. Wash. Pub.*, 114.

CHIAO, J. W. and DRAY, S. (1969). Identification and genetic control of rabbit haptoglobin allotypes. *Biochem. Genet.*, 3, 1-13.

CHIARELLI, B., CARLI, L. de and NUZZO, F. (1962). Analisi morfometrica dei cromosomi del coniglio. *Caryologica*, 15, 565-568.

CLAUSEN, J. J. and SYVERTON, J. T. (1962). Comparative chromosomal study of 31 cultured mammalian cell lines. *J. Nat. Cancer Inst.*, 28, 117-145.

CONWAY, T. P., DRAY, S. and LICHTER, E. A. (1969). Identification and genetic control of three rabbit gamma-A immunoglobin allotypes. *J. Immunol.*, 102, 544-554.

DAVE, M. J., TAKAGI, N., OISHI, H. and KIKUCHI, Y. (1965).

Chromosome studies on the hare and rabbit. *Proc. Japan Acad.*, 41, 244-248.

DRAY, S. (1963). Genetics of gamma-globulin. *Proc. XI Inter. Congr. Genet.*, 2, 165-180.

DRAY, S., YOUNG, G. O. and GERALD, L. (1963). Immunological identification and genetics of rabbit allotypes. *J. Immunol.*, 91, 403-415.

DUBISKI, S. (1969). Immunochemistry and genetics of a 'new' allotypic specificity Ae^{14} of rabbit gamma-G immunoglobins: recombination in somatic cells. *J. Immunol.*, 103, 120-128.

DUBISKI, S., RAPACZ, J. and DUBINSKA, A. (1962). Heredity of rabbit gamma-globilun iso-antigens. *Acta Genet. Stat. Med.*, 12, 136-155.

GILMAN-SACHS, A., MAGE, R. G., YOUNG, G. O., ALEXANDER, C. and DRAY, S. (1969). Identification and genetic control of two rabbit immunoglobulin allotypes at a second light chain locus, the *c* locus. *J. Immunol.*, 103, 1159-1167.

HAMERS, R., HAMERS-CASTERMAN, C. and LAGNAUX, S. (1966). A new allotype in the rabbit linked with Asl which may characterize a new class of IgC. *Immunology*, 10, 399-408.

HURST, C. C. (1905). Experimental studies on heredity in rabbits. *Linn. Soc. J. Zool.*, 29, 283-324.

ISSA, M., ATHERTON, G. W. and BLANK, C. E. (1968). The chromosomes of the domestic rabbit. *Cytogenetics*, 7, 361-375.

KELUS, A. S. and GELL, P. G. H. (1963). Immunogenetics of rabbit gammaglobulin. *Proc. XI Inter. Congr. Genet.*, 1, 194.

KELUS, A. S. and GELL, P. G. H. (1967). Immunoglobulin allotypes of experimental animals. *Progr. Allergy*, 11, 141-184.

KNIGHT, K. L. and DRAY, S. (1968). Identification and genetic control of two rabbit a_2-macroglobulin allotypes. *Biochemistry* 7, 1165-1171.

KNOPFMACHER, H. P. (1942). Thesis: Brown University.

MELANDER, Y. (1956). The chromosome complement of the rabbit. *Hereditas*, 42, 432-435.

MYERS, L. B., O'LEARY, J. L. and FOX, R. R. (1964). Classification of chromosomes in normal and ataxic rabbits. *Neurology*, 14, 1058-1065.

NACHTSHEIM, H. (1929). Das Rexkaninchen und seine Genetik. *Z. Ind. Abst. Vererbgsl.*, 52, 1-52.

NICHOLS, W. W., LEVAN, A., HANSEN-MELANDER, E. and MELANDER, Y. (1965). The idiogram of the rabbit. *Hereditas*, 35, 63-76.

PAP, E. (1921). Vererbung von Farbe und Zeichnung bei dem Kaninchen. *Z. Ind. Abst. Vererbgsl.*, 26, 185-270.

PARKHURST, R. T. and WILSON, W. K. (1933). Rexing the lilac rabbit. *J. Hered.*, 24, 35-39.

PEASE, M. S. (1928). Yellow fat in rabbits, a linked character? *Verh. V. Intern. Kongr. Vererbungswissenschaft*, 2, 1153-1156.

PICKARD, J. N. (1941). Waved – a new coat type in rabbits. *J. Genet.*, 42, 215-222.

PRUNERIAS, M., JACQUEMONT, C. and MATHIVON, M. F. (1965).

Etudes sur les relations virus-chromosomes. V. Le caryotype du lapin domestique. *Ann. Inst. Pasteur*, 109, 465-471.

PUNNETT, R. C. (1912). Inheritance of coat colour in rabbits. *J. Genet.*, 2, 221-238.

PUNNETT, R. C. (1915). Further experiments on the inheritance of coat colour in rabbits. *J. Genet.*, 5, 38-50.

PUNNETT, R. C. (1924). On the Japanese rabbit. *J. Genet.*, 14, 225-240.

PUNNETT, R. C. (1926). Note on a Chinchilla-Japanese cross in rabbits. *J. Genet.*, 17, 217-220.

PUNNETT, R. C. (1928). Further notes on Dutch and English rabbits. *J. Genet.*, 20, 247-260.

PUNNETT, R. C. (1930). On the series of allelomorphs connected with the production of black pigment in rabbits. *J. Genet.*, 23, 265-274.

RAY, M. and WILLIAMS, T. W. (1966). Karyotype of rabbit chromosomes from leucocyte cultures. *Canad. J. Genet. Cytol.*, 8, 393-397.

RIFAAT, O. M. (1954). A revised map of the fifth chromosome of the domestic rabbit. *Heredity*, 8, 107-116.

ROBINSON, R. (1951). Dutch-type white-spotting in rabbits. *Nature*, 168, 300.

ROBINSON, R. (1953). Segregation of the satin gene in the rabbit. *J. Hered.*, 44, 95-96.

ROBINSON, R. (1956). A review of independent and linked segregation in the rabbit. *J. Genet.*, 54, 358-369.

ROBINSON, R. (1958). Genetic studies of the rabbit. *Bibliogr. Genet.*, 17, 229-558.

SAWIN, P. B. (1934). Linkage of wide-band and agouti genes. *J. Hered.*, 25, 477-481.

SAWIN, P. B. (1955). Recent genetics of the domestic rabbit. *Advanc. Genet.*, 7, 183-226.

SAWIN, P. B. and GLICK, D. (1943). Atropinesterase, a genetically determined enzyme in the rabbit. *Proc. Nat. Acad. Sci. U.S.A.*, 29, 55-59.

SAWIN, P. B., GRIFFIN, M. A. and STUART, C. A. (1944). Genetic linkage of blood types in the rabbit. *Proc. Nat. Acad. Sci. U.S.A.*, 30, 217-221.

SCHIFF, R. and STORMONT, C. (1968). The genetic control of rabbit red cell esterase variation. *Proc. XII Inter. Congr. Genet.*, 1, 127.

SCHIFF, R. and STORMONT, C. (1970). The biochemical genetics of rabbit erythrocyte esterases: two new esterase loci. *Biochem. Genet.*, 4, 11-23.

SHARMA, G. P., PARSHAD, R. and GHUMAN, S. K. (1963). On the meiotic and somatic chromosomes of the rabbit. *Res. Bull. Panjab Univ. Sci.*, 14, 171-173.

SHAVER, E. L. and CARR, D. H. (1969). The chromosome complement of rabbit blastocysts in relation to the time of mating and ovulation. *Canad. J. Genet. Cytol.*, 11, 287-293.

SLEEPER, H. A. (1940). Thesis: Brown University.

SPENDLOVE, W. and ROBINSON, R. (1970). A linkage test with satin in the rabbit. *Genetica,* 41, 635-637.

TEPLITZ, R. and OHNO, S. (1963). Postnatal induction of ovogenesis in the rabbit. *Exp. Cell Res.,* 31, 183-189.

ZELNIK, J. (1965). [The inheritance of coat colour and wool characters in Angora crossed with brown rex (Havana) rabbits.] *Pol'nohospodarstvo,* 11, 533-543.

CHAPTER 14

Guinea-Pig

Cavia cobaya $(n = 31 + X + Y)$
In the early days of mammalian genetics, the domestic guinea-pig featured almost as prominently as other laboratory rodents but, latterly, the animal has receded from favour. In spite of several systematic searches for possible linkage among the known mutants, the results have been largely negative. However, even negative results are not devoid of value since these imply that a variety of chromosome segments are tagged by genes. The relatively large number of chromosomes is probably the main determinant in the absence of linkage. Some 25 mutant genes have been described for the guinea-pig if the majority of genes described by Ibsen (1932) are regarded merely as provisional. Out of the positively recognized mutants, 15 have been employed in investigations for linkage. A recent review is that of Robinson (1970).

TABLE 14.1. Genes/loci of the guinea-pig which have featured in studies on linkage: assigned linkage group, symbol, and conventional designation.

Linkage Group	Symbol	Designation	Prime Characteristic
	a	Non-agouti	Coat colour
	b	Brown	Coat colour
	c	Albino	Coat colour
	e	Yellow	Coat colour
	f	Faded	Coat colour
	H	Histocompatibility	Immunogenetics
	l	Long hair	Hair texture
II	*m*	Rough modifier	Hair texture
	p	Pink-eye	Coat colour
	Pll	PLL response	Immunogenetics
I	*Px*	Polydactyly	Feet
I	*R*	Rough	Hair texture
	s	Piebald	White spotting
II	*si*	Silver	Coat colour
	sk	Sticky	Lipid secretion
	sm	Salmon-eye	Coat colour
	St	Star	Hair texture

LINKAGE GROUP

Merely two pairs of genes have been observed to display linkage. The paucity of linked segregation is remarkable considering the fair number of mutants available and, above all, the systematic manner with which each gene has been tested against most of the others (Fig. 14.1). A summary of the linkage data is provided by Table 14.2. There is a tendency for crossingover to occur more frequently in the female although none of the differences emerge as significant.

Two forms of polydactyly occur in the guinea-pig. That due to the heterozygosity of the dominant *Px* gene (*PxPx* is monstrous and lethal) and another form which is due to the accumulative effects of three or four mainly recessive genes (Wright, 1934). The results presented by Wright (1941) are suggestive that one of these may be linked to *a*. Wright states that if recessive polydactyly is assumed to be due to a single gene, a recombination value of 44.3 ± 2.9 would be consistent with the data.

Green *et al.* (1969) have shown that the ability to respond immunologically to poly-L-lysine (PLL) and certain other compounds and to dinitro-phenyl (DNP)-poly-L-arginine and certain other compounds is due to a dominant autosomal complex locus or to two linked loci. The responses to PLL and DNP are highly specific and this fact lends colour to the concept of linked loci, rather than that of a complex. However, overt data on the strength of the linkage has not been published at this time. For the moment, only the *Pll* locus will be recognized as a distinct entity.

Ellman *et al.* (1970) have since shown that *Pll* is linked to a major histocompatibility locus. In a testcross of 18 animals, the linkage is complete but the small sample can only exclude weak to moderate linkage. The authors have compared the present results with those for the *H-2* locus of the mouse. In that species, *H-2* is a complex locus, mediating not only the nature of histological antigens but also the immune response. A similar complex may exist for the guinea-pig. The histocompatibility locus has not been precisely designated. In this review, it will be simply denoted as *H*.

Group I (Px - R)

The mean crossover value for the *Px - R* intercept is 45.7 ± 1.6 crossover units. Formally, this value is significant and linkage between the two genes seems to be reasonably established. However, there are complications. The major one is that the expression of *Px* is variable and some impenetrance is evident for certain crosses. The extent of the impenetrance and the effect this is likely to have on detection of linkage is discussed by Wright (1941, 1949). Wright (1941) divides the early data into three categories: the last one being regarded as the least reliable for linkage estimation. This group of data has not been used in the present calculations. It is unfortunate that the impenetrance is such that the effect cannot be readily removed by statistical manipulation.

The second complication is the presence of significant heterogeneity between the four segregations for *Px - R*. This stems from the excessively high crossover value for females of the repulsion backcross. No obvious reason is apparent for this unless, of course, it is a manifestation of the variable impenetrance. This batch of data is the smallest in the total sample and rejection of it gives the mean crossover value of 44.7 ± 1.7.

A sex difference of crossingover (3.5 ± 3.2) can be observed from Table 14.2 although it is not of statistical importance. Its magnitude is due mainly to the high crossingover value for the female repulsion data. Without these data, the sex difference almost disappears.

Group II (m, si)

The mean crossover value is 21.7 ± 4.3. The expression of *si* is variable and impenetrance can be a serious problem upon certain genetic backgrounds (Wright, 1959). Allowance for this can effect the magnitude of the crossover value but not the existence of linkage. A difference of crossingover occurs between the sexes but fails to attain significance.

SEX-LINKAGE

No instances of sex-linked genes are known and it may be

mentioned that many of the genes have been examined for partial sex-linkage with negative results. No Y chromosome associated histocompatible genes were found by Bauer (1960); employing skin transplants between two of the surviving inbred stocks developed by Sewall Wright (Wright 2 and 13).

TABLE 14.2 Established linkage groups in the guinea-pig.

Group Loci	Sex	Mating type	Crossover value	References		
I $Px - R$	♂	CBB	42.4±2.5	Wright (1941, 1949)		
	♂	RBB	45.3±2.8	,,	,,	,,
	♀	CBB	48.4±3.6	,,	,,	,,
	♀	RBB	56.3±5.6	,,	,,	,,
II $m - si$	♂	RBB	20.3±5.2	Wright (1959)		
	♀	RBB	25.0±7.8	,,	,,	

INDEPENDENT ASSORTMENT

The extent of testing for linked segregation is summarized by Table 14.3 and Fig. 14.1. These data are remarkable on two accounts, (1) for their coverage (in that so many of the genes have been tested in all combinations with each other) and (2) the quantity of data for most gene pairs is such as to exclude all but weak linkage. Both of these attributes are due to the sustained work of Wright.

Most of the gene combinations of Table 14.3 are undoubtedly inherited independently. Loose linkage cannot be ruled out *a priori* and Fig. 14.1 conveys an estimate of the closest linkage which would be compatible with the data to hand. This is approximated by multiplying the standard error by 196 and subtracting the product from the observed recombination value. For most gene pairs, the derived value borders on formal independency.

Ibsen (1922) reported significant excess recombination between the c and p genes. However, the excessive recombination is only apparent for one of the two crosses outlined and, if the results are pooled, the results become

TABLE 14.3 Quasi-independent gene assortment in the guinea-pig.

Loci	Recombination value	Phase balance	References
a - b	49±2	98	Sollas (1909), Castle (1916), Ibsen (1923), Gregory (1928.) Wright (1941)
a - c	47±2	100	Castle (1916), Ibsen (1923), Wright (1941)
a - e	49±1	98	Sollas (1909), Castle (1913, 1916), Ibsen (1923), Wright (1941)
a - f	45±3	100	Wright (1941)
a - m	48±2	99	Wright (1916, 1941)
a - p	49±2	96	Ibsen (1923), Gregory (1928), Wright (1941)
a - Px	46±3	100	Wright (1941)
a - R	51±1	98	Wright (1916, 1941), Ibsen (1923)
a - s	51±2	100	Wright (1941)
a - si	44±3	89	Wright (1959)
a - sm	51±5	60	Gregory (1928)
a - St	52±2	84	Wright (1949)
a - ♂	50±2	32	Wright (1941)
b - c	50±3	100	Castle (1916), Ibsen (1923), Wright (1941)
b - e	50±2	95	Sollas (1909), Castle (1916), Ibsen (1923), Wright (1941)
b - f	54±3	100	Wright (1941)
b - m	49±3	100	,, ,,
b - p	50±2	100	Ibsen (1923), Gregory (1928), Wright (1941)
b - Px	56±3	100	Wright (1941)
b - R	49±2	100	Ibsen (1923), Wright (1941)
b - s	50±2	100	Wright (1941)
b - si	54±5	79	Wright (1959)
b - sm	50±5	45	Gregory (1928)
b - St	46±3	68	Wright (1949)
b - ♂	52±2	49	Wright (1941)
c - e	50±2	100	Castle (1916), Ibsen (1923), Wright (1941)
c - f	50±4	19	Wright (1941)
c - l	125±50	100	Castle (1913)
c - m	49±3	100	Wright (1941)
c - p	50±3	98	Ibsen (1922, 1923), Wright (1941)
c - Px	48±4	100	Wright (1941)
c - R	48±2	100	Castle (1913), Ibsen (1923)
c - s	47±2	100	Wright (1941)

Loci	Recombination value	Phase balance	References
c - si	36±11	55	Wright (1959)
c - St	51±2	54	Wright (1949)
c - ♂	49±2	60	Wright (1941)
e - f	47±4	100	,, ,,
e - m	48±2	100	,, ,,
e - p	50±2	100	Ibsen (1923), Wright (1941)
e - Px	48±3	100	Wright (1941)
e - R	49±1	100	Ibsen (1923), Wright (1941)
e - s	50±2	100	Wright (1941)
e - si	41±2	88	Wright (1959)
e - St	51±2	95	Wright (1949)
e - ♂	50±2	37	Wright (1941)
f - m	57±4	100	,, ,,
f - p	50±4	85	,, ,,
f - Px	46±4	100	,, ,,
f - R	52±3	100	,, ,,
f - s	49±3	100	,, ,,
f - si	50±9	0	Wright (1959)
f - St	52±4	100	Wright (1949)
f - ♂	46±3	67	Wright (1941)
l - R	38±27	100	Castle (1913)
m - p	49±3	100	Wright (1941)
m - Px	50±4	100	Wright (1941)
m - R	48±2	97	Wright (1916, 1941)
m - s	53±3	100	Wright (1941)
m - St	48±3	7	Wright (1949)
m - ♂	49±3	38	Wright (1941)
p - Px	55±4	100	,, ,,
P - R	49±2	100	Ibsen (1923), Wright (1941)
p - s	49±2	100	Wright (1941)
p - si	54±4	67	Wright (1959)
p - sm	50±5	57	Gregory (1928)
p - St	52±2	84	Wright (1949)
p - ♂	49±2	64	Wright (1941)
Px - s	52±5	100	,, ,,
Px - si	51±5	63	Wright (1959)
Px - St	47±2	51	Wright (1949)
Px - ♂	53±3	100	Wright (1941)
R - s	53±2	100	,, ,,
R - si	49±2	100	Wright (1959)
R - St	47±2	59	Wright (1949)
R - ♂	51±2	100	Wright (1941)
s - si	50±10	0	Wright (1959)
s - St	50±3	89	,, ,,

Loci	Recombination value	Phase balance	References
s - ♂	47±2	86	Wright (1941)
si - St	46±3	100	Wright (1959)
si - ♂	53±2	63	,, ,,
St - ♂	48±2	65	Wright (1949)

insignificant. The crosses engender some complicated phenotypic interactions and contribute only a small amount of statistical information. Subsequent results by Ibsen (1923) and Wright (1941) on the simultaneous segregation of the two genes showed fully normal and independent assortment.

Wright (1941) discussed the significance of much of the independence data; in particular, incorporating the results of his 1927 and 1928 papers. The *Px* gene has variable expression, including impenetrance, of a type which cannot be easily corrected statistically. The *si* gene behaves erratically displaying variable impenetrance and phenotypic interactions with other genes, notably *f* and *s*. The *sssisi* phenotype acts as a crossover class in the assortment of *s* and *si* and occurs significantly in excess of expectation, thereby producing the curious recombination value of 64.8 ± 6.2. Estimation based upon the other three classes gives the value of 49.6 ± 10.1 and this has been adopted as the more realistic of the two.

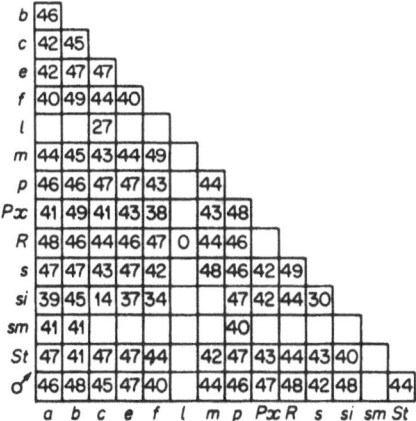

Fig. 14.1. Summary of the extent of independent assortment in the guinea-pig.

The pooled data on the e - f pair produce the significant recombination value of 43.9 ± 2.9. However, Wright (1941) is cautious in accepting the result as evidence for linkage because the cross in which the linkage is most obvious involves phenotypes where f represents difficulties of classification. In the cross where misclassification is not a problem, the recombination value is 47.4 ± 3.7, *prima facie* not significant. Linkage is possible, therefore, but apparently improbable. The latter determination is used for Table 14.3 and Fig. 14.1.

Herbertson *et al.* (1959) state that sk is inherited autosomally and independently of several colour genes and of rough. However, it is admitted that loose linkage cannot be excluded. No linkage data are presented nor are the colour genes specified.

CHROMOSOME MORPHOLOGY

The karyotype of the guinea-pig has a relatively large number of elements for a rodent. Apart from early techniques tending to produce inconsistent results in any event, the presence of numerous small chromosomes was a further confusing factor. However, Makino (1947) definitely established that the karyotype consists of 32 elements and this determination has since been adequately confirmed. The more pertinent recent papers are as follows: Sachs (1952, 1953), Hsu and Pomerat (1953), Awa, Sasaki, and Takayama (1959), Ohno, Weiler, and Stenius (1961), Sharma, Parshad, and Ghuman (1963), Prunerias, Mathivon, Leung, and Gazzolo (1965), Watson, Blumenthal, and Hutton (1966), Dobrijanov and Goljdman (1967a, b), and Jagiello (1969).

It is especially difficult to characterize the chromosomes of the guinea-pig. The large number of small elements could be ascribed to an exceptional amount of fragmentation (or its alternative, lack of fusion). If fragmentation has occurred, it has effectively destroyed any individuality which the chromosomes may have possessed. Published karyotypes indicate very clearly the existence of a single large subtelocentric chromosome, easily identifiable, and a large telocentric which is almost certainly the X. Beyond this, recognition of individual chromosomes becomes increasingly difficult. Even categorizing the elements

into groups presents problems because, except for one medium
sized metacentric (represented as chromosome 5 by Cohen and
Pinsky, 1966), all (or most) of the others are either telocentrics
or subtelocentrics varying continuously in size from the largest
to the smallest. The Y chromosome is almost certainly a small
acrocentric, not easily separable from the small autosomes.

Ohno *et al.* have described a curious condition for the large
subtelocentric pair of chromosomes in mitotic cells. The small
second arm of one chromosome appeared to be more condensed
than the corresponding arm of the other and, consequently,
gave the impression of being abbreviated. The longer arm
stained differentially to give the chromosome the appearance of
being satellited. The difference could also be seen in meiosis.
The heterochromia area was depicted as a nucleolus organizer.
The marked condensation could be indicative of a localized and
differentiated 'inactivation' between the two chromosomes, the
exact nature of which has yet to be determined.

Dobrijanov and Goljdman (1967a) feature a similar condition
for the large subtelocentric element which could be readily
identified in their material. In this case, however, the authors
attribute the heteromorphism to a structural difference in size
of the smaller arm, rather than to differential condensation.
This hypothesis seems to be supported by the observation of
individuals with both chromosomes either of one form or of the
other — as opposed to individuals in which only one of the pair
of chromosomes is invariably longer (or less condensed) than
the other. This discovery would seem worthy of verification by
examination of animals from a variety of sources, since a
chromosomal polymorphism is conceivable.

Cohen and Pinsky (1966) have investigated the karyotype for
a stock of guinea-pigs which they refer to as an albino
short-haired tailless *Cavia porcellus*. The haploid chromosome
number is 32, as for *C. cobaya*. However, they feel that the
karyotypes differ in several respects. Whereas *cobaya* possesses
one medium size metacentric (chromosome 5), *porcellus* does
not but has more reputed submetacentrics than does *cobaya*. *C.
porcellus* is portrayed as possessing about seven small
telocentrics while *cobaya* has up to 15.

A single large subtelocentric element could be readily
identified for the *porcellus* karyotype. Cohen and Pinsky note

that the appearance of the chromosome was variable. In some individuals, the small second arm seemed to be unusually long while, in others, the arm seemed to be lacking. The authors propose that the difference could be due to a translocation. The chromosome with the unusually long arm is represented as having a duplicated region while the apparently abbreviated chromosome is deficient. Six possible combinations of the two chromosomes could exist and, so far, three have been identified. No obvious phenotypic differences were detected.

At present, the number of animals examined is quite small and it will be interesting to ascertain if all of the chromosomal combinations are viable. The homozygous deficient form, for instance, could be lethal. The authors speculate that the condition described by Ohno *et al.* for the large subtelocentric (chromosome 1) of *cobaya* might be similar to the above. However, it is not clear for the Ohno case whether or not the chromosomes differed in relative size or if the second arm was displaying differential condensation.

CHIASMA FREQUENCY

No systematic analysis of the occurrence of chiasma has been published thus far. However, Sharma, Parshad, and Ghuman (1963) briefly commented that each of the autosomal bivalents in spermatogenetic material had one or two chiasmata. Just as cursorily, Jagiello (1969) noted a mean frequency of 1.04 chiasmata per bivalent in 50 metaphase cells specially selected for clarity. In neither instance was a frequency distribution presented. It is conceivable that a low mean frequency may be typical for the guinea-pig because of the large number of small chromosomes.

TABLE 14.4. Summary of karyological investigations of the guinea-pig

Defined autosomal groups	No. of metacentrics			No. of acrocentrics			Sex chromosomes		References
	Large	Medium	Small	Large	Medium	Small	X	Y	
—	0	1	0	1	14	15	Lm	Sa	Awa et al. (1959)
3	1	7	0	1	9	13	Lm	Sa	Ohno et al. (1961)
—	1	10	3	1	0	16	Lm	Sa	Watson et al. (1966)
4	1	7	1	3	9	10	Mm	Sa	Dobrijanov and Goljdman (1967b)
4	0	12	0	2	10	7	Mm	Sa	Cohen and Pinsky (1966)
4	0	10	0	1	13	7	Lm	—	Jagiello (1970)

L = large, M = medium, S = small, m = metacentric, a = acrocentric; acrocentric embraces subtelocentric and telocentric elements.

BIBLIOGRAPHY

AWA, A., SASAKI, M. and TAKAYAMA, S. (1959). An *in vitro* study of the somatic chromosomes in several mammals. *Jap. J. Zool.*, 12, 257-265.

BAUER, J. A. (1960). Genetics of skin transplantation and an estimate of the number of histocompatibility genes in inbred guinea-pigs. *Annals N.Y. Acad. Sci.*, 87, 78-92.

CASTLE, W. E. (1905). Heredity of coat characters in guinea-pigs and rabbits. *Car. Inst. Wash. Pub.*, 23, 78pp.

CASTLE, W. E. (1913). Reversion in guinea-pigs and its explanation. *Car. Inst. Wash. Pub.*, 179, 1-10.

CASTLE, W. E. (1916). An expedition to the home of the guinea-pig and some breeding experiments with material there obtained. *Car. Inst. Wash. Pub.*, 241, 1-55.

COHEN, M. M. and PINSKY, L. (1966). Autosomal polymorphism via a translocation in the guinea-pig, *Cavia porcellus*. *Cytogenics*, 5, 120-132.

DOBRIJANOV, D. S. and GOLJDMAN, I. L. (1967a). [Chromosomes of the guinea-pig.] *Byull. Exsp. Biol. Med.*, 63 (4), 100-104.

DOBRIJANOV, D. S. and GOLJDMAN, I. L. (1967b). [The normal karyotype of the guinea-pig.] *Tsitol. Genet.* 1 (5), 78-82.

ELLMAN, L., GREEN, I., MARTIN, W. J. and BENACERRAF, B. (1970). Linkage between the poly-l-lysine gene and the locus controlling the major histocompatibility antigens in strain 2 guinea-pigs. *Proc. Nat. Acad. Sci. U.S.A.*, 66, 322-328.

GREEN, I., PAUL, W. E. and BENACERRAF, B. (1969). Genetic control of immunological responsiveness in guinea-pigs to 2-4-dinitrophenyl conjugates of poly-L-arginine, protamine and poly-L-ornithine. *Proc. Nat. Acad. Sci. U.S.A.*, 64, 1095-1102.

GREGORY, P. W. (1928). Some new genetic types of eyes in the guinea-pig. *J. Exp. Zool.*, 52, 159-181.

HERBERTSON, B. M., SKINNER, M. E. and TATCHELL, J. A. H. (1959). Sticky: a new mutant in the guinea pig. *J. Genet.*, 56, 315-324.

HSU, T. C. and POMERAT, C. M. (1953). Mammalian chromosomes *in vitro*. II. A method for spreading the chromosomes of cells in tissue culture. *J. Hered.*, 44, 23-29.

IBSEN, H. L. (1922). A cross in guinea-pigs best explained by assuming 75 per cent crossing over. *Anat. Rec.*, 23, 96.

IBSEN, H. L. (1923). Evidence of the independent inheritance of six pairs of allelomorphs in guinea-pigs. *Anat. Rec.*, 26, 392-393.

IBSEN, H. L. (1932). Modifying factors in guinea-pigs. *Proc. 6th Inter. Congr. Genet.*, 2, 97-101.

JAGIELLO, G. M. (1969). Some cytologic aspects of meiosis in female guinea-pig. *Chromosoma*, 27, 95-101.

MAKINO, S. (1947). Notes on the chromosomes of four species of small mammals. *J. Fac. Sci., Hokkaido Univ.*, IV, 9, 345-357.

OHNO, S., WEILER, C. and STENIUS, C. (1961). A dormant nucleolus organizer in the guinea pig, *Cavia cobaya*. *Exp. Cell Res.*, 25, 498-503.

PRUNERIAS, M., MATHIVON, M.F., LEUNG, T. K. and GAZZOLO, L. (1965). Culture euploidie de cellules epidermiques adultes en couche manocellulaire. *Ann. Inst. Pasteur,* 108, 149-165.

ROBINSON, R. (1970). Genetic linkage in the guinea-pig. *Ann. Genet. Select. Anim.,* 2, 241-248.

SACHS, L. (1952). Polyploid evolution and mammalian chromosomes. *Heredity,* 6, 358-364.

SACHS, L. (1953). Simple methods for mammalian chromosomes. *Stain Tech.,* 28, 169-172.

SHARMA, G. P., PARSHAD, R. and GHUMAN, S. K. (1963). On the meiotic and somatic chromosomes of the guinea-pig. *Res. Bull. Panjab Univ. Sci.,* 14, 167-169.

SOLLAS, I. B. J. (1909). Inheritance of color and of supernumary mammae in guinea-pigs, with a note on the occurrence of a dwarf form. *Rep. Evol. Com. Roy. Soc.,* 5, 51-79.

TUSCANY, R. (1963). [Contribution to the characteristics of the chromosome set in somatic cells of the guinea-pig.] *Czech. Morph.,* 2, 124-130.

WATSON, E. D., BLUMENTHAL, H. T. and HUTTON, W. E. (1966). A method for the culture of leucocytes of the guinea-pig with karyotypic analysis. *Cytogenetics,* 5, 179-185.

WRIGHT, S. (1916). An intensive study of the inheritance of color and of other coat characters in guinea-pigs. *Car. Inst. Wash. Pub.,* 241, 57-160.

WRIGHT, S. (1927). The effects in combination of the major colour factors of the guinea-pig. *Genetics,* 12, 530-569.

WRIGHT, S. (1928). An eight factor cross in the guinea-pig. *Genetics,* 13, 508-531.

WRIGHT, S. (1934). The results of crosses between inbred strains of guinea-pigs differing in number of digits. *Genetics,* 19, 537-551.

WRIGHT, S. (1941). Tests for linkage in the guinea-pig. *Genetics,* 26, 650-669.

WRIGHT, S. (1949). On the genetics of hair direction in the guinea-pig. II. Evidence for a new dominant gene, star, and tests for linkage with eleven other loci. *J. Exp. Zool.,* 122, 325-340.

WRIGHT, S. (1959). On the genetics of silvering in the guinea-pig, with special reference to interaction and linkage. *Genetics,* 44, 387-405.

CHAPTER 15

Deermouse

Peromyscus maniculatus ($n = 23 + X + Y$)
This new-world cricetid species (or group of species) of
continental North America could serve as an excellent foil to
the ubiquitous old-world house mouse. However, for a number
of reasons, some technical but mostly arising from lack of
opportunity, the deermouse has not been utilized as extensively
as its innate reproductive and genetic qualities would merit.
Some 40 - 46 mutant gene/loci are known, of which 21 have
been involved in studies of linkage. A detailed review of early
research on linkage is provided by Robinson (1964). A general
review of *Peromyscus* heredity has been recently contributed by
Rasmussen (1968).

TABLE 15.1. Genes/loci of the deermouse which have featured in studies
on linkage: assigned linkage group, symbol, and conventional designation.

Linkage Group	Symbol	Designation	Prime Characteristic
II	*b*	Brown	Coat colour
	bg	Boggler	Behaviour
I	*c*	Albino	Coat colour
II	*d*	Dilute	Coat colour
	ep	Epilepsy	Behaviour
	Es-1	Erythrocyte esterase-1	Electrophoretic variant
I	*f*	Flex-tail	Tail
	g	Grey-band	Coat colour
	hr	Hairless	Coat colour
	i	Ivory	Coat colour
	n	Nude	Hypotrichosis
III	*Nb*	Wide-band	Coat colour
	p	Pink-eye	Coat colour
	Pm	Erythrocyte antigen	Immunogenetics
	S	White face	White spotting
	sb	Snub-nose	Skeleton
I	*si*	Silver	Coat colour
	sph	Spherocytosis	Blood
III	*v*	Waltzer	Behaviour
	Wc	Whitecheek	Coat colour
	y	Yellow	Coat colour

LINKAGE GROUPS

Five pairs of linked genes have so far been discovered, between them determining three linkage groups. The complete data are summarized in Table 15.2. None of the observed differences in crossingover between the sexes are individually significant although it is interesting that the amount is greater for the female in four cases. In this respect, the deermouse conforms with observations in other species where a sex difference in crossingover has been established. A separate table for the sexes is omitted because it would virtually be a duplication of Table 15.2. Randerson (1965) briefly mentions that the *Es-1* and *Pm* loci may be closely linked but the relevant data have yet to be published.

Group I (c, f, p, si)

Linkage between *c* and *p* was first observed by Sumner (1922) but he gave no firm data from which a crossover value could be obtained. Clark (1936, 1938) has supplied the data producing the values shown in Table 15.2. It may be noted that Feldman (1937) failed to observe an association between *c* and *p*. According to his figures, 82 crossovers were observed out of 183 animals, yielding a crossover percentage of 44.8 ± 3.7.

Some inconsistency is evident for the data on the *f - si* intercept. Two batches of data show a significantly lower amount of crossingover than the others. Both of these genes display impenetrance and allowance has been made for the phenomenon. Apart from this, the reason for the low values is obscure (Huestis and Lindstedt, 1946). The mean crossover value for the intercept may be taken to be 25 ± 1.7.

Strictly, it is not possible to arrange the four loci of this group in linear order but, on the principle of minimizing the linkage map, the order may be provisionally taken to be:

$$\frac{\begin{array}{cccc} c & si & p & f \end{array}}{\begin{array}{ccc} 11 & 8 & 19 \end{array}}$$

However, Rasmussen (1968) cites a personal communication from Huestis and Silliman which proposes that the order of the four loci is: *p - c - sb - si - f*. Alas, the data in support of the proposal have not been published. It will be seen that this order would not be incompatible with those data which have been published except that the map length is now longer than given above. No information is available for *sb* but, on this basis and from the known distances between *c* and *f* and between *f* and *si*, it would seem that *sb* may be fairly closely linked to either *c* or *si* or perhaps to both.

Group II (b, d)

McIntosh's (1956) data on the *b* - *d* intercept in the female consisted of four groups, one of which was inconsistent with the other three. The crossover value was 19.4 ± 4.8 for the anomalous group, as opposed to 43.1 ± 2.9 for the other three. The difference is significantly large. McIntosh speculated that the females of the group may have carried an inversion in the pertinent chromosome.

Group III (Nb, v)

The two genes display a mean crossover value of 11 ± 1.

SEX-LINKAGE

No mutants belonging to loci on the *X* chromosome are known but the observations of Shaw and Barto (1965) and Shaw (1966) indicate the presence of a form of glucose-6-phosphate (G6PD) which appears to be homologous to that of several other species (man, horse, donkey, and two hares; Ohno, 1967). The form is termed type A for the deermouse. No electrophoretic variants have yet been isolated among about 400 animals, hence the mode of genetic determination is unknown. However, the comparable form in the various species cited above is controlled by an *X*-linked gene and it is conjectured that a similar situation may hold for the deermouse.

A conspicuous feature of the deermouse data is the absence of data on partial sex-linkage. Though it can be argued that this sort of linkage may not occur or be rare, the degree of sex-linkage could be such that it could pass undetected unless a specific test is applied. Sex segregates in every cross, hence information on partial sex-linkage can usually be collected incidentally to other observations.

TABLE 15.2. Established linkage groups in the deermouse.

Group Loci	Sex	Mating type	Crossover value	References
I c - f	♂	RBB	30.0 ± 5.5	Huestis and Lindsted (1946)
	♀	RBB	40.7 ± 4.4	,, ,, ,, ,,
	-	RII	25.9 ± 9.3	,, ,, ,, ,,
c - p	♂	RBB	13.5 ± 5.6	Clark (1938)
	♀	RBB	18.4 ± 2.8	,, ,,
	?	?BB	44.8 ± 3.7	Feldman (1937)
f - si	♂	CBB	9.1 ± 2.9	Huestis and Lindsted (1946)
	♂	RBB	22.5 ± 3.8	,, ,, ,, ,,
	♀	CBB	21.4 ± 2.3	,, ,, ,, ,,
	♀	RBB	27.7 ± 3.5	,, ,, ,, ,,
	-	RII	8.2 ± 2.9	Huestis and Piestrak (1942)
II b - d	♂	CBB	39.7 ± 3.5	McIntosh (1956)
	♀	CBB	43.1 ± 2.9	,, ,,
	♀	CBB	19.4 ± 4.8	,, ,,
III Nb - v	♂	CBB	11.4 ± 1.5	,, ,,
	♀	CBB	10.5 ± 1.4	,, ,,

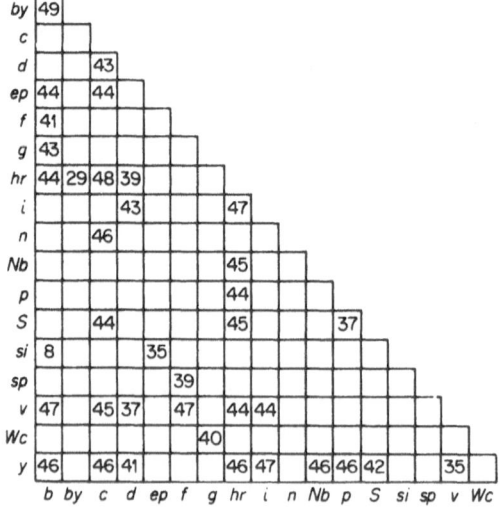

Fig. 15.1. Summary of extent of independent assortment in the deermouse.

INDEPENDENT ASSORTMENT

The extent of tests which have given no indication of linkage are shown by Table 15.3 and Fig. 15.1. The entries in these are calculated in the usual manner as described earlier for other species. It is apparent that loose linkage cannot be excluded for a number of pairs of genes.

With data of this nature, it is to be expected that several deviations could exceed the conventional level of significance by chance alone. One such fortuitous event may be the data of Feldman (1937) for the p and S genes. These two genes display a recombination value of 44 ± 3. This value is just significant but the result is probably spurious. Gene c is linked to p but, despite this, c shows no linkage with S. However, c and p are not closely linked and should c and S lie on opposite sides of p, a loose linkage similar to that observed would be the outcome. The matter can only be settled with additional breeding data.

CHROMOSOME MORPHOLOGY

The haploid number of chromosomes is 28. As yet, no studies have been made on frequency distribution of chiasmata. The early reports on deermouse karyology are briefly summarized by Robinson (1964). Much of this early work has been superseded by later studies using modern techniques and, more significantly, sampling the various subspecies on a greater scale than hitherto. The pertinent literature is as follows: Singh and McMillan (1966), Sparkes and Arakaki (1966). Ohno, Weiler, Poole, Christian, and Stenius (1966), Arakaki and Sparkes (1967), Hsu and Arrighi (1966, 1968), and Kreizinger and Shaw (1970).

A summary of the main features of deermouse karyology is provided by Table 15.4. It will be appreciated that such a classification cannot entirely avoid being arbitrary. For this reason, not too much should be read into the apparent differences between investigations. On the other hand, Ohno et al. (1966), Hsu and Arrighi (1968), Arakaki et al. (1970), and Kreisinger and Shaw (1970) have all emphasized the extraordinary variability of karyotype between individuals

TABLE 15.3. Quasi-independent gene assortment in the deermouse.

Loci	Recombination value	Phase balance	References
b - *bg*	50±11	100	Barto (1955)
b - *ep*	50±3	26	Barto (1956)
b - *f*	50±4	0	Huestis and Barto (1936)
b - *g*	49±3	100	Blair (1947)
b - *hr*	50±3	0	McIntosh (1956)
b - *si*	52±23	0	Huestis and Barto (1934)
b - *v*	55±4	100	McIntosh (1956)
b - *y*	52±3	0	,, ,,
bg - *hr*	50±11	0	Barto (1955)
c - *d*	52±4	0	Barto (1942)
c - *ep*	50±3	0	Barto (1956)
c - *hr*	53±2	?	Sumner (1922), Feldman (1937), Clark (1938)
c - *n*	52±3	0	Barto (1942)
c - *S*	54±5	?	Feldman (1937)
c - *v*	51±3	0	Barto (1942)
c - *y*	48±1	?	Sumner (1922), Feldman (1937), Clark (1938)
d - *hr*	45±3	0	Barto (1942)
d - *i*	51±4	0	,, ,,
d - *v*	50±7	83	Barto (1956)
d - *y*	53±6	0	Clark (1938)
ep - *si*	47±6	0	Barto (1956)
f - *sph*	49±5	100	Huestis, Anderson, and Motulsky (1956)
f - *v*	52±3	41	Barto (1956)
g - *Wc*	47±3	100	Blair (1944)
hr - *i*	52±3	0	Barto (1942)
hr - *Nb*	52±3	100	McIntosh (1956)
hr - *p*	51±3	?	Feldman (1937)
hr - *s*	51±3	?	,, ,,
hr - *v*	52±4	0	Barto (1956)
hr - *y*	49±3	?	Feldman (1937), Clark (1938)
i - *v*	57±6	29	Barto (1956)
i - *y*	54±3	0	McIntosh (1956)
Nb - *y*	54±4	0	,, ,,
p - *S*	44±3	?	Sumner and Collins (1922), Feldman (1937)
p - *y*	55±5	?	Feldman (1937)
S - *y*	48±3	?	,, ,,
v - *y*	44±4	37	Barto (1956)

belonging to different subspecies. A certain amount of inter-subspecies variation is conceivable (perhaps, even to be expected) but the variation has also been observed within subspecies. However, the extent and taxonomic significance of the variability has yet to be fully evaluated. The variation is not simply that of relative sizes of chromosomes or confusion over the status of the very small elements but appears to involve such major features as number of metacentric versus acrocentrics. A single large metacentric and large acrocentric seem prominent in many of the recently prepared karyotypes, otherwise there is considerable intergrading of elements.

It is difficult to dismiss the variation as due to artefacts, as is possible for older material, and the simplest explanation is that some of the differences are existing as polymorphisms. Ohno *et al.* (1966) have depicted two cases for the *bairdii* subspecies. The first involved one of the largest submetacentrics with a large acrocentric and the second involved a small metacentric with a small acrocentric. Despite the differential positioning of the centromere, the disparate homologues seem able to synapse. In one mating, the differences were transmitted from parent to offspring simultaneously and showed independent assortment.

It is probably too early for profitable speculation on the effects of the above variation upon the field of genetics. An obvious effect could be otherwise inexplicable variation of linkage between some gene pairs. The odd differences in strength of linkage (or absence in some crosses but presence in others) noted earlier for the c - p, f - si and b - d intercepts would indeed be the sort of variation to be anticipated. However, it may not be possible to place the phenomenon on a reliable basis, until a great many loci are represented by mutant genes than are known at present. Even then, these may have to be evenly distributed both along and between the chromosomes.

This *sine qua non* is remote at the present time but if karyotyping of the *maniculatus* species complex is energetically pursued, eventually a combined genetic and karyological approach may be possible. The former should be able to provide independent confirmation of the movement of chromatin material between chromosomes revealed by the latter. Some preliminary work should be possible, however, on the fertility and detection of peculiar chromosome configuration at meiosis of hybrids between animals with different karyotypes. A few karyological observations of this nature have been presented by Ohno *et al.* (1966) and Arakaki *et al.* (1970).

The *X* chromosome seems to be relatively stable for most subspecies, being a large metacentric or submetacentric. On the other hand, the *Y* chromosome emerges as exceptionally variable (Hsu and Arrighi, 1968). The element ranges from a large submetacentric to a small acrocentric. While much of the variation occurs between subspecies, some variation can occur within subspecies. The apparently wide variation shown by the *Y* has been investigated by Kreizinger and Shaw (1970). Microscopic examination was supplemented by tritiated thymidine labelling. The latter procedure was the most informative in that it was possible to identify a putative *Y* in all cases, whereas this was not possible with the former. In particular, a search for unmatched chromosomes, as a means of separating the *X* and *Y* was not wholly reliable. The size of the presumptive *Y* was variable but was consistently within the 'medium sized' range. The position of the centromere varied from metacentric to telocentric. The *X* was always among the largest of the complement although it was metacentric in one individual and subtelocentric in another three. Almost all of the confusion surrounding the ready identification of the sex chromosomes was due to variability of karyotype displayed by the animals examined. In this respect, Kreizinger and Shaw's paper deserves to be read in order to appreciate the extraordinary diversity which has been revealed by the studies of Hsu and Arrighi (1968) and the authors themselves.

BIBLIOGRAPHY

ARAKAKI, D. T. and SPARKES, R. S. (1967). The chromosomes of *Peromyscus maniculatus hollesteri. Cytologia*, 32, 180-183.

ARAKAKI, D. T., VEOMETT, I. and SPARKES, R. S. (1970). Chromosome polymorphism in deermouse siblings. *Experientia*, 26, 425-426.

BARTO, E. (1942). Independent inheritance of certain characters in the deermouse, *Peromyscus maniculatus. Papers Mich. Acad. Sci. Arts Lettr.* 27, 195-213.

BARTO, E. (1955). Boggler, an inherited abnormality of the deermouse *(Peromyscus maniculatus)* characterised by a tremor and a staggering gait. *Contr. Lab. Vert. Biol. Univ. Mich.*, 71, 1-26.

BARTO, E. (1956). Tests for independence of waltzer and ep sonogenic convulsive from certain other genes in the deer mouse. *Contr. Lab. Vert. Biol. Univ. Mich.*, 74, 1-16.

BLAIR, W. F. (1944). Inheritance of the white-cheek character in mice of the genus *Peromyscus. Contr. Lab. Vert. Biol. Univ. Mich.*, 25, 1-7.

BLAIR, W. F. (1947). An analysis of certain genetic variations in pelage color of the Chihuahua deermouse *(Peromyscus maniculatus blandus). Cont. Lab. Vert. Biol. Univ. Mich.*, 35, 1-18.

BRADSHAW, W. N. and GEORGE, W. A. (1969). The karyotype in *Peromyscus maniculatus nubiterrae. J. Mammal.*, 50, 822-824.

CLARK, F. H. (1936). Linkage of pink-eye and albinism in the deermouse. *J. Hered.*, 27, 256-260.

CLARK, F. H. (1938). Inheritance and linkage relations of mutant characters in the deermouse. *Contrib. Lab. Vert. Biol. Univ. Mich.*, 7, 1-11.

FELDMAN, H. W. (1937). Segregation of mutant characters of deermice. *Amer. Nat.*, 71, 426-429.

HUESTIS, R. R. (1940). Tests for linkage in *Peromyscus. Genetics*, 25, 121.

HUESTIS, R. R. (1942). Gene interaction in *Peromyscus. Genetics*, 27, 146-147.

HUESTIS, E. E. (1944). Tests for linkage of silver pelage and flexed-tail in *Peromyscus. Yearb. Amer. Phil. Soc.* (1953), 160-161.

HUESTIS, R. R. (1946). Linkage of flexed tail and albinism in *Peromyscus. Genetics*, 31, 219.

HUESTIS, R. R., ANDERSON, R. S. and MOTULSKY, A. G. (1956). Hereditary spherocytosis in *Peromyscus. J. Hered.*, 47, 225-228.

HUESTIS, R. R. and BARTO, E. (1934). Brown and silver deermice. *J. Hered.*, 25, 219-223.

HUESTIS, R. R. and BARTO, E. (1936). Flex-tailed *Peromyscus. J. Hered.*, 27, 73-75.

HUESTIS, R. R. and LINDSTEDT, G. (1946). Linkage relations of flexed-tail in *Peromyscus. Amer. Nat.*, 80, 85-90.

HUESTIS, R. R. and PIESTRAK, V. (1942). An aberrant ratio in *Peromyscus. J. Hered.*, 33, 289-291.

HSU, T. C. and ARRIGHI, F. S. (1966). Chromosomal evolution in the genus *Peromyscus (Cricetidae, Rodentia). Cytogenetics*, 5, 355-359.

HSU, T. C. and ARRIGHI, F. S. (1968). Chromosomes of *Peromyscus* (Rodentia, Cricetidae). *Cytogenetics*, 7, 417-446.

KREIZINGER, J. D. and SHAW, M. W. (1970). Chromosomes of *Peromyscus* (Rodentia, Cricetidae). II. The Y chromosome of *Peromyscus maniculatus. Cytogenetics*, 9, 52-70.

McINTOSH, W. B. (1956). Linkage in *Peromyscus* and sequential tests for independent assortment. *Contr. Lab. Vert. Biol. Univ. Mich.*, 73, 1-27.

OHNO, S. (1967). *Sex chromosomes and sex-linked genes.* Berlin:

Springer-Verlag.

OHNO, S., WEILER, C., POOLE, J., CHRISTIAN, L. and STENIUS, C. (1966). Autosomal polymorphism due to pericentric inversions in the deermouse *(Peromyscus maniculatus)* and some evidence of somatic segregation. *Chromosoma*, 18, 177-187.

RANDERSON, S. (1965). Erythrocyte esterase forms controlled by multiple alleles in the deermouse. *Genetics*, 52, 999-1005.

RASMUSSEN, D. I. (1968). Genetics. In King, J. A. (Editor). *Biology of Peromyscus. Amer. Soc. Mammalogists Spec. Pub.*, 2, 340-372.

ROBINSON, R. (1964). Linkage in *Peromyscus. Heredity*, 19, 701-709.

SHAW, C. R. (1966). Glucose-phosphate dehydrogensae: homoglous molecures in deermouse and man. *Science*, 153, 1013-1015.

SHAW, C. R. and BARTO, E. (1965). Autosomally determined polymorphism of glucose-6-phosphate dehydrogenase in *Peromyscus. Science*, 148, 1099-1100.

SINGH, R. P. and McMILLAN, D. B. (1966). Karyotypes of three subspecies of *Peromyscus. J. Mammal.*, 47, 261-266.

SPARKES, R. S. and ARAKAKI, D. T. (1966). Intrasubspecific and intersubspecific chromosomal polymorphism in *Peromyscus maniculatus. Cytogenetics*, 5, 411-418.

SUMNER, F. B. (1922). Linkage in *Peromyscus. Amer. Nat.*, 56, 412-417.

SUMNER, F. B. and COLLINS, H. H. (1922). Further studies of color mutations in mice of the genus *Peromyscus. J. Exp. Zool.*, 36, 289-316.

TABLE 15.4. Summary of investigations on deermouse karyology (*Peromyscus maniculatus* subspecies)

Subspecies	Defined autosomal groups	No. of metacentrics			No. of acrocentrics			Sex Chromosomes		References
		Large	Medium	Small	Large	Medium	Small	X	Y	
austerius	3	1	4	9	2	3	4	Lm	Sm	Hsu and Arrighi (1968)
bairdii	4	3	4	7	3	4	2	Lm	Sa	Singh and McMillan (1966)
bairdii	3	3	3	12	3	2	0	Lm	Sm	Sparkes and Arakaki (1966)
bairdii	4	2	4	11	1	3	2	La	Sa	Ohno et al. (1966)
bairdii	4	2	3	11	2	3	2	La	Sa	Ohno et al. (1966)
gambelii	3	1	3	11	1	2	5	La	?	Kreizinger and Shaw (1970)
gracilis	4	3	4	9	2	2	3	Lm	Sa	Singh and McMillan (1966)
gracilis	3	3	2	4	1	3	0	Lm	Mm	Sparkes and Arakaki (1966)
hollisteri	4	1	3	10	2	3	4	La	Sa	Ohno et al. (1966)
hollisteri	4	1	3	8	2	3	6	La	Sa	Ohno et al. (1966)
hollisteri	3	2	3	9	3	2	4	Lm	Sm	Arakaki and Sparkes (1967)
noveboracensis	4	3	3	4	2	5	6	Lm	Sa	Singh and McMillan (1966)
nubiterrae	4	2	9	8	2	1	1	Lm	Sa	Bradshaw and George (1969)
oreas	4	1	3	15	2	1	1	?	Mm	Kreizinger and Shaw (1970)
rubidus	3	3	2	9	2	2	5	Lm	Sa	Sparkes and Arakaki (1966)
rufinus	3	1	4	3	2	6	7	Lm	Lm	Hsu and Arakaki (1968)
rufinus	3	1	4	13	1	2	2	Lm	Sa	Hsu and Arakaki (1968)
rufinus	4	1	4	3	4	5	6	La	Mm	Kreizinger and Shaw (1970)
rufinus	4	1	4	13	1	2	2	La	Mm	Kreizinger and Shaw (1970)
rufinus	4	1	5	3	1	8	5	?	Mm	Kreizinger and Shaw (1970)
sonoriensis	4	1	3	14	1	3	1	?	Mm	Kreizinger and Shaw (1970)

L = large; M = medium; S = small; a = acrocentric; m = metacentric; acrocentric embraces subtelocentric and acrocentric elements.

CHAPTER 16

Golden Hamster

Mesocricetus auratus $(n = 21 + X + Y)$
The golden hamster is one of the more recent additions to the range of laboratory animals. The species is proving to be an excellent laboratory subject and, already, a number of mutant genes have been described. *In toto*, approximately 16 have been reported to date and those which have been utilized in linkage research are presented in Table 16.1. A general review of hamster genetics may be found in Robinson (1968).

TABLE 16.1. Genes/loci of the golden hamster which have featured in studies on linkage; assigned linkage group, symbol, and conventional designation.

Linkage Group	Symbol	Designation	Prime Characteristic
I	b	Brown	Coat colour
	Ba	White band	White spotting
I	c^d	Acromelanic albino	Coat colour
	e	Yellow	Coat colour
XXII	Mo	Mottle-white	Coat colour
XXII	pa	Hind-leg paralysis	Neuromuscular
	ru	Ruby-eye	Coat colour
	s	Piebald	White spotting
XXII	To	Tortoiseshell	Coat colour
	Wh	Anophthalmic white	Coat colour

LINKAGE GROUP

Group I (b, c^d)
Only one case of linkage has so far been discovered. This is between b and c^d, and has the value of 36 ± 4, based largely on female gametogenesis (Robinson, unpublished observations). This constitutes the first autosomal group. The apparent association between ru and s is almost certainly due to inviability interaction (Robinson, 1959c, 1971).

SEX-LINKAGE

Three loci are known to be carried by the X chromosome but, at this time, no information is available on their serial order. The genes are *Mo* (Magalhaes, 1954), *To* (Robinson, 1966), and *pa* (Nixon and Connelly, 1968). This number means that the hamster has the distinction of falling into second place to the house mouse for number of sex-linked genes. If the suggestion is taken up that the sex-linked group of genes is conventionally designated as the final linkage group of the species, the XXII linkage group may be regarded as the first to be established for the hamster, requiring only numerical values for the group to have substance.

INDEPENDENT ASSORTMENT

A number of the known genes have been tested among themselves for possible linkage and the extent of this work is summarized by Table 16.2 and Fig. 16.1. It is by no means certain, of course, that all of these genes are truly independent. One or more could still be found to be loosely linked. Furthermore, not all of the possible combinations have been investigated. Fig. 16.1 shows the closest linkage which would not be in serious disagreement with the published information. These values are approximated by multiplying the standard error by 196 and subtracting the product from the recombination value.

There are two cases in which linkage might be indicated. These are between *ru* and *s*, and *ru* and the sex segment of chromosome. However, in both of these cases, the evidence points to an inviability interaction rather than to chromosomal linkage. The apparent association between *s* and *ru* came into being because, in the early experiments, the genes *ru* and *s* entered the crosses in repulsion and the doubly homozygous *russ* consequently was a recombination class. Both of these genes are associated with inviability effects (small size, sterility, high mortality) and these effects are more than proportionally magnified for the *russ* animal. Consequently, there was a deficiency of recombinants. When data were secured in which

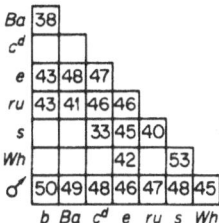

Fig. 16.1. Summary of extent of independent assortment in the hamster.

the two linkage phases were roughly equally balanced, the spurious linkage disappeared (Robinson, 1959c, 1971).

TABLE 16.2. Quasi-independent gene assortment in the golden hamster.

Loci	Recombination value	Phase balance	References
b - Ba	46 ± 4	0	Robinson (1971)
b - e	51 ± 4	0	Robinson (1962a)
b - ru	51 ± 4	0	,, ,,
b - $\male$	56 ± 3	0	Robinson (1962a, 1971)
Ba - e	53 ± 3	100	Robinson (1962b)
Ba - ru	55 ± 7	100	,, ;,
Ba - $\male$	54 ± 2	100	Robinson (1962b, 1971)
c^d - e	49 ± 3	0	Robinson (1959a, 1971)
c^d - ru	50 ± 2	0	,, ,, ,,
c^d - s	43 ± 5	24	,, ,, ,,
c^d - $\male$	50 ± 2	0	,, ,, ,,
e - ru	48 ± 1	68	Robinson (1958, 1959a, b, 1962a, 1971)
e - s	48 ± 2	90	Robinson (1958, 1959a, 1962b, 1964, 1971)
e - Wh	47 ± 3	100	Robinson (1964)
e - $\male$	49 ± 1	82	Robinson (1958, 1959a, b, 1962a, b, 1964, 1971)
ru - s	46 ± 3	51	Robinson (1958, 1959a, 1962b, 1971)
ru - $\male$	50 ± 1	94	Robinson (1958, 1959a, b, 1962a, b, 1971)
s - Wh	55 ± 3	100	Robinson (1964)
s - $\male$	51 ± 2	66	Robinson (1958, 1959a, 1962b, 1964, 1971)
Wh - $\male$	50 ± 3	100	Robinson (1964)

CHROMOSOME MORPHOLOGY

The hamster has a haploid complement of 22 chromosomes (Husted, Hopkins, and Moore, 1945; Matthey, 1951, 1952; Sachs, 1952). These determinations have been amply confirmed by other workers. The more informative reports are as follows: Koller (1938), Sheaffer (1955), Ohno, Becak, and Becak (1964), Awa, Sasaki, and Takayama (1959), Jordan (1959), Emmons and Husted (1962), Ohno and Weiler (1961), Saksela and Moorhead (1962), Lehman et al. (1963), Walkanowska (1964), and Senin and Pogosiantz (1967).

An interesting feature of the karyotype is the even grading in size of the chromosomes from the largest to the smallest. There are no outstandingly large elements, such as characterize the karyotype of other species (e.g., the Norway rat and guinea-pig). This implies that the summary classification of Table 16.3 tends to be arbitrary.

The sex chromosomes have been repeatedly examined. This is due in part to the early observations of Koller (1938), who concluded that both prereduction and post-reduction separation may occur for the XY bivalent. This conclusion is disputed by Matthey (1952) but upheld by Emmons and Husted (1962). The X is certainly one of the largest (if not the largest) chromosome of the karyotype. It is probably relatively large as a mammalian X element per se. According to Ohno et al. (1964), to compensate for the large size and to maintain the appropriate X: autosome balance for somatic nuclei, the whole of one X and one arm of the other X is heterochromatic in the female and one arm of the X is heterochromatic in the male. The Y chromosome has been described as large by some observers but as small by others. This difference may be simply an artefact although it could be indicative of size variation (Lehman et al. 1963).

CHIASMA FREQUENCY

Several investigations have contributed information on the distribution of chiasmata among the chromosomes. The main results are presented by Table 16.4. The observations of

TABLE 16.3. Summary of karyological investigations of the golden hamster.

Defined autosomal groups	No. of metacentrics			No. of acrocentrics			Sex chromosomes		References
	Large	Medium	Small	Large	Medium	Small	X	Y	
—	2	11	4	0	0	4	Lm	Lm	Matthey (1952)
—	1	14	2	0	0	4	Lm	Sm	Awa et al. (1959)
—	2	13	2	0	0	4	Lm	Lm	Ohno and Weiler (1961)
13	2	13	1	0	0	5	Lm	Mm	Lehman et al. (1963)
6	2	13	2	0	0	4	Lm	Lm	Senin and Pogosiantz (1967)

L = large, M = medium, S = small, m = metacentric; acrocentric embraces subtelocentric and telocentric elements.

Emmons and Husted (1962) have been approximated from details given in their paper but it is felt that any error due to this will be offset by other factors tending to underestimation of the real frequency. For instance, the zero frequency may represent bivalents with one chiasma (perhaps terminally situated) which have precociously separated. By this reasoning, the mean frequency may in fact be about 1.17 and 1.03 for the two XY entries of the table. Chiasmata seem to occur more frequently in the female than in the male, a conclusion which is in general agreement with the pattern observed for the house mouse and Norway rat.

TABLE 16.4. Chiasmata frequency in the golden hamster.

Bivalent	Sex	No. of bivalents	No. of chiasmata				Mean	References
			0	1	2	3		
Autosome	♂	180	0	121	58	1	1.33	Koller (1938)
	♀	132	-	-	-	-	1.46	Sheaffer (1955)
XY	♂	276	23	206	47	0	1.08	Koller (1938)
	♂	1734	104	1583	47	0	0.97	Emmons and Husted (1962)
XX	♀	40	-	-	-	-	1.38	Sheaffer (1955)

BIBLIOGRAPHY

AWA, A., SASAKI, M. and TAKAYAMA, S. (1959). An *in vitro* study of the somatic chromosomes in several mammals. *Jap. J. Zool.*, 12, 257-265.
EMMONS, L. R. and HUSTED, L. (1962). The sex bivalent of the golden hamster. *J. Hered.*, 53, 227-232.
HUSTED, L., HOPKINS, J. T. and MOORE, M. B. (1945). The *X* bivalent of the golden hamster. *J. Hered.*, 36, 93-96.
JORDAN, M. (1959). Les heterochromosomes chez le golden hamster *(Mesocricetus auratus). Folia Biol. (Krakow)*, 7, 73-81.
KOLLER, P. C. (1938). The genetical and mechanical properties of the sex chromosomes. IV, The golden hamster. *J. Genet.*, 36, 177-195.
LEHMAN, J. M., MacPHERSON, I. and MOORHEAD, P. S. (1963). Karyotype of the Syrian hamster. *J. Nat. Cancer Inst.*, 31, 639-650.
MAGALHAES, H. (1954). Mottle-white, a sex-linked lethal mutation in the golden hamster. *Anat. Rec.*, 120, 752.
MATTHEY, R. (1951). Chromosomes de Muridae. *Experientia*, 7, 340-341.

MATTHEY, R. (1952). Chromosomes de Muridae (Microtinae et Cricetinae). *Chromosoma*, 5, 113-138.

NIXON, C. W. and CONNELLY, M. E. (1968). Hind leg paralysis: a new sex-linked mutation in the Syrian hamster. *J. Hered.*, 59, 276-278.

OHNO, S., BECAK, W. and BECAK, M. L. (1964). *X*-autosome ratio and the behaviour pattern of individual *X* chromosomes in placental mammals. *Chromosoma*, 15, 14-30.

OHNO, S. and WEILER, C. (1961). Sex chromosome behaviour pattern in germ and somatic cells of *Mesocricetus auratus*. *Chromosoma*, 12, 362-373.

ROBINSON, R. (1958). Genetic studies of the Syrian hamster. I. The mutant genes cream, ruby-eye and piebald. *J. Genet.*, 56, 85-102.

ROBINSON, R. (1959a). Genetic studies of the Syrian hamster. II. Partial albinism. *Heredity*, 13, 165-177.

ROBINSON, R. (1959b). Genetic studies of the Syrian hamster. III. Variation of dermal pigmentation. *Genetica*, 30, 393-411.

ROBINSON, R. (1959c). Genetic independency of four mutants in the Syrian hamster. *Nature*, 183, 125-126.

ROBINSON, R. (1962a). Genetic studies of the Syrian hamster. IV. Brown pigmentation. *Genetica*, 33, 81-87.

ROBINSON, R. (1962b). Genetic studies of the Syrian hamster. V. White band. *Heredity*, 17, 477-486.

ROBINSON, R. (1964). Genetic studies of the Syrian hamster. VI. Anophthalmic white. *Genetica*, 35, 241-250.

ROBINSON, R. (1966). Sex-linked yellow in the Syrian hamster. *Nature*, 212, 824-825.

ROBINSON, R. (1968). Genetics and karyology. In Hoffman, R. A., Robinson, P. F. and Magalhaes, H. (Editors). *The golden hamster: its biology and use in medical research*. Ames: Iowa State University Press.

ROBINSON, R. (1971). Genetic studies of the Syrian hamster. VII. Independence data. *Heredity*, 26, 65-71.

SACHS, L. (1952). Polyploid evolution and mammalian chromosomes. *Heredity*, 6, 357-364.

SAKSELA, E. and MOORHEAD, P. S. (1962). Enhancement of secondary constrictions and the heterochromatic *X* in human cells. *Cytogenetics*, 1, 225-244.

SENIN, V. M. and POGOSIANTZ, E. E. (1967). On the normal karyotype of golden hamster. *Genetika (Mosk.)*, 1967 (4), 81-88.

SHEAFFER, C. I. (1955). The *X* bivalent of the golden hamster. *Virginia J. Sci.*, 6, 46-52.

WALKNOWSKA, J. (1964). The chromosomes in ontogenesis of golden hamster. *Folia Biol (Krakow)*, 12, 321-346.

CHAPTER 17

Cat

Felis catus $(n = 18 + X + Y)$
The domestic cat is coming more into its own of recent years as
a laboratory animal and it is hoped that greater attention will be
paid to the more formal aspects of the species' genetics. About
20 mutant genes have been described but few of these have
featured in studies on their possible linkage relationships. A
detailed account of these mutants and of the animal's genetics
may be found in Robinson (1959, 1971).

TABLE 17.1. Genes/loci of the cat which have featured in studies on
linkage; assigned linkage group, symbol, and conventional designation

Linkage group	Symbol	Designation	Prime characteristic
	a	Non-agouti	Coat colour
	c^s	Siamese	Coat colour
	d	Dilution	Coat colour
	l	Long hair	Hair texture
	M	Manx	Skeleton
XIX	O	Orange	Coat colour
	Pd	Polydactyly	Feet
	r	Cornish rex	Hair texture
	W	White	Coat colour

LINKAGE GROUPS

No autosomal groups are known.

SEX-LINKAGE

A gene borne by the X chromosome is responsible for the
orange and tortoiseshell colours. This case is well established
with abundant data (Doncaster, 1913; Bamber and Herdman,
1927, 1932; Robinson, 1970).
The sex linkage of O would seem to be complete and the

451

gene may be regarded as located in the X differentiated segment. On the other hand, males of the $O+$ phenotype (tortoiseshell) do occur rarely and it has been suggested that these could result from infrequent crossingover (Komai, 1952; Jude and Searle, 1957). However, it seems more likely that the rare tortoiseshell males arise from a variety of karyological chimeras which have involved the sex chromosomes (Thuline and Norby, 1961; Chu *et al.*, 1964; Biggers and McFeely, 1966; Smith and Jones, 1966; McFeely *et al.*, 1967; Malouf *et al.*, 1967; and Loughman *et al.*, 1970). The typical chimera is of the constitution XY/XXY but other combinations (some complex) could theroretically exist and in fact do. An animal of constitution XY/XY, with O and $+$ genes on the different X chromosomes, could occur and be karyologically undetectable as a chimera. The majority of tortoiseshell males are sterile but the XY/XY individual could conceivably be fertile. In these cases, the tortoiseshell phenotype has behaved as a screen to pick out males of unusual chromosome constitutions.

INDEPENDENT ASSORTMENT

The amount of data available on random assortment is so scanty that a quantitative analysis is not worthwhile. The greatest amount is provided by Searle and Jude (1956) for the genes d and r. Backcross data in coupling show a recombination value of 44 ± 10 for 25 animals. Todd (1963) contributed data which shows mutually independent inheritance of a, b, c^s and M (eight animals) and again in 1966 for the independence of Pd and W (22 animals). Tjebbes' (1924) data were less informative but, as far as these go, independent assortment of c^s, l, W and sex is indicated. The comments of Keeler and Cobb (1936) could be brought forward in favour of independent segregation of c^s and l. It is obvious, however, that loose linkage is not excluded for any of these genes.

CHROMOSOME MORPHOLOGY

Many of the early determinations of the haploid chromosome

TABLE 17.2. Summary of karyological investigations of the cat

Defined autosomal groups	No. of metacentrics			No. of acrocentrics			Sex chromosomes		References
	Large	Medium	Small	Large	Medium	Small	X	Y	
—	3	7	1	2	3	2	Mm	Sa	Awa et al. (1959)
—	2	6	4	1	3	2	Sm	Sm	Ohno et al. (1962)
6	2	4	6	1	3	2	Mm	Sa	Matano (1963)
9	4	6	2	1	3	2	Mm	Sa	Cranmoor and Alpen (1964)
—	2	2	6	1	5	2	Mm	Sa	Chu et al. (1964)
6	2	3	3	4	3	2	Mm	Sm	Gustavsson (1965)
6	2	3	3	4	4	2	Mm	Sa	Hsu and Rearden (1965)
5	5	4	3	1	3	2	Mm	Sm	Hare et al. (1966)
3	2	5	5	1	3	2	Mm	Sa	O'Reilly and Whitaker (1969)
6	2	3	3	4	4	2	Mm	Sa	Loughman et al. (1970)

M = medium, S = small, a = acrocentric, m = metacentric; acrocentric embraces subtelocentric and telocentric elements.

number proved to be inaccurate. The correct amount of 19 was found by Minouchi (1928), Minouchi and Ohta (1934), and Matthey (1934). This number has been confirmed by many later studies, among which the following are the most important: Koller (1941), Awa, Sasaki, and Takayama (1959), Ohno, Stenius, Weiler, Trujillo, Kaplan, and Kinosita (1962), Matano (1963), Chu, Thuline, and Norby (1964), Cranmore and Alpen (1964), Gustavsson (1965), Hsu and Rearden (1965), Hare, Weber, McFeely, and Yang (1966), Wurster and Benirschke (1968), and O'Reilly and Whitaker (1969).

A generalized description of the karyotype is shown by Table 17.2. The individual chromosomes of the cat seem to be relatively distinctive and there is greater agreement between the various investigations than the table would imply. The X is a medium sized metacentric while the Y is small, sometimes described as metacentric, and, at other times, as acrocentric.

Some workers claim to be able to separate almost all of the individual elements (e.g. Chu *et al.*, 1964; Hsu and Rearden, 1965). A satellited chromosome is of regular recurrence (Cranmore and Alpen, 1964; Chu *et al.*, 1964; Gustavsson, 1965); while Wurster and Benirschke (1968) depict the element as the 'carnivore chromosome'. Chu *et al.* (1964) and O'Reilly and Whitaker (1964) have tabulated numerical data on arm length ratios for many of the larger chromosomes. It is of interest that Wurster and Benirschke (1968) have compared the karyotype as a whole with many other Felid and carnivore species, with special emphasis on their evolutionary modification and significance.

CHIASMA FREQUENCY

A report by Koller (1941) has given information on the distribution of chiasmata among the chromosomes. Table 17.3 presents details of the findings. There is a small decrease in mean frequency from diakinesis to metaphase, such as might be expected if a proportion of the chiasmata had terminized or 'slipped off' the ends of the bivalent. A rough, and hence tentative, estimate of the chiasma frequency for the *XY* gave a mean of 1.03 for 134 bivalents.

TABLE 17.3. Chiasma frequency in autosomes of the male cat

Cell stage	No. of bivalents	No. of chiasmata				Mean	Reference
		0	1	3	4		
Diakinesis	126	0	55	53	18	2.71	Koller (1941)
Metaphase	108	1	72	26	9	2.40	Koller (1941)

BIBLIOGRAPHY

AWA, A., SASAKI, M. and TAKAYAMA, S. (1959). An in *vitro* study of the chromosomes in several mammals. *Jap. J. Zool.*, 12, 257-265.

BAMBER, R. C. and HERDMAN, E. C. (1927). The inheritance of black, yellow and Tortoiseshell coat colour in cats. *J. Genet.*, 18, 87-97.

BAMBER, R. C. and HERDMAN, E. C. (1932). A report on the progeny of a tortoiseshell male cat, together with a discussion of his genetic constitution. *J. Genet.*, 26, 115-128.

BIGGERS, J. D. and McFEELY, R. A. (1966). Intersexuality in domestic animals. *Advanc. Reprod. Physiol.*, 1, 29-59.

CHU, E. H. Y., THULINE, H. C. and NORBY, D. E. (1964). Triploid-diploid chimerism in a male tortoiseshell cat. *Cytogenetics*, 3, 1-18.

CRANMORE, D. and ALPEN, E. L. (1964). Chromosomes of the cat. *Nature*, 204, 99-100.

DONCASTER, L. (1913). On sex-limited inheritance in cats and its bearing on the sex-limited transmission of certain human abnormalities. *J. Genet.*, 3, 11-23.

GUSTAVSSON, I. (1965). Somatic chromosomes of the cat. *Acta Vet. Scand.*, 6, 274-285.

HARE, W. C. D., WEBER, W. T., McFEELY, R. A. and YANG, T. J. (1966). Cytogenetics in the dog and cat. *J. Small Anim. Pract.*, 7, 575-592.

HSU, T. C. and REARDEN, H. H. (1965). Further karyological studies in Felidae. *Chromosoma*, 16, 365-371.

JUDE, A. C. and SEARLE, A. G. (1957). A fertile tortoiseshell tomcat. *Nature*, 79, 1087-1088.

KEELER, C. E. and COBB, V. (1936). Siamese-Persian cats. *J. Hered.*, 27, 339-340.

KOLLER, P. C. (1941). The genetical and mechanical properties of the sex chromosomes. VIII. The cat *(Felis domestica). Proc. Roy. Soc. Edin, B.*, 61, 78-94.

KOMAI, T. (1952). On the origin of the tortoiseshell male cat — a correction. *Proc. Jap. Acad.*, 28, 150-155.

LOUGHMAN, W. D., FRYE, F. L. and CONDON, T. B. (1970). *XY/XXY* bone marrow mosaicism in three male tricolor cats. *Amer. J. Vet. Res.*, 31, 307-314.

MALOUF, N., BENIRSCHKE, K. and HOEFNAGEL, D. (1967). *XX/XY* chimerism in a tricolour male cat. *Cytogenetics,* **6,** 228-241.

MATANO, Y. (1963). A study of the chromosomes in the cat. *Jap. J. Genet.,* **38,** 147-156.

MATTHEY, R. (1934). La formule chromosomiale du chat domestique. *C.R.Soc. Biol., Paris,* **117,** 435-436.

McFEELY, R. A., HARE, W. C. D. and BIGGERS, J. D. (1967). Chromosome studies in 14 cases of intersexes in domestic animals. *Cytogenetics,* **6,** 242-253.

MINOUCHI, O. (1928). On the chromosomes of the cat. *Proc. Imp. Acad. Tokyo,* **4,** 128-130.

MINOUCHI, O. and OHTA, T. (1934). On the chromosome number and the sex-chromosomes in the germ-cells of male and female cats. *Cytologia,* **5,** 355-362.

OHNO, S., STENIUS, C., WEILER, C. P., TRUJILLO, J. M., KAPLAN, W. D. and KINOSITA, R. (1962). Early meiosis of male germ cells in fetal testis of *Felis domestica. Exp. Cell Res.,* **27,** 401-404.

O'REILLY, K. J. and WHITAKER, A. M. (1969). The development of feline cell lines for the growth of feline enteritis (panleucopaenia) virus. *J. Hyg.,* **67,** 115-124.

ROBINSON, R. (1959). Genetics of the domestic cat. *Bibliogr. Genet.,* **18,** 273-362.

ROBINSON, R. (1970). Gene assortment and preferential mating in the breeding of German fancy cats. *Heredity,* **25,** 207-216.

ROBINSON, R. (1971). *Genetics for cat breeders.* Oxford: Pergamon Press.

SEARLE, A. G. and JUDE, A. C. (1956). The 'rex' type of coat in the domestic cat. *J. Genet.,* **54,** 506-512.

SMITH, H. A. and JONES, T. C. (1966). *Veterinary pathology,* London: Bailliere, Tindall and Cassell.

THULINE, H. C. and NORBY, D. E. (1961). Spontaneous occurrence of chromosome abnormality in cat. *Science,* **134,** 554-555.

TJEBBES, K. (1924). Crosses with Siamese cats. *J. Genet.,* **14,** 355-366.

TODD, N. B. (1963). Independent assortment of Manx and three coat colour mutants in the domestic cat. *J. Hered.,* **54,** 266.

TODD, N. B. (1966). The independent assortment of dominant white and polydactyly in the cat. *J. Hered.,* **57,** 17-18.

WURSTER, D. H. and BENIRSCHKE, K. (1968). Comparative cytogenetic studies in the order carnivora. *Chromosoma,* **24,** 336-382.

CHAPTER 18

Dog

Canis familiaris (n = 38 + X + Y)
In spite of its undoubted popularity, very little work appears to have been accomplished with the formal genetics of the dog. The heredity of coat colour has received considerable attention and a number of loci have been defined. However, there is controversy over the status of certain colour variations and recognition of several genes (Winge, 1950; Little, 1957; Burns and Fraser, 1966). A large number of abnormalities have been shown to be monogenically determined. The most comprehensive survey of canine genetics is that of Burns and Fraser (1966); while a remarkable compilation of breeding results is provided by Little (1957).

TABLE 18.1. Gene/loci of the dog which have featured in studies on linkage: assigned linkage group, symbol, and conventional designation.

Linkage group	Symbol	Designation	Prime characteristic
XXXIX	*ha*	Haemophila A	Blood
XXXIX	*hb*	Haemophila B	Blood
XXXIX	*sc*	Carpal subluxation	Skeleton

LINKAGE GROUPS

No cases of autosomal linkage have been reported at this time.

SEX-LINKAGE

At least three instances of *X* sex-linkage have been described. The classic case is that of haemophila A, well known because it appears to be homologous to the similar disease in man (Merkens, 1938; Hutt *et al.*, 1948; Graham *et al.*, 1949;

Brinkhous and Graham, 1950; Graham, 1952). A clinically distinct haemophila B, again resembling the human defect, is also sex-linked (Mustard *et al.*, 1960; Rowsell *et al.*, 1960). A carpal subluxation is described by Pick *et al.* (1967), which is sex-linked. A form of cystinuria is also considered to be due to an *X* borne gene (Knox, 1966).

Pick *et al.* (1967) found evidence of close linkage between *ha* and *sc*. No crossover young were observed among 49 testcross progeny. By Stevens' method, these figures are indicative of an upper limit of 7.5 per cent of crossingover.

CHROMOSOME MORPHOLOGY

The karyotype is composed of a relatively large number of chromosomes. This aspect may have been the cause of incorrect counts by several early studies. However, Minouchi (1928) established that the diploid number was 78 and this determination has since been repeatedly confirmed (Ahmed, 1941; Makino, 1949, 1956; Hsu and Pomerat, 1953; Matthey, 1954; Takayama, 1958; Awa *et al.*, 1959; Takayama and Makino, 1961; Ohno *et al.*, 1964; Newham and Davidson, 1966; Eliasson *et al.*, 1968; Kakpakova *et al.*, 1968).

The more recent investigations have concentrated on obtaining detailed karyotypes. There appears to be a large measure of agreement between the various reports. All of the autosomes are either telocentric or acrocentric. One of the autosomes may be sufficiently larger than the others to be regularly distinguishable but this is not indisputable. Apart from this element, the autosomes graduate from the largest to the smallest with very few distinguishing features. The *X* is either submetacentric or metacentric and is one of the largest chromosomes of the complement. In contrast, the *Y* is one of the smallest, if not the smallest, and appears to be submetacentric. The following papers should be consulted: Brown *et al.* (1963), Moore and Lambert (1963), Sofuni and Makino (1963), Fraccaro *et al.* (1964), Gustavsson (1964), Barberis *et al.* (1964), Ford (1965), Gustavsson and Sundt (1965), Pakes and Griesener (1965), Basrur and Gilman (1966), Brown *et al.* (1966), Hare *et al.* (1966), Hsu and Benirschke

(1967), Weiss *et al.* (1967), Borgaonker *et al.* (1968), Tzessarskaya (1968), and Clough *et al.* (1970).

Several accounts of individuals with abnormal chromosomal elements occur in the literature. These abnormal conditions range from a karyotype with 78 normal chromosomes, plus a minute extra element, to a karyotype with 77 normal chromosomes and a large metacentric body. The minute element could arise from a variety of causes; whereas the large metacentric is indicative of a translocation or centric fusion of two of the normal acrocentric autosomes (Shive *et al.*, 1965; Hare *et al.*, 1966; Patterson *et al.*, 1966; Hare *et al.*, 1967).

Clough *et al.* (1970) have reported the discovery of a *XXY* dog, male in appearance, with a penis capable of erection but with hypoplastic and aspermatogenetic testes.

CHIASMA FREQUENCY

Counts of chiasma frequency have been undertaken by Ahmed (1941). Table 18.2 summarizes the observations for three samples of animals (reconstituted from the data of the paper). The means for each sample differ and may be a reflection of breed differences. In addition, it was observed that the Sealyham material displayed a higher rate of terminalization for bivalents with two chiasmata than the other samples. A comparison of chiasma frequency for diakinesis and metaphase for the Spaniel gave means of 2.13 ± 0.19 and 1.93 ± 0.07, respectively. The decline in mean frequency is in the expected direction but the difference is insignificant.

Eliasson *et al.* (1968) have briefly commented on the frequency of chiasmata. All of the autosomes usually showed one chiasma but the larger bivalents very often exhibited two and sometimes three chiasmata. The smaller bivalents always revealed a greater amount of terminalization than the larger.

BIBLIOGRAPHY

AHMED. I. A. (1941). Cytological analysis of chromosome behaviour in three breeds of dogs. *Proc. Roy. Soc. Edin., B*, **61**, 107-118.

TABLE 18.2. Frequency distribution per bivalent at meiotic metaphase in the male dog.

Breed	No. of bivalents	Chiasma frequency				Mean	Reference
		1	2	3	4		
Sealyham	380	—	326	64	1	2.06 ± 0.03	Ahmed (1941)
Spaniel	380	—	360	30	—	2.13 ± 0.01	Ahmed (1941)
Spaniel × Manchester Terrier	380	—	356	34	—	1.62 ± 0.02	Ahmed (1941)
Total	1140	—	1042	128	1	1.94 ± 0.02	

AWA, A., SASAKI, M. and TAKAYAMA, S. (1959). An *in vitro* study of the somatic chromosomes in several mammals. *Jap. J. Zool.*, 12, 257-265.

BARBERIS, L., SARTI, M. and SORRENTINO, R. (1964). The chromosomes of *Canis familiaris:* systematic and evolutionary interest. *Natura (Milano)*, 55, 234-240.

BASRUR, P. K. and GILMAN, J. P. W. (1966). Chromosome studies in canine lymphosarcoma. *Cornell Vet.*, 56, 541-469.

BORGAONKAR, D. S., ELLIOT, O. S., WONG, M. and SCOTT, J. P. (1968). Chromosome study of four breeds of dog. *J. Hered.*, 59, 157-160.

BRINKHOUS, K. M. and GRAHAM, J. B. (1950). Haemophilia in the female dog. *Science*, 111, 723-724.

BROWN, R. C., SWANTON, M. C. and BRINKHOUS, K. M. (1963). Canine haemophilia and male pseudohermaphroditism. Cytogenetic studies. *Lab. Invest.*, 12, 961-967.

BROWN, R. C., CASTLE, W. L. K., HUFFINES, W. H. and GRAHAM, J. B. (1966). Pattern of DNA replication in chromosomes of the dog. *Cytogenetics*, 5, 206-222.

BURNS, M. and FRASER, M. N. (1966). *Genetics of the dog.* Edinburgh: Oliver and Boyd.

CLOUGH, E., PYLE, R. L., HARE, W. C. D., KELLY, D. F. and PATTERSON, D. F. (1970). An *XXY* sex chromosome constitution in a dog with testicular hypoplasia and congenital heart disease. *Cytogenetics*, 9, 71-77.

ELIASSON, K., GUSTAVSSON, I., HULTEN, M. and LINDSTEN, J. (1968). The meiotic chromosomes of the male dog. *Hereditas*, 58, 135-137.

FORD, L. (1965). Leukocyte culture and chromosome preparations from adult dog blood. *Stain Tech.*, 40, 317-320.

FORD, L. (1969). Identification and chromosome interpretation of pachytene bivalents from *Canis familiaris. Canad. J. Genet. Cytol.*, 11, 389-402.

FRACCARO, M. I., GUSTAVSSON, I., HULTEN, M., LINDSTEN, J., MANNINI, A. and TIEPOLO, L. (1964). DNA replication patterns of canine chromosomes *in vivo* and *in vitro. Hereditas*, 52, 265-270.

GRAHAM, J. B. (1952). Further observations on canine hemophila. *J. Elisha Mitchell Sci. Soc.*, 68, 153.

GRAHAM, J. B., BUCKWALTER, J. A., HARTLEY, L. J. and BRINKHOUS, K. M. (1949). Canine hemophila: observations on the course, the clotting anomaly and the effect of blood transfusions. *J. Exp. Med.*, 90, 97-111.

GUSTAVSSON, I. (1964). The chromosomes of the dog. *Hereditas*, 51, 187-189.

GUSTAVSSON, I. and SUNDT, C. O. (1965). Chromosome complex of the family Canidae. *Hereditas*, 54, 248-254.

HARE, W. C. D., WEBER, W. T., McFEELY, R. A. and YANG, T. J. (1966). Cytogenetics in the dog and cat. *J. Small Anim. Pract.*, 7,

575-592.

HARE, W. C. D., WILKINSON, J. S. and McFEELY, R. A. (1967). Bone chondroplasia and a chromosome abnormality in the same dog. *Amer. J. Vet. Res.*, **28**, 583-587.

HSU, T. C. and BENIRSCHKE, K. (1967). *An atlas of mammalian chromosomes. Vol. I.* Berlin: Springer Verlag.

HSU, T. C. and POMERAT, C. M. (1953). Mammalian chromosomes *in vitro.* II. A method of spreading the chromosomes of cells in tissue culture. *J. Hered.*, **43**, 167-172.

HUTT, F. B., RICKARD, C. G. and FIELD, R. A. (1948). Sex-linked hemophila in dogs. *J. Hered.*, **39**, 2-9.

KAKPAKOVA, E. S., POGOSYANZ, E. E. and PONOMARKOV, V. I. (1968). Peculiarities of the karyotype of the tranmissible sarcoma cells in the dog. *Vop. Onkol.*, **14** (11), 43-50.

KNOX, W. E. (1966). Cystinuria. In Stanbury, J. B., Wyngaarden, J. B. and Frederickson, D. S. (Editors). *The metabolic basis of inherited disease.* New York: McGraw-Hill Book Co.

LITTLE, C. C. (1957). *The inheritance of coat colour in dogs.* New York: Comstock Publishing Associates.

MAKINO, S. (1949). A review of the chromosomes of domestic mammals. *Jap. J. Zootech. Sci.*, **19**, 5-15.

MAKINO, S. (1956). *Review of the chromosome numbers in mammals.* Tokyo: Hokuryukan.

MATTHEY, M. R. (1954). Chromosomes et systematique des canides. *Mammalia*, **18**, 225-230.

MERKENS, J. (1938). Haemophilie bij honden. *Ned. Ind. Bl. Diergeneesk*, **50**, 149-151.

MINOUCHI, O. (1928). The spermatogenesis of the dog with special reference to meiosis. *Jap. J. Zool.*, **1**, 255-268.

MOORE, W. and LAMBERT, P. D. (1963). The chromosomes of the Beagle dog. *J. Hered.*, **54**, 273-276.

MUSTARD, J. F., ROWSELL, H. C., ROBINSON, G. A., HOEKSEMA, T. D. and DOWNIE, H. G. (1960). Canine haemophilia B (Christmas disease). *Brit. J. Haematol.*, **6**, 259-266.

NEWNHAM, R. E. and DAVIDSON, W. M. (1966). Comparative study of the karyotypes of several species in carnivora. *Cytogenetics*, **5**, 152-163.

OHNO, S., BECAK, W. and BECAK, M. L. (1964). X-autosome ratio and the behaviour pattern of individual X chromosomes in placental mammals. *Chromosoma*, **15**, 14-30.

PAKES, S. P. and GRIESEMER, R. A. (1965). Current status of chromosome analysis in veterinary medicine. *J. Amer. Vet. Med. Assoc.*, **146**, 138-145.

PATTERSON, D. F., HARE, W. C. D., SHIVE, R. J. and LUGINBUHL, H. (1966). Congenital malformations of the cardiovascular system associated with chromosomal abnormalities. *Zbl. Vet. Med.*, **13**, 669-686.

PICK, J. R., GOYER, R. A., GRAHAM, J. B. and RENWICK, J. H. (1967). Subluxation of the carpus in dogs. *Lab. Invest.*, **17**, 243-248.

ROWSELL, H. C., DOWNIE, H. G., MUSTARD, J. F., LEESON, J. E. and ARCHIBALD, J. A. (1960). A disorder resembling hemophilia B (Christmas disease) in dogs. *J. Amer. Vet. Med. Assoc.*, 137, 247.

SHIVE, R. J., HARE, W. C. D. and PATTERSON, D. F. (1965). Chromosome anomalies in dogs with congenital heart disease. *Cytogenetics*, 4, 340-348.

SOFUNI, T. and MAKINO, S. (1963). A supplementary study on the chromosomes of venereal tumours of the dog. *Gann*, 54, 149.

TAKAYAMA, S. (1958). Existence of a stem-lineage in an infectious venereal tumour of the dog. *Jap. J. Genet.*, 33, 56-64.

TAKAYAMA, S. and MAKINO, S. (1961). A study of the chromosomes in venereal tumours of the dog. *Z. Krebsforschung*, 64, 253-261.

TZESSARSKAYA, T. P. (1968). Somatic chromosomes of the dog. *Genetika, (Mosc.)*, 10, 158.

WEISS, E., HOFFMAN, R. and TOTHEMUND, I. (1967). Die Karyogramme der haussaugetiere. *Zuchthyg.*, 2, 152-155.

WINGE, O. (1950). *Inheritance in dogs.* New York: Comstock Publishing Co.

CHAPTER 19

American Mink

Mustela vison $(n = 14 + X + Y)$

The mink is apparently of greater commercial, rather than laboratory, value at this time. However, considerable interest is taken in the retention and propagation of colour mutants and a large number have been described. In most cases, sufficient breeding data have been published to establish the mode of inheritance of individual mutants but little on the possible linkage relationships between the mutants. A general account of mink genetics is provided by Shackleford (1957). The value of mink as a laboratory animal is discussed by Padget, Gorham, and Henson (1968).

TABLE 19.1. Gene/loci of the American mink which have featured in studies on linkage: assigned linkage group, symbol, and conventional designation.

Linkage group	Symbol	Designation	Prime characteristic
	al	Aleutan	Coat colour
I	*b*	Brown-eyed pastel	Coat colour
I	*Eb*	Ebony	Coat colour
	F	Blufrost	Coat colour
	p	Silverblu	Coat colour
	S	Black cross	Coat colour

LINKAGE GROUP

Group I (b, Eb)
Shackleford (1949) has reported data which are very suggestive of linkage of 24.2 ± 4.9 for the genes *b* and *Eb*.

INDEPENDENT ASSORTMENT

A problem in assessing independent heredity of the genes of the

465

mink is that little or no data have been obtained from systematic experiments. The collation of breeding results from breeders are fine for determining the type of monogenic heredity of a mutant but less so for random assortment. The difficulty is that of deciding whether or not the data are of a single linkage phase or a mixture. If the latter, the manifestation of linkage could be obscured. With this provision in mind, the data of Table 19.2 are worth noting.

TABLE 19.2. Quasi-independent gene assortment in the American mink

Loci	Recombination value	Phase balance	Closest linkage	References
al - p	71 ± 2	0	38	Shackleford (1949)
F - S	42 ± 6	0	30	Shackleford (1949)
p - S	50 ± 5	0	40	Nes (1963)

CHROMOSOME MORPHOLOGY

The count of 30 chromosomes for the mink was found by Lande (1957) and Humphrey and Spencer (1959). Later work has confirmed this determination and has provided detailed descriptions of the karyotype. In good preparations, the karyotype presents an interesting picture. Almost all of the chromosomes can be distinguished by virtue of relative size, position of the centromere and, on occasion, by the presence of secondary constrictions (Fredga, 1961; Nes, 1962). Table 19.3 provides a generalized summary of the main conclusions. The X is a medium sized submetacentric while the Y is a small submetacentric. The following papers should be consulted for finer details: Fredga (1961), Nes (1962, 1966), Shida and Sasaki (1962), Basrur et al. (1963), Evsikov and Isakova (1968), Hsu and Benirschke (1968), Itoh et al. (1968), and Chang et al. (1969).

TABLE 19.3. Summary of karyological investigations of the mink.

Defined autosomal groups	No. of metacentrics			No. of acrocentrics			Sex chromosomes		References
	Large	Medium	Small	Large	Medium	Small	X	Y	
11	7	4	2	—	—	1	Mm	Sm	Fredga (1961)
11	5	3	2	—	3	1	Mm	Sm	Nes (1962)
6	7	4	2	—	—	1	Mm	Sm	Shida and Sasaki (1962)
11	7	4	2	—	—	1	Mm	Sm	Itoh et al. (1968)
—	8	3	2	—	—	1	Mm	Sm	Chang et al. (1969)

M = medium, S = small, m = metacentric; acrocentric embraces subtelocentric and telocentric elements.

BIBLIOGRAPHY

BASRUR, P. K., GRAY, D. P. and GILMAN, J. P. W. (1963). Somatic chromosomes of mink. *Canad. J. Genet Cytol.*, 5, 96-97.

CHANG, M. C., PICKWORTH, S. and McGAUGHEY, R. W. (1969). Experimental hybridization and chromosomes of hybrids. In Benirschke, K. (Editor) *Comparative Mammalian Cytogenetics.* Berlin: Springer Verlag.

EVSIKOV, V. I. and ISAKOVA, G. K. (1968). [Some results of karyological studies in minks of various genotypes.] *Genetika (Mosc.)*, 4, 34-47.

FREDGA, K. (1961). The chromosomes of the mink. *J. Hered.*, 52, 90-94.

HSU, T. C. and BENIRSCHKE, K. (1968). *An atlas of mammalian chromosomes. Vol. II.* Berlin: Springer Verlag.

HUMPHREY, D. G. and SPENCER, N. (1959). Chromosome number in the mink. *J. Hered.*, 50, 245-247.

ITOH, M., SASAKI, M., SHINBA, H. and SHIOTA, Y. (1968). The chromosomes of four mutant strains of the mink. *Zool. Mag.*, 77, 374-378.

LANDE, O. (1957). The chromosomes of the mink. *Hereditas*, 43, 578-582.

NES, N. (1962). Chromosome studies in Heggedal and standard dark mink. *Acta Vet. Scand.*, 3, 275-294.

NES, N. (1963). An investigation of the relation of the Heggedal factor to the black cross factor. *Acta Agric. Scand.*, 13, 359-370.

NES, N. (1966). Diploid-triploid chimerism in a true hermaphrodite mink. *Hereditas*, 56, 159-170.

PADGETT, G. A., GORHAM, J. R. and HENSON, J. B. (1968). Mink as a biomedical model. *Lab. Anim. Care*, 18, 258-266.

SHACKLEFORD, R. M. (1949). Six mutations affecting coat color in ranch bred mink. *Amer. Nat.*, 83, 49-68.

SHACKLEFORD, R. M. (1957). *Genetics of the ranch mink.* Black Fox Magazine Publication.

SHIDA, G. and SASAKI, M. (1962). An *in vitro* study of the somatic chromosomes in the mink. *Zool. Mag.*, 71, 98-101.

Index to Part B

Consolidated Index for Parts A and B

473